Topics in Applied Physics Volume 44

Topics in Applied Physics Founded by Helmut K. V. Lotsch

1 **Dye Lasers** 2nd Ed. Editor: F. P. Schäfer
2 **Laser Spectroscopy** of Atoms and Molecules. Editor: H. Walther
3 **Numerical and Asymptotic Techniques in Electromagnetics** Editor: R. Mittra
4 **Interactions on Metal Surfaces** Editor: R. Gomer
5 **Mössbauer Spectroscopy** Editor: U. Gonser
6 **Picture Processing and Digital Filtering** 2nd Edition. Editor: T. S. Huang
7 **Integrated Optics** 2nd Ed. Editor: T. Tamir
8 **Light Scattering in Solids** Editor: M. Cardona
9 **Laser Speckle** and Related Phenomena Editor: J. C. Dainty
10 **Transient Electromagnetic Fields** Editor: L. B. Felsen
11 **Digital Picture Analysis** Editor: A. Rosenfeld
12 **Turbulence** 2nd Ed. Editor: P. Bradshaw
13 **High-Resolution Laser Spectroscopy** Editor: K. Shimoda
14 **Laser Monitoring of the Atmosphere** Editor: E. D. Hinkley
15 **Radiationless Processes** in Molecules and Condensed Phases. Editor: F. K. Fong
16 **Nonlinear Infrared Generation** Editor: Y.-R. Shen
17 **Electroluminescence** Editor: J. I. Pankove
18 **Ultrashort Light Pulses** Picosecond Techniques and Applications Editor: S. L. Shapiro
19 **Optical and Infrared Detectors** 2nd Ed. Editor: R. J. Keyes
20 **Holographic Recording Materials** Editor: H. M. Smith
21 **Solid Electrolytes** Editor: S. Geller
22 **X-Ray Optics.** Applications to Solids Editor: H.-J. Queisser
23 **Optical Data Processing.** Applications Editor: D. Casasent
24 **Acoustic Surface Waves** Editor: A. A. Oliner
25 **Laser Beam Propagation in the Atmosphere** Editor: J. W. Strohbehn
26 **Photoemission in Solids I.** General Principles Editors: M. Cardona and L. Ley
27 **Photoemission in Solids II.** Case Studies Editors: L. Ley and M. Cardona
28 **Hydrogen in Metals I.** Basic Properties Editors: G. Alefeld and J. Völkl
29 **Hydrogen in Metals II** Application-Oriented Properties Editors: G. Alefeld and J. Völkl
30 **Excimer Lasers** Editor: Ch. K. Rhodes
31 **Solar Energy Conversion.** Solid-State Physics Aspects. Editor: B. O. Seraphin
32 **Image Reconstruction from Projections** Implementation and Applications Editor: G. T. Herman
33 **Electrets** Editor: G. M. Sessler
34 **Nonlinear Methods of Spectral Analysis** Editor: S. Haykin
35 **Uranium Enrichment** Editor: S. Villani
36 **Amorphous Semiconductors** Editor: M. H. Brodsky
37 **Thermally Stimulated Relaxation in Solids** Editor: P. Bräunlich
38 **Charge-Coupled Devices** Editor: D. F. Barbe
39 **Semiconductor Devices** for Optical Communication Editor: H. Kressel
40 **Display Devices** Editor: J. I. Pankove
41 **Computer Application in Optical Research** Editor: B. R. Frieden
42 **Two-Dimensional Digital Signal Processing I.** Linear Filters Editor: T. S. Huang
43 **Two-Dimensional Digital Signal Processing II.** Transforms and Median Filters. Editor: T. S. Huang
44 **Turbulent Reacting Flows** Editors: P. A. Libby and F. A. Williams
45 **Hydrodynamic Instabilities and the Transition to Turbulence** Editors: H. L. Swinney and J. P. Gollub
46 **Glassy Metals I** Editors: H.-J. Güntherodt and H. Beck
47 **Sputtering by Particle Bombardment I** Editor: R. Behrisch
48 **Optical Information Processing** Fundamentals Editor: S. H. Lee

Turbulent Reacting Flows

Edited by P. A. Libby and F. A. Williams

With Contributions by
R. W. Bilger   K. N. C. Bray   E. E. O'Brien
C. R. Ferguson   P. A. Libby   A. M. Mellor
F. A. Williams

With 38 Figures

Springer-Verlag Berlin Heidelberg New York 1980

Professor *Paul A. Libby*, PhD
Professor *Forman A. Williams*, PhD

Department of Applied Mechanics and Engineering Sciences, University of California,
La Jolla, CA 92093, USA

ISBN 3-540-10192-6 Springer-Verlag Berlin Heidelberg New York
ISBN 0-387-10192-6 Springer-Verlag New York Heidelberg Berlin

Library of Congress Cataloging in Publication Data. Main entry under title: Turbulent reacting flows. (Topics in applied physics; 44). Bibliography: p. Includes index. 1. Turbulence. 2. Combustion. 3. Thermochemistry. I. Bilger, R. W. II. Libby, Paul A. III. Williams, Forman Arthur, 1934- IV. Series. QA913.T87 532'.517 80-18381

Monophoto typesetting, offset printing and bookbinding: Brühlsche Universitätsdruckerei, Giessen
2153/3130-543210

Preface

Fluid mechanical turbulence is one of the most challenging fields of engineering science, rich in problems involving theory and experiment and having fundamental and applied significance. When chemical reactions occur in turbulent flows as is the case in many devices involving combustion, the challenge takes on a new dimension; the broad field of fluid mechanical turbulence is joined with another area of fluid mechanics, aerothermochemistry, which relates to the coupling between fluid dynamics, thermodynamics, and chemistry. There result sublimely complex problems involving the interaction of turbulent motions with the chemical and thermodynamic behavior associated with combustion. It is the purpose of this volume to provide background material that is needed to unravel the essentials of these complexities and interactions and to summarize recent theoretical approaches that have been developed in the field.

Turbulent reacting flows have attracted increased interest in recent years. The requirements of increased combustion efficiency and decreased pollutant emissions from a variety of devices, from power plants to jet engines, and the introduction of new devices such as the gas dynamic laser have led to the need for improved methods of prediction and calculation for turbulent flows involving chemical reactions. This increase in interest has led to a series of specialized meetings, workshops, publications, etc. References to the literature resulting from this activity will appear in the various chapters which follow. However, the field of turbulent reacting flows is in such a state of flux that it is difficult for a research worker, familiar to some extent with the essential background ingredients, fluid mechanics, turbulence, and aerothermochemistry, to gain perspective on our understanding of turbulent combustion and on our current capabilities to predict the structure of turbulent reacting flows. There thus appears a need for a tutorial treatment of the field.

In a book devoted to such a specialized field as turbulent reacting flows appropriate background knowledge must be assumed. Thus our exposition presumes that the reader has some knowledge of fluid mechanical turbulence and of aerothermochemistry; in Chap. 1 we suggest a few books providing this requisite background. With this knowledge as a foundation the present volume is self-contained and may be used as a textbook for a graduate course in mechanics, applied physics, or engineering or as an introduction to a new research area for the reader. One of our goals is to set forth the variety, challenge, and importance of the problems in turbulent reactant flows and thus to encourage research workers to enter the field.

The contents of the book are proscribed in another fashion. Many problems in turbulent combustion involve the injection of liquid or solid fuels into oxidizing gas streams. In these situations the combustion processes are accompanied by vaporization and/or sublimation and thus by additional complications. Similar remarks pertain to radiative transfer in turbulent reacting flows; in many flows involving combustion radiation from particles arising from incomplete combustion of injected solid fuel, from condensed-phase products of combustion, from refractory impurities contained in fuels, or from the nucleation of carbon in portions of the flow with excess fuel can provide significant energy transfer. With respect to both of these complications we shall limit discussion to a presentation in Chap. 6 of current entries into the literature on these largely untouched problems.

Throughout, our exposition emphasizes the fundamental bases and theoretical aspects of each topic. Experimental results and occasionally, in an idealized sense, the means to obtain them are introduced in order to reinforce the physical interpretation of a theoretical notion, to guide theoretical developments, e.g., in the introduction of models to close the describing equations, and to indicate the accuracy of theoretical prediction. We note that the application of experimental techniques to turbulent combustion is a separate discipline outside the scope of the present volume.

Chapter 1 sets the foundations of subsequent, more specialized discussions and reviews the general theoretical basis for the study of turbulent reacting flows in the context of combustion. It also establishes the notation used in the remainder of the book. Chapter 2 also has an introductory nature; it relates to the problems of an applied and practical nature which are of technological interest and which conceivably can be clarified and elucidated by the study of turbulent combustion. At this juncture the reader should have gained a sense of the complexities involved in the most general cases of turbulent combustion and will be prepared in the best tradition of the engineering sciences to deal with limited cases which, because of their nature, are amenable to facilitated treatment. Accordingly, Chaps. 3 and 4 will discuss the cases of nonpremixed and premixed reactants, respectively. Since there are practical situations which closely correspond to these two limits, their consideration is not of only academic interest. Throughout these chapters repeated use will be made of the notion of probability density functions in guiding the reader's perception of the phenomena involved and in developing suitable models and approximations. These functions relating to one or more fluid mechanical variables at each spatial location can be considered to play a more fundamental role than is suggested by such uses. In fact if a direct attack on the problem of describing turbulent reacting flows via the probability density function were feasible, many of the problems we discuss would be readily solvable; this is far from true, but certain limited problems can be attacked in this direct fashion, and Chap. 5 describes this approach and the problems involved in its extension.

Of necessity the flows treated in Chaps. 3–5 will be relatively idealized. Although many useful results are obtained from analyses of these simplified situations, there are flows of applied interest which do not lead themselves to these indications but rather involve in a significant way some of the effects which have been neglected. In addition, the current state of flux of the subject of turbulent reacting flows is manifest by several novel approaches whose thorough discussion would be inappropriate in a tutorial volume. However, attention should be called to them because of the difficulties and limitations associated with the more developed approaches emphasized in the various chapters. Finally, there are topics excluded from our exposition since they lie beyond the possible scope of this volume even though they are relevant to the study of turbulent reacting flows. In Chap. 6 we offer comments on aspects of turbulent reacting flows generally disregarded in Chaps. 3–5, on novel and developing approaches to the subject and on outstanding problems. References providing entries to the relevant literature of these topics are given in order to stimulate further interest and the additional research required to overcome current difficulties and limitations.

Because of the variety and complexity of the problems associated with turbulent reacting flows, their study can involve sophisticated techniques of applied mathematics and of experimental methods. Our approach is to emphasize the engineering and physical aspects of the subject, referring wherever indicated by exposition to experimental results and experimental methods and to problems of interest to applied mathematicians. Our hope is to provide a perspective of, and to stimulate interest in, a challenging and important field.

Our research on turbulent reacting flows has been supported over a period of several years by the Office of Naval Research as part of Project SQUID and by the Air Force Office of Scientific Research. This volume is a consequence of our research interests and we therefore gratefully acknowledge the contribution of these agencies. We also note with thanks the contributions of Ms. Barbara Hanson in the preparation of the manuscript of the Preface and of Chaps. 1 and 6.

La Jolla, July 1980

Paul A. Libby
Forman A. Williams

Contents

Contributors

Bilger, Robert W.
Department of Mechanical Engineering, University of Sydney,
New South Wales 2006, Australia

Bray, Kenneth N. C.
Department of Aeronautics and Astronautics, The University,
Southampton, SO9 5NH, England

O'Brien, Edward E.
Department of Mechanics, State University of New York at Stony Brook,
Stony Brook, NY 11794, USA

Ferguson, Colin R.
School of Mechanical Engineering, Purdue University,
West Lafayette, IN 47907, USA

Libby, Paul A.
Department of Applied Mechanics and Engineering Sciences,
University of California, San Diego, La Jolla, CA 92093, USA

Mellor, Arthur M.
School of Mechanical Engineering, Purdue University,
West Lafayette, IN 47907, USA

Williams, Forman A.
Department of Applied Mechanics and Engineering Sciences,
University of California, La Jolla, CA 92093, USA

1. Fundamental Aspects

P. A. Libby and F. A. Williams

With 2 Figures

The basic equations that describe turbulent reacting flows usually are accepted to be the Navier-Stokes equations, suitably amplified to include chemical reactions, species conservation, and variable-property effects. This chapter offers an exposition of these equations. It opens with suggestions for background reading that may serve as a point of departure. After preliminary delimitation of the types of problems covered within this volume, it continues with a review of elements of fluid mechanical turbulence and of aerothermochemistry necessary for an understanding of the phenomena. This review includes establishment of notation; discussions of transport properties and of chemical kinetics; alternative approaches to probabilistic descriptions, such as time averaging, Favre averaging, and equations for evolution of probability-density functionals; specific turbulence topics of potential relevance to reacting flows, such as intermittency and spectral decompositions; nondimensionalizations resulting in similarity parameters that help to quantify limiting behaviors; methods for treatment of chemical reactions in turbulent reacting flows; and some discussion of methods of closure. The intent in this chapter is to set the stage for more detailed considerations that are given in subsequent chapters.

1.1 Background Literature

A variety of books are available for obtaining the appropriate background in fluid-mechanical turbulence. We especially recommend *Bradshaw* [1.1, 2] and *Tennekes* and *Lumley* [1.3] as providing a useful and practical focus. More extensive treatments are offered by *Hinze* [1.4] and by *Monin* and *Yaglom* [1.5]. Background information in aerothermochemistry and combustion may be obtained from the books by *Penner* [1.6] and by *Williams* [1.7], for example. More recent texts on combustion include those of *Strehlow* [1.8] and *Glassman* [1.9].

A current review of fluid mechanical turbulence is given by *Reynolds* [1.10]. Reviews on turbulent reacting flows have been prepared by *Hill* [1.11] and by *Libby* and *Williams* [1.12]. *Launder* and *Spalding* [1.13] and *Spalding* [1.14] provided entries into the current literature on the more application-oriented aspects of turbulence phenomenology.

1.2 Turbulence in Reacting Liquids and Gases

Liquids and gases constitute two limiting states of fluids which fundamentally exhibit very different properties. Although the present volume concerns chemical reactions in gases, the results are applicable to liquids in certain respects. Therefore an introductory discussion of the differences between liquids and gases seems warranted.

The most convincing justification for the use of the Navier-Stokes level of description of flows differs for liquids and gases. Most gases of interest are "dilute" in the kinetic-theory sense; the duration of molecular collisions is short compared with the time between collisions. By contrast, a molecule in a liquid experiences collisions most of the time. For dilute gases the kinetic theory is relatively well developed and provides the most convincing foundation from which the Navier-Stokes description can be derived. For liquids, complexities of dense-gas kinetic theory cause a phenomenological, continuum-mechanical derivation of the Navier-Stokes equations to be more satisfying than kinetic-theory derivations. Thus, the origins of the describing equations may be viewed as differing for the two limiting states.

When the starting point is phenomenological, no avenue is open for questioning the validity of the describing equations. In contrast, a derivation from kinetic theory can be challenged by questioning specific assumptions, such as molecular chaos, that are introduced during the derivation. Recently a few investigators have raised such challenges to Navier-Stokes turbulence for dilute gases (see, for example, [1.15]). The possibility of the existence of associated influences on turbulent reacting flows recently has been brought out by *Tsugé* and *Sagara* [1.16], who achieved agreement with anomalous experimental results on hydrogen–oxygen ignition by means of an analysis that is based on kinetic theory and that denies applicability of a Navier-Stokes level of description to certain experiments. Although the possibilities for failure of a Navier-Stokes description are interesting, this topic is too new and much too incomplete to be covered within the present volume. Therefore we adopt the classical viewpoint that, for gases as well as liquids, the basic equations underlying the description of turbulent reacting flows are the Navier-Stokes equations, and the probabilistic structures appropriate for such flows are to be erected upon those equations.

Given this identical level of description, there remain significant differences between liquids and gases experiencing chemical reactions. Aside from subsidiary influences such as buoyant convection, the density is essentially constant in a reacting liquid but changes substantially as a result of heat release for most reactions in gases (notably for all combustion reactions). This simplifies the dynamics of turbulent reacting flows for liquids, in that the turbulent behavior of the velocity field is unaffected by the chemistry. In contrast, for the turbulent reacting flows of interest in gases, the full interaction between turbulence and chemistry occurs: The turbulence influences the effective chemical behavior and the heat release associated with the exothermic chemical reactions alters the turbulence. In this respect, the problems addressed in the present volume are more complex than those of chemical reactions in turbulent liquids.

Offsetting this relative simplicity of liquids is a complication associated with molecular transport properties. In gases, diffusivities of heat, chemical species, and momentum all are of the same order of magnitude (with the possible exception of hydrogen which diffuses an order of magnitude faster than other species). However, in liquids the diffusivity of heat is large compared with that of momentum, which in turn is large compared with diffusivities of species. This introduces into the dynamics of turbulent mixing of liquids elements not found in turbulent mixing of gases (see for example, [1.17, 18]) and correspondingly may influence reaction dynamics in turbulent liquids. In many problems this complexity may be relatively insignificant; often it is thought that for sufficiently intense turbulence molecular transport is unimportant in comparison with turbulent transport. However, in general, studies of effects of large differences in molecular transport coefficients are well motivated primarily for liquids. Such effects are excluded from the present volume.

A final way in which turbulent reacting liquids behave more simply than turbulent reacting gases stems from the fact that reaction rates usually tend to be relatively high in liquids. This difference ultimately rests on the continual intermolecular collisions occurring in liquids, in comparison with the relatively rare intermolecular encounters of gases. The rapid reaction in liquids allows certain simplifications associated with the notion of "fast chemistry" to be applied more widely in liquids than in gases. This and the aforementioned constant-density property of liquids allow many of our considerations to apply to chemical reactions in turbulent liquids. In particular, one limiting case of reactions in gases, that involving highly dilute reactants undergoing fast chemistry, is directly applicable to reactions in liquids when large differences in coefficients of molecular transport are unimportant.

1.3 The Eulerian Viewpoint and Notation

Although there are specific processes, such as turbulent diffusion, in which clarification can be achieved through use of a Lagrangian viewpoint [1.3], complexities that arise in attempting to tie the coordinate system to fluid parcels in turbulent flows of general types usually lead to the adoption of the Eulerian viewpoint. Following this latter convention, we consider the behavior with time of the various quantities characterizing the flow at a fixed spatial point. These quantities are the components of velocity and the state variables, pressure, density, and concentrations of chemical species. Each of these quantities is considered to have associated with it an ensemble average or mean value, which may vary in space and time, and a fluctuation, which always depends on space and time and which typically varies much more rapidly than the mean.

For purposes of exposition it will be convenient to employ Cartesian tensor notation. Thus we consider a Cartesian coordinate system with coordinates x_i, $i=1,2,3$ and associated velocity components u_i, $i=1,2,3$. Furthermore, we shall reserve the subscripts k and l to indicate summation; thus, e.g., $u_k u_k = u_1^2 + u_2^2 + u_3^2$. The subscripts i and j will indicate either quantities associated with the coordinate directions, e.g., $u_i u_j$, the product of the ith velocity component with the jth velocity component; or quantities identifying chemical species or reactions, e.g., Y_i, the mass fraction of species i. In subsequent chapters when specialized applications are discussed it will be more convenient to introduce x, y, z and u, v, w as notation for spatial coordinates and velocity components, but the subscripts i and j will continue to identify the various species.

1.4 Transport Properties

The starting point for our considerations will be the conservation equations for mass, momentum, and energy for a multicomponent, reacting, gas mixture. Establishment of these equations and their exploitation is fundamental to aerothermochemistry and combustion theory. The transport properties of the gas mixture enter these equations and are the subject of separate study; *Hirschfelder* et al. [1.19] provided the most detailed discussion of transport properties in gases.

As discussed in Sect. 1.2, we assume that momentum conservation for gas mixtures of interest is described by the Navier-Stokes equations, i.e., that the gas mixture may be treated as a Newtonian fluid with the viscosity coefficient μ taken to be a given function of temperature and composition. Effects of bulk viscosity seem always to be unimportant and are neglected herein. Mass conservation arises at several levels: overall conservation of mass for the gas mixture leads to the usual "continuity equation". Conservation of individual species includes accumulation, convection, diffusion, and creation or destruction by chemical activity. Treatment of species conservation can sometimes be simplified by invoking conservation of elements. Energy conservation contains the same types of terms as species conservation; we exclude Soret and Dufour phenomena since it is probable that these effects are usually small.

Under the assumptions stated above, the transport of momentum, exclusive of effects of gradients in hydrostatic pressure, is described by the viscous stress tensor,

$$\tau_{ij} = \mu\left(\frac{\partial u_i}{\partial x_j} + \frac{\partial u_j}{\partial x_i}\right) - \frac{2}{3}\mu \frac{\partial u_k}{\partial x_k}\delta_{ij}, \tag{1.1}$$

where $\delta_{ij}=1$ if $i=j$, 0 if $i \neq j$. This tensor has the property that the viscous force in the ith coordinate direction is $(\partial/\partial x_k)\tau_{ik}$.

In turbulent flows molecular transport of chemical species enters in indirect, subtle ways which generally call for special treatment. To avoid excessive complication and to maintain consistency with the current state of development of the theory it is satisfactory to describe the flux of species due to molecular transport in terms of Fick's law with a single diffusion coefficient for all species; thus

$$\varrho_i V_{ij} = -\varrho \mathscr{D} \partial Y_i / \partial x_j , \tag{1.2}$$

where ϱ_i is the partial density of species i, ϱ is the mixture density, V_{ij} is the diffusional velocity of species i in the jth coordinate direction, $\mathscr{D}$ is the diffusion coefficient, and Y_i is the mass fraction of species i. The product $\varrho\mathscr{D}$, like μ, is taken to be a given function of temperature and composition. The advantage of the Fick approximation is that the equations for energy, species, and element conservation are simplified considerably. The disadvantage is that in gas mixtures involving a significant spread in the molecular weights of the participating species the selection of the single diffusion coefficient to describe accurately the diffusional characteristics of all species cannot be readily rationalized.

In energy conservation the Fourier law for molecular conduction of heat describes the transport, subject to the stated assumptions. Thus, the heat flux in the ith coordinate direction is

$$q_i = -k \partial T / \partial x_i , \tag{1.3}$$

where k, the coefficient of thermal conductivity of the gas mixture, is a known function of the temperature T and of the chemical composition.

There are nondimensional combinations of μ, $\mathscr{D}$, and k which are frequently approximated to simplify further the description of turbulent gas flows and to permit thereby the analysis of idealized systems. The Schmidt number, $\mathrm{Sc} = \mu/\varrho\mathscr{D}$, is a nondimensional measure of the relative importance of viscous and diffusional properties of the gas and often is taken equal to unity, although its value may be as low as 0.25 for binary mixtures of gases with large and small molecular weights. The corresponding quantity involving the thermal conductivity is the Prandtl number, $\sigma = \mu c_\mathrm{p}/k$, where c_p is the specific heat at constant pressure of the gas mixture. Frequently σ is set equal to one despite exact values measurably different therefrom for some gases and gas mixtures. The Lewis number, which is the ratio of the Schmidt number to the Prandtl number, tends to be closer to unity than either of the other two nondimensional ratios.

1.5 Chemical Kinetics

Chemistry makes itself felt in reacting turbulent flows through the chemical source terms that appear in the equations for species and energy conservation. Quantities $\dot{w}_i$, defined as the mass rate of production per unit volume of species i (g/cm^3 s), occur in these conservation equations. To obtain an expression for

$\dot{w}_i$, consider a general system involving N chemical species and R reaction steps of the form

$$\sum_{i=1}^{N} \nu'_{ij}\mathcal{M}_i \underset{k_{b_j}}{\overset{k_{f_j}}{\rightleftarrows}} \sum_{i=1}^{N} \nu''_{ij}\mathcal{M}_i, \quad j=1,2,\ldots,R, \tag{1.4}$$

where $\mathcal{M}_i$ is the chemical symbol for species i, ν'_{ij} and ν''_{ij} are the (constant) stoichiometric coefficients for species i appearing as reactant and product, respectively, in reaction j, and the k's denote rate constants that depend on temperature for reactions in ideal gas mixtures. Note that the ν's are integers if the equation describes a fundamental chemical step occurring at the molecular level through collisions between ν'_{ij} molecules of each type i. Note also that the subscripts f and b on k identify rates for the reaction occurring in the "forward" and "backward" directions, i.e., in the directions indicated by arrows. If the reaction is a fundamental step, then the rate constants k_{f_j} and k_{b_j} will be expressed in the units $(\mathrm{g/cm^3})^{m_j-1}\,\mathrm{s}^{-1}$ and $(\mathrm{g/cm^3})^{n_j-1}\,\mathrm{s}^{-1}$, respectively, where $m_j=\sum_{i=1}^{N}\nu'_{ij}$ and $n_j=\sum_{i=1}^{N}\nu''_{ij}$ are the overall orders of the forward and backward reactions, respectively. The mass-based units adopted here differ from the mole-based units common in the chemical literature, but the conversion is straightforward, merely involving a molecular-weight factor.

The chemical source term must account for the production or destruction of each species due to each reaction step. For species i, the contributions of all reactions may be expressed compactly as $\dot{w}_i=\sum_{j=1}^{R}\dot{w}_{ij}$, where

$$\dot{w}_{ij}=(\nu''_{ij}-\nu'_{ij})k_{f_j}\varrho^{\hat{m}_j}\prod_{i=1}^{N} Y_i^{\hat{\nu}'_{ij}}\left(1-\frac{k_{b_j}}{k_{f_j}}\varrho^{\hat{n}_j-\hat{m}_j}\prod_{i=1}^{N} Y_i^{\hat{\nu}''_{ij}-\hat{\nu}'_{ij}}\right) \tag{1.5}$$

is the mass rate of production per unit volume for the ith species by the jth reaction step. Observe that the sign of the factor $\nu''_{ij}-\nu'_{ij}$ automatically accounts for creation of products and destruction of reactants, zeros being possible values for the coefficients, so that the formulation is entirely general. The carets on the various exponents may be ignored for the present.

The parenthesis on the right-hand side of (1.5) has a special significance. If the factors multiplying the parenthesis are "large" for a particular value of j, then the parenthesis must be "small" for that j. The limiting condition

$$1-\frac{k_{b_j}}{k_{f_j}}\varrho^{n_j-m_j}\prod_{i=1}^{N} Y_i^{\nu''_{ij}-\nu'_{ij}}=0 \tag{1.6}$$

is the statement of chemical equilibrium for the jth reaction step and is seen from this development to be related in some way to large values of chemical reaction-rate factors. Thus, (1.6) need not imply that $\dot{w}_{ij}=0$; instead it means that the rate of production of any species by the jth reaction step is indeterminate and must be found by special means. Further discussion of this loose wordage will be given later.

The rate constants k_{f_j} and k_{b_j} are expressible, for example, as

$$k_{f_j}=B_jT^{\alpha_j}e^{-T_{aj}/T}, \tag{1.7}$$

where B_j, α_j, and T_{aj} are constants characterizing the rate of the jth reaction. The exponential term is the Arrhenius factor, containing the activation temperature T_{aj} (K), which may be obtained from tabulations of the activation energy E_j (kcal/mole), used more conventionally, through $T_{aj}=E_j/R_0$, where $R_0=1.9865$ cal/mole K is the universal gas constant. The product $B_jT^{\alpha_j}$ is the preexponential factor in the reaction-rate constant and has a weak temperature dependence if $\alpha_j \neq 0$. Generally $-2<\alpha_j<2$; the units for the constant B_j in the preexponential factor are those cited above for k_{f_j} divided by $(K)^{\alpha_j}$. There are reactions, notably recombination processes, for which $T_{aj}=0$, so that the entire temperature dependence of k_{f_j} is described by T^{α_j}. For most bimolecular reactions $T_{aj} \neq 0$, and in these cases it is sufficiently large (typically of order 10,000 K) so that the temperature dependence in the preexponential factor usually is negligible in comparison, $\alpha_j=0$ being an accurate approximation.

In studying reacting flows the fluid mechanician needs to know the chemical species involved, the chemical reaction steps, and the forward and backward rates associated with these steps. Unfortunately, some chemical systems involve a large number of significant species and a large number of reaction steps. Even the transformation of a simple pair of reactants to one product may in reality involve a variety of intermediates. As a consequence the information provided to the fluid mechanician may be imprecise and incomplete.

Such lack of information is circumvented by the introduction of reactions (1.4) that do not represent fundamental steps. The rates still are expressed in the form (1.5), but the exponents appearing therein, as well as the constants in (1.7), are determined empirically and do not represent parameters describing specific reactive collisions between molecules. Consequently the overall orders ($\hat{m}_j$ and $\hat{n}_j$) and the orders with respect to species ($\hat{\nu}'_{ij}$ and $\hat{\nu}''_{ij}$) need not be integers, the summation relationships for m_j and n_j need not apply to quantities with carets, and T_{aj} in (1.7) becomes an overall activation temperature, not derivable through theoretical analysis of elementary reaction steps. For many applications such phenomenological descriptions of the chemical behavior are adequate, and in fact there are situations in which rate parameters for all of the fundamental reactions of the form (1.4) are known, but for simplicity and enhanced understanding the large number of reactions is replaced by a smaller number, often one or two, with empirical exponents. It should be kept in mind that in turbulent flows such procedures may introduce errors. In contrast to the situation which prevails in many problems in mechanics, namely that approximations can lead to clearly identifiable and assessible errors, the errors in the description of chemical behavior can be difficult to establish and can lead to orders-of-magnitude errors in product concentrations if justifications for the approximations are not obtained.

1.6 Conservation Equations

These preliminaries enable us to set forth the conservation equations for a multicomponent gas mixture involving N species and M elements. The simplest is the equation for overall mass conservation, namely

$$\frac{\partial \varrho}{\partial t}+\frac{\partial}{\partial x_k}(\varrho u_k)=0. \tag{1.8}$$

Conservation of momentum in the three coordinate directions is described by

$$\frac{\partial}{\partial t}(\varrho u_i)+\frac{\partial}{\partial x_k}(\varrho u_k u_i)=-\frac{\partial p}{\partial x_i}+g_i+\frac{\partial}{\partial x_k}\tau_{ik}, \tag{1.9}$$

where p is the hydrostatic pressure, a thermodynamic variable, and g_i is the body force per unit volume in the ith coordinate direction. In (1.9) the viscous stress tensor τ_{ik} is given by (1.1). If gravity is responsible for g_i and the axis corresponding to $i=3$ is vertical, then $g_i=0$, $i=1$, 2 and $g_3=-\varrho g$, with g being the acceleration of gravity.

The next equation to be considered relates to conservation of energy; it can take on a variety of forms with static temperature, static enthalpy, stagnation enthalpy, or internal energy as the principal variable. If we confine our attention to low-speed flows, i.e., to flows with a characteristic Mach number small compared to unity, then the energy equation in terms of static enthalpy is the most compact and useful. Thus we have

$$\frac{\partial}{\partial t}(\varrho h)+\frac{\partial}{\partial x_k}(\varrho u_k h)=\frac{\partial p}{\partial t}+\frac{\partial}{\partial x_k}\left[\frac{\mu}{\sigma}\frac{\partial h}{\partial x_k}+\mu\left(\frac{1}{\mathrm{Sc}}-\frac{1}{\sigma}\right)\sum_{i=1}^{N} h_i\frac{\partial Y_i}{\partial x_k}\right], \tag{1.10}$$

where h is the mixture enthalpy and h_i is the enthalpy of species i considered as a known function of the temperature T. Neglected in (1.10) are the kinetic energy per unit mass of mixture compared to the mixture enthalpy per unity mass and the viscous dissipation, negligible in low speed flows. Also neglected are radiant energy transfer and work associated with the action of body forces.

Consequences of the approximations in (1.10) deserve further discussion. Dufour terms, excluded earlier, seem always negligible, not only through order-of-magnitude estimates but also because these indirect effects of species diffusion on energy flux have never been documented to be significant in chemically reacting flows of multicomponent gases. The inverse effect of heat conduction on species diffusion could have measurable influences on very light or very heavy species, and therefore the Soret effect is omitted below in the species conservation equation only because this effect has not been studied for turbulent reacting flows. The restriction in (1.10) to low-speed flows excludes, for example, the important application to supersonic combustion. The restriction is not intended to imply that turbulent reacting flows at high Mach numbers are unworthy of study, or even that analysis of such flows is much more difficult than that of low speed flows. Instead, the current status of the field is such that relatively little detailed study has been given to turbulent reacting flows at high Mach number, and therefore the more general equation will not be needed in this volume. It might be noted that in turbulent combustion high velocities accentuate problems of ignition and extinction, so that in a practical sense requisite theoretical analyses may be complex. It is a simple matter to reinstate kinetic-energy and viscous-dissipation terms in (1.10) if desired.

It is also straightforward to include the work associated with body forces, but numerical estimates suggest that if the body forces are produced by earth's gravity then this work term always will be negligible in comparison with rates of thermal energy change associated with reactions. Effective body forces from high rates of swirl may not be negligible and are included in (1.10) implicitly; they appear explicitly under transformation to a rotating frame of reference. The neglect of radiant energy terms in (1.10) can introduce measurable inaccuracies in analyses of some furnace flows or particle-laden flows, for example, and radiation effects are excluded mainly because insufficient attention has been paid to them. It is easy to include in (1.10) the term describing the most important of these effects, the radiant energy loss, but complications arise in the analysis when this is done.

Finally, we observe that the term $\partial p/\partial t$ often is omitted in (1.10). This term plays a role in the propagation of acoustic waves, but other acoustic aspects have been excluded through the approximation of low Mach number. Low Mach number, however, is not a sufficient excuse for eliminating $\partial p/\partial t$; this term is important, for example, in describing engine cycles of internal combustion engines. It is retained here to emphasize that some further study of its influence on turbulent reacting flows of all types is warranted. It may be noted that pressure gradients always should be kept in momentum conservation (1.9) for turbulent flows, so that even if $\partial p/\partial t$ is discarded, many effects of pressure variations remain. In many problems, the body-force and pressure-gradient terms in (1.9) alone incorporate with reasonable accuracy the influences of gravity and pressure variations.

The mixture enthalpy in (1.10) is related implicitly to the temperature by its definition in terms of the species enthalpies, namely by

$$h = \sum_{i=1}^{N} Y_i h_i . \tag{1.11}$$

For most purposes in the present context it is sufficiently accurate to approximate the species enthalpy by

$$h_i = c_{pi} T + \Delta_i , \tag{1.12}$$

where c_{pi} and Δ_i are constants selected to approximate the actual $h_i(T)$ variation in the temperature range of interest and the heat liberated in chemical transformations. The combination of (1.11) and (1.12) leads to

$$h = c_p T + \sum_{i=1}^{N} Y_i \Delta_i, \tag{1.13}$$

where $c_p = \sum_{i=1}^{N} Y_i c_{pi}$ is the specific heat at constant pressure for the mixture, introduced earlier. Note from the second term in (1.13) and from the absence of h in undifferentiated form in its conservation equation (1.10) that an arbitrary constant can be added to all the species constants Δ_i (the heats of formation) without altering the results. This feature sometimes can be used to advantage in simplifying (1.13); for example, in a model chemical system involving two reactants and a single product it is frequently possible to replace (1.13) by

$$h = c_p T - Y_p \Delta_p, \tag{1.14}$$

where Y_p denotes the mass fraction of product and Δ_p the heat of combustion (the chemical heat released per unit mass of product formed).

In (1.10) the simplification which results from the approximation $\mathrm{Sc} = \sigma = 1$ is evident. In this case we have for energy conservation

$$\frac{\partial}{\partial t}(\varrho h) + \frac{\partial}{\partial x_k}(\varrho u_k h) = \frac{\partial p}{\partial t} + \frac{\partial}{\partial x_k}\left(\mu \frac{\partial h}{\partial x_k}\right). \tag{1.15}$$

In complex problems (1.15) often is considered adequate although the correct Schmidt and Prandtl numbers may differ from unity.

In order to facilitate either comparison with experiment or incorporation of experimental results in the development of theory, it may be desirable to express energy conservation in terms of the physically observable quantity, temperature, namely as

$$c_p \left[\frac{\partial}{\partial t}(\varrho T) + \frac{\partial}{\partial x_k}(\varrho u_k T)\right] = \frac{\partial p}{\partial t} - \sum_{i=1}^{N} h_i \dot{w}_i + \frac{\partial}{\partial x_k}\left(k \frac{\partial T}{\partial x_k}\right) + \varrho \mathscr{D} \frac{\partial T}{\partial x_k} \sum_{i=1}^{N} c_{pi} \frac{\partial Y_i}{\partial x_k}, \tag{1.16}$$

which may be derived from (1.10) by use of species conservation equations. Here the volumetric rate of chemical heat release, $-\sum_{i=1}^{N} h_i \dot{w}_i$, the divergence of conductive heat transport, $-\partial q_k/\partial x_k$, see (1.3), and convection of enthalpy by diffusion velocities, $-\sum_{i=1}^{N} \varrho_i V_{ik} \partial h_i/\partial x_k$, see (1.1), appear sequentially on the right-hand side. In this form the assumption of equal specific heats at constant pressure for all species, i.e., $c_{pi}=c_p$, eliminates the last term, producing a simpler equation. If, in addition, $\sigma=1$, then k may be replaced by μc_p in the next-to-last term.

The equation for conservation of individual species is

$$\frac{\partial}{\partial t}(\varrho Y_i)+\frac{\partial}{\partial x_k}(\varrho u_k Y_i)=\frac{\partial}{\partial x_k}\left(\varrho \mathscr{D}\frac{\partial Y_i}{\partial x_k}\right)+\dot{w}_i, \qquad i=1,2,\ldots,N, \tag{1.17}$$

where all of the quantities have been previously defined. The last term on the right side of (1.17) is the chemical source term that has been discussed in Sect. 1.5.

It is sometimes convenient to utilize the concept of element mass fraction Z_i since these are conserved quantities; as a starting definition we have

$$Z_i=\sum_{j=1}^{N}\mu_{ij}Y_j, \qquad i=1,2,\ldots,M, \tag{1.18}$$

where the μ_{ij} coefficients are the number of grams of the ith element in a gram of the jth species. Thus, for example, if we consider water identified with the subscript 3 involving the elements oxygen and hydrogen identified with $i=1$, 2, respectively, we have $\mu_{13}=8/9$ and $\mu_{23}=1/9$. Provided that these coefficients are consistently calculated, it is irrelevant whether the element is considered to be the atomic or molecular form, e.g., whether atomic or molecular oxygen is used in the computation of the coefficients for species involving oxygen.

One of the advantages of the simplification of a single diffusion coefficient is now realized; if the appropriate equations of species conservation given by (1.17) are multiplied by μ_{ij} and summed over j, we find the equation for conservation of element i, namely

$$\frac{\partial}{\partial t}(\varrho Z_i)+\frac{\partial}{\partial x_k}(\varrho u_k Z_i)=\frac{\partial}{\partial x_k}\left(\varrho \mathscr{D}\frac{\partial Z_i}{\partial x_k}\right), \qquad i=1,2,\ldots,M. \tag{1.19}$$

The chemical source term vanishes because in chemical reactions elements are conserved; mathematically we can state $\sum_{i=1}^{N}\mu_{ij}\dot{w}_j=0$, $i=1, 2, \ldots, M$.

With the concept of element conservation established, we have two alternative paths to describe the composition of a gas mixture. In both paths (1.8) is included and assures overall mass conservation. In one path $N-1$ species equations (1.17) are explicitly considered. The alternative is to consider $M-1$ element conservation equations and $N-M$ species conservation equations. If N is large relative to M, i.e., there are a large number of species involving only a few elements, there is relatively little advantage of the second path. However, if there are involved only a few more species than elements, the analysis is facilitated by dealing with the simpler equations of element conservation and with a small number of equations of species conservation.

The equation of state constitutes a relationship among the pressure, density, temperature, and species mass fractions; we have

$$p=\varrho R_0 T \sum_{i=1}^{N} \frac{Y_i}{W_i}, \tag{1.20}$$

where R_0 is the universal gas constant, e.g., 82 atmos. cm^3/mole K. If the molecular weights of the various species do not greatly differ from one another, an approximate form of (1.20) is appropriate, namely

$$p=\varrho RT, \tag{1.21}$$

where R is the mass-based gas constant specific to the chemical system in question.

1.7 Probabilistic Structure

Turbulence is characterized by rapid fluctuations of velocities and state variables in time and space. These fluctuations occur when the inertial forces are sufficiently large in comparison with the viscous forces (1.1) that appear in the last term in (1.9). Typically the relevant parameter is the ratio of dynamic to viscous forces, the Reynolds number, $\mathrm{Re}=\varrho UL/\mu$, where L is a characteristic length and U a characteristic velocity. Viscosity provides damping of fluctuations, and if Re is sufficiently large, then the damping is too small and flow instabilities occur. There have been many analyses of the hydrodynamic instability that develops in the vicinity of the critical value of Re above which self-induced fluctuations amplify. Especially at values of Re well above the critical value, this instability leads rapidly to complex fluctuations having a highly random character.

It is commonly accepted that the equations in Sect. 1.6 comprise a deterministic set, in the sense that, when proper initial and boundary conditions are included, a unique solution exists. However, at high Re this solution is believed to be highly sensitive to these initial and boundary conditions, at points sufficiently far removed from the boundaries and for sufficiently late times. A small change in an applied condition can produce a large change in the solution. Even the fastest and most accurate computers available today cannot obtain solutions with adequate space and time resolution for the full set of equations in these circumstances. It then becomes both practical and philosophically attractive to introduce statistical methods. Since the initial and boundary conditions cannot be controlled accurately enough to produce unique solutions, it may be assumed that these conditions are given only in a probabilistic sense, and the solutions to the equations then themselves become probabilistic quantities, the reacting turbulent flow being viewed as a stochastic process. Attention is then focused upon obtaining probabilistic aspects of the solutions.

Since the dependent variables, such as u_i, are functions of x_i and t, the probabilistic description fundamentally involves probability-density functionals, such as $\mathscr{P}(u_i)$, defined by stating that the probability of finding the ith component of velocity in a small range δu_i about a particular function $u_i(x_i, t)$ is $\mathscr{P}(u_i)\,\delta u_i$, the symbol δu_i denoting the volume element in the function space. In addition to $\mathscr{P}(u_i)$, $\mathscr{P}(\varrho)$, $\mathscr{P}(T)$, etc., there are joint probability-density functionals, such as $\mathscr{P}(\varrho, u_i)$, defined so that $\mathscr{P}(\varrho, u_i)\delta\varrho\delta u_i$ represents the probability of finding the fluid density in a small range $\delta\varrho$ about $\varrho(x_i, t)$ and the ith component of velocity in a small range δu_i about $u_i(x, t)$. From the equations in Sect. 1.6, a functional differential equation may be written for the joint probability-density functional $\mathscr{P}(\varrho, u_1, u_2, u_3, Y_1, Y_2, \ldots, Y_N, T)$, in a manner resembling that indicated by *Dopazo* and *O'Brien* [1.20]. In principle, this equation contains all of the statistical information about the flow, and the problem of describing the turbulent reacting flow reduces to the problem of solving this equation. Unfortunately, methods for solving functional differential equations are not highly developed, and therefore this approach, while direct, is intractable. However, approaches along these lines will be discussed in Chap. 5.

From probability-density functionals can be calculated probability-density functions. A representative probability-density function is $P(u_i)$, defined so that $P(u_i)du_i$ is the probability that the ith component of velocity will lie in a small range du_i about the value u_i at particular, fixed values of x_i and t. In notation that will be employed later, the fact that x_i and/or t are assigned fixed values will be indicated explicitly by placing one or both of them inside the parenthesis of the symbol P, separated from u_i by a semicolon. There are joint probability-density functions, such as $P(\varrho, u_i)$, having $P(\varrho, u_i)d\varrho du_i$ equal to the probability that the fluid density lies in a small range $d\varrho$ about ϱ and the ith component of velocity lies in a small range du_i about u_i, at particular values x_i and t, and $P(u_{i1}, u_{j2})$, having $P(u_{i1}, u_{j2})du_{i1}du_{j2}$ equal to the probability that the ith component of velocity lies in the small range du_{i1} about u_{i1} at particular values x_{i1}, t_1 and the jth component of velocity lies in the small range du_{j2} about u_{j2} at other particular values x_{i2}, t_2. Note that, with these definitions, $P(\varrho, u_i)$ is a one-point, one-time, probability-density function, while $P(u_{i1}, u_{j2})$ is a two-point, two-time, probability-density function. Aspects of behavior in time at a fixed point can be investigated by putting $x_{i1}=x_{i2}$, and aspects of spatial structure at a fixed time can be studied by putting $t_1=t_2$. Multi-point, multi-time, probability-density functions of any order can be generated from the probability-density functional, but none of these can provide complete statistical information about the flow. Lower-order probability-density functions can be derived from higher-order probability-density functions; for example, $P(u_i)=\int_0^\infty P(\varrho, u_i)d\varrho$. Although equations can be derived for any multi-point, multi-time, probability-density function from the equation for the probability-density functional, no one or group of these derived equations is closed without the introduction of additional, ad hoc, statistical hypotheses; that is, there are more unknowns than equations. There is a close analogy here to the Liouville equation of mechanics. Statistical hypotheses that lead to closed equations for probability density functions will be introduced in Chap. 5.

The rest of this chapter is devoted to the study of properties of turbulent reacting flows that can be defined in terms of probability-density functions, to further explanation of the meaning and attributes of probability-density functions, and to investigation of statistical averages of variables defined in Sect. 1.6. It may be remarked here and will be shown more thoroughly later that all averages, including multi-point and/or multi-time averages, can be defined in terms of suitable probability-density functions. These averages usually are the quantities of greatest practical interest, and descriptions of turbulent reacting flows often have been based on conservation equations for these averaged quantities. Just as hypotheses are required to close equations for probability-density functions, so are hypotheses needed to achieve closure of systems of equations for averaged quantities. In turbulence, hypotheses effecting closure are termed models. It is primarily the need for these modeling hypotheses that makes the subject so perplexing; only in limiting cases of relatively little interest, such as flows having low turbulence intensities or low gradients of fluctuating quantities, can closures be justified through perturbation methods rather than methods exhibiting ad hoc character. Since different closure hypotheses lead to different predictions, appreciable uncertainties remain concerning most turbulent reacting flows.

1.8 Time Averaging and Favre Averaging

In many flows, a simplified and more easily accepted view of statistical averages can be developed from the idea of time averaging. In most flow problems involving turbulence and having engineering and applied significance, the information of interest is limited; it is usually sufficient to determine the mean values of the various fluid-mechanical quantities, some measure of the extent of the variations about those means, and some measures of how correlated are the various quantities. Thus, it is customary to decompose the various quantities into mean and fluctuating components. For example, for the velocity components, $u_i(\boldsymbol{x}, t)=\bar{u}_i(\boldsymbol{x}, t)+u_i'(\boldsymbol{x}, t)$, where $\boldsymbol{x}=(x_1, x_2, x_3)$. If the mean $\bar{u}_i$ varies sufficiently slowly with time, then the statistical structure of the turbulence may be nearly stationary, i.e., independent of time. For stationary stochastic processes, means may be calculated as ordinary time averages of the quantities of interest. For purposes of exposition here, we shall assume that the mean is time independent and shall thus write $\bar{u}_i(\boldsymbol{x})$; there is no formal difficulty in introducing slowly varying mean quantities into our considerations.

When the density in the conservation equations described in Sect. 1.6 is constant, time averaging applied to those equations yields a simplified description of turbulent flows. These equations are of interest in the study of turbulent reacting flows under the special circumstances noted earlier and will in fact arise in Chap. 3.

In turbulent combustion the overall effects of the chemical reactions of interest are exothermic, with the consequence that the temperature of the products is generally considerably higher than that of the reactants. Thus there are usually significant inhomogeneities in the density in the flow, and it is therefore necessary to consider the density in the same statistical fashion as the other fluid-mechanical quantities. If this is done, it is unfortunately found that a wide variety of quantities involving density fluctuations occur in the averaged equations. The implication from this situation is that the density plays an important and pervasive role in the dynamics of turbulence; such a conclusion is consistent with experience from gas dynamic phenomena in which density variations on the same order as those arising in combustion play dominant roles in determining flow behavior.

Several modifications of the usual averaging techniques have been suggested in order to simplify the equations applicable to turbulent flows with density variations. The most general is that due to *Favre* [1.21]; high-speed turbulent flows in which the density variations are due to viscous heating provide the original motivation for this averaging technique, but it appears to have utility in any high-Re flow with such variations whether they be due to inhomogeneities in composition or temperature.

In Favre averaging all fluid-mechanical quantities except the pressure are mass averaged. Thus, for example, the mean velocity components are

$$\tilde{u}_i(\boldsymbol{x}) = \overline{\varrho u_i}(\boldsymbol{x})/\bar{\varrho}(\boldsymbol{x}) . \tag{1.22}$$

Note that two symbols are introduced to denote averages; the bar indicates the conventional time average whereas the tilde denotes a mass-averaged quantity. We use a double prime to denote the fluctuations about the mass-averaged mean; thus

$$u_i(\boldsymbol{x}) = \tilde{u}_i(\boldsymbol{x}) + u_i''(\boldsymbol{x}, t) . \tag{1.23}$$

Because of the nature of these fluctuations, their mean value is not zero whereas their mass-averaged value is zero; thus $\overline{\varrho u_i''} = 0$.

It is illuminating to consider the consequence of applying the two means of averaging to a typical nonlinear term arising from the convective terms in the conservation equations. Consider

$$\overline{\varrho u_i u_j} = \bar{\varrho}\bar{u}_i\bar{u}_j + \bar{\varrho}\overline{u_i' u_j'} + \bar{u}_i\,\overline{\varrho' u_j'} + \bar{u}_i\,\overline{\varrho' u_j'} + \overline{\varrho' u_i' u_j'} = \bar{\varrho}\tilde{u}_i\tilde{u}_j + \overline{\varrho u_i'' u_j''} . \tag{1.24}$$

The first terms on the right sides are the mean quantities, usually retained on the left side of the conservation equations. The term $\overline{\varrho u_i'' u_j''}$ resembles $\bar{\varrho}\overline{u_i' u_j'}$, and in the Favre formulation the three terms involving the density fluctuations do not appear. The terms $\bar{u}_i \overline{\varrho' u'_j}$ and $\bar{u}_j \overline{\varrho' u'_i}$ are said by *Morkovin* [1.22], in the course of his discussion of high-speed turbulent boundary layers, to be mass transfer terms since they describe a momentum exchange due to an interaction between the mean velocity and volumetric changes. The triple correlation $\overline{\varrho' u'_i u'_j}$ represents such an exchange due to fluctuation-fluctuation interaction. The most important implication to be derived from (1.24) is that Favre averaging automatically incorporates these various modes of momentum exchange into a smaller number of terms.

Similar considerations apply to other nonlinear terms arising in the conservation equations. A potential disadvantage of Favre averaging arises from the molecular transport terms, e.g., from the stress tensor τ_{ij}, which becomes more complicated, rather than simpler, in the Favre notation. In effecting closure, modeling hypotheses often are introduced for these molecular terms, and the modeling may differ according to the two means of averaging. Therefore, the modeling of molecular transport provides no good basis for selecting one method of averaging over another, and the relative simplicity of the turbulent transport in the Favre formulation makes this mass-average approach preferable. Problems also may arise in the chemical production term (see Chap. 5).

The experimental determination of mass-averaged quantities requires in principle the measurement at a given point in space of the instantaneous density as well as the variable being averaged. This usually means that the pressure must be measured or assumed constant for purpose of the density determination; then from (1.20) measurement of the instantaneous temperature $T(\boldsymbol{x}, t)$ and mass fractions $Y_i(\boldsymbol{x}, t)$, $i=1, 2, \ldots, N$ yield $\varrho(\boldsymbol{x}, t)$. Correlation of the density with the variable being averaged and subsequent division by $\bar{\varrho}(\boldsymbol{x})$ yield the desired average. The implication from this discussion is that the experimental determination of Favre-averaged variables often is difficult. Accordingly, it may be necessary in order to carry out comparison between the predictions of a calculation based on Favre averaging and an experiment to relate the two averages. This is always possible but leads to correlations of density fluctuations and the conventional fluctuations of the variable in question; for example,

$$\tilde{u}_i(\boldsymbol{x}) = \bar{u}_i(1 + \overline{\varrho' u'_i}/\bar{\varrho}\bar{u}_i)\,. \tag{1.25}$$

Thus we see that before the comparison can be made some estimate or a measurement of the second term on the right side must be made.

1.9 Probability-Density Functions

The restriction to statistical stationarity may be employed to present a more intuitive view of probability-density functions than that introduced in Sect. 1.7. Suppose that the pressure is a known constant and that suitable instrumentation[1] is available for measurement, with a certain repetition rate, of the instantaneous values of the three velocity components u_i, $i=1, 2, 3$, the temperature, and the mass fractions Y_i, $i=1, 2, \ldots, N$. We then have $N+4$ time series, and from (1.20) we may generate an additional time series in the density. Time series of data obtained in this fashion can be used to develop statistics of the various variables and of their correlations. Thus, for example, we can directly compute the conventional and Favre-averaged velocity components $\bar{u}_i$ and $\tilde{u}_i$ and the correlations $\overline{u_i'u_j'}$ and $\overline{\varrho u_i''u_j''}$. In addition, from these data certain probability-density functions also can be obtained.

An example is the probability-density function for the ith component of velocity, $P(u_i;\boldsymbol{x})$, which has the properties

$$\int_{-\infty}^{\infty} P(u_i;\boldsymbol{x})du_i=1\,, \tag{1.26}$$

$$\int_{-\infty}^{\infty} u_i P(u_i;\boldsymbol{x})du_i=\bar{u}_i(\boldsymbol{x})\,, \tag{1.27}$$

and

$$\int_{-\infty}^{\infty} u_i^2 P(u_i;\boldsymbol{x})du_i=\overline{u_i^2}(\boldsymbol{x})=\bar{u}_i^2(\boldsymbol{x})+\overline{u_i'^2}(\boldsymbol{x})\,. \tag{1.28}$$

Observe that, from formulas of this type, the nth moment of the fluctuation of the ith component of velocity, $\overline{u_i'^n}(\boldsymbol{x})$, can be calculated from $P(u_i;\boldsymbol{x})$ at the point $\boldsymbol{x}$. In particular, the variance $\overline{u_i'^2}(\boldsymbol{x})$, the skewness $\overline{u_i'^3}(\boldsymbol{x})/[\overline{u_i'^2}(\boldsymbol{x})]^{3/2}$, and the kurtosis $\overline{u_i'^4}(\boldsymbol{x})/[\overline{u_i'^2}(\boldsymbol{x})]^2$ (a measure of the curvature or flatness of the probability-density function) may be obtained. In this regard we note that the extent of the time series used to develop $P(u_i;\boldsymbol{x})$ determines the statistical accuracy of the results; in particular if the higher moments are desired, sufficient entries must be available to provide accurately the "tails" of the distribution.

The physical significance of $P(u_i;\boldsymbol{x})$ can be understood as follows: Consider a narrow range of velocities δ centered about a nomial velocity denoted by u_i^*; then $P(u_i^*;\boldsymbol{x})\delta$ is the fraction of time the velocity component u_i is within the range $u_i^*-\frac{1}{2}\delta\leqq u_i\leqq u_i^*+\frac{1}{2}\delta$. This interpretation suggests the basis for determining experimentally $P(u_i;\boldsymbol{x})$ from the time series in u_i; the entire range of experimentally observed values of u_i is divided into equal intervals, termed "bins", and the time series is scanned. The number of entries in each bin, divided by the total number of entries in the time series, yields a discretized probability-density function. The selection of a bin size involves a compromise between resolution on the one hand and the number of entries in each bin on the other.

1 Although somewhat beyond the capability of present instrumentation, the experiment we consider is within the spirit of present methods in experimental turbulence and is the goal of much current research in combustion diagnostics.

Since Favre-averaged statistics are of importance, it is useful to extend these ideas to multivariable or joint probability density functions. From the time series for u_i and for the density ϱ, the function $P(\varrho, u_i; \boldsymbol{x})$ may be formed, with the properties

$$\int_0^\infty \int_{-\infty}^\infty P(\varrho, u_i; \boldsymbol{x}) du_i d\varrho = 1, \tag{1.29}$$

$$\int_0^\infty \int_{-\infty}^\infty \varrho P(\varrho, u_i; \boldsymbol{x}) du_i d\varrho = \bar{\varrho}, \tag{1.30}$$

$$\int_0^\infty \int_{-\infty}^\infty \varrho u_i P(\varrho, u_i; \boldsymbol{x}) du_i d\varrho = \overline{\varrho u_i}(\boldsymbol{x}) = \bar{\varrho}\tilde{u}(\boldsymbol{x}), \tag{1.31}$$

and

$$\int_0^\infty \int_{-\infty}^\infty \varrho u_i^2 P(\varrho, u_i; \boldsymbol{x}) du_i d\varrho = \overline{\varrho u_i^2}(\boldsymbol{x}) = \bar{\varrho}\tilde{u}_i^2(\boldsymbol{x}) + \overline{\varrho u_i''^2}(\boldsymbol{x}). \tag{1.32}$$

Again, repeated weighing of the two integrals leads to all Favre-averaged statistics involving u_i and the density at $\boldsymbol{x}$. The earlier physical interpretation and the discussion of the means for obtaining this one-variable probability-density function are readily extended to this two-variable case. However, the extent of the time series required to achieve statistical reliability is significantly increased in going from one to two variables.

Bilger [1.23] introduced the idea of a Favre probability-density function defined, for example, as $\tilde{P}(u_1; \boldsymbol{x}) = \int_0^\infty \varrho P(\varrho, u_1; \boldsymbol{x}) d\varrho / \bar{\varrho}$. Application to (1.29–32) reduces the determination of Favre-averaged statistics of u_i to the same form as that of the conventional averages given in (1.26–28).

The probability-density functions discussed here are special cases of more general, multivariable probability-density functions. From the time series we have imagined to be collected at the particular spatial point in the flow, we can in principle form the multivariable probability-density function $P(u_1, u_2, u_3, \varrho, T, Y_1, Y_2, \ldots, Y_N; \boldsymbol{x})$. This function provides most of the statistical information of interest regarding the flow at $\boldsymbol{x}$, although because of the method of data collection, spectral information, discussed in the following section, is lost. As an example of the application of this function, consider the determination of the mean flux $\overline{\varrho u_1'' u_2''}$; from the definitions,

$$\overline{\varrho u_1'' u_2''} = \int_0^1 \ldots \int_0^1 \int_0^1 \int_0^\infty \int_0^\infty \int_{-\infty}^\infty \int_{-\infty}^\infty \int_{-\infty}^\infty \varrho(u_1 - \tilde{u}_1)(u_2 - \tilde{u}_2)$$
$$\cdot P(u_1, u_2, u_3, \varrho, T, Y_1, Y_2, \ldots, Y_N) du_1 du_2 du_3 d\varrho dT dY_1 dY_2 \ldots dY_N. \tag{1.33}$$

Because of the utility of multivariable probability-density functions, there has been considerable effort to develop theories for their direct calculation; the present status of these efforts is reviewed in Chap. 5. At present it must be admitted that we are far from a theory permitting direct calculation of the probability-density function used formally in (1.33). Nevertheless, notions based on probability-density considerations are frequently invoked to guide the reasoning and modeling used in the development of approximate methods of analysis for turbulent reacting flows. We shall see examples of such notions later in this chapter and in the chapters which follow.

It is important to recognize an essential difference in the ranges applicable to the various variables appearing in the multivariable probability-density functions; for example, the velocity components do not involve sharp bounds, but by their nature the mass fractions are restricted to the range zero to unity. A consequence of these different behaviors is that for some variables Gaussian distributions, with their attractive analytic properties, are appropriate, while they are generally inappropriate for quantities such as mass fractions and other variables with sharp bounds.

1.10 Intermittency

Sharp limits on the range of values for some of the variables of interest have important consequences on the probability-density functions in reacting flows involving fast chemistry and in flows involving intermittency. Here we focus on intermittency since it is a feature in many turbulent flows and since it plays an important role in the behavior of such flows, whether they involve chemical reactions or not.

The phenomenon of intermittency is illustrated in Fig. 1.1, which is taken from *Fernholz* [1.24]; shown is a turbulent boundary layer with the turbulent fluid tagged with smoke and with the contiguous, irrotational fluid clear. We see that the boundary between the two fluids is highly convoluted. The low Reynolds number of the experiment in this photograph leads to a more elaborate structure than found in flows of more practical interest, but the three-dimensional nature of the oscillating boundary and the sharp edges of that boundary are found in all flows involving intermittency.

Measurements of the properties of the flow at a point involving intermittency result in probability-density functions which reflect the intermittency. The possibility of discriminating between passages of turbulent and irrotational fluid has led to a technique of experimental turbulence termed conditioned sampling and has resulted in significant increases in our understanding of the physics of turbulence. *Van Atta* [1.25], *Gibson* [1.26], and *Kovasznay* et al. [1.27] should be consulted in this regard.

To understand the significance of intermittency relative to probability-density functions, consider a simple turbulent flow without reaction, namely a

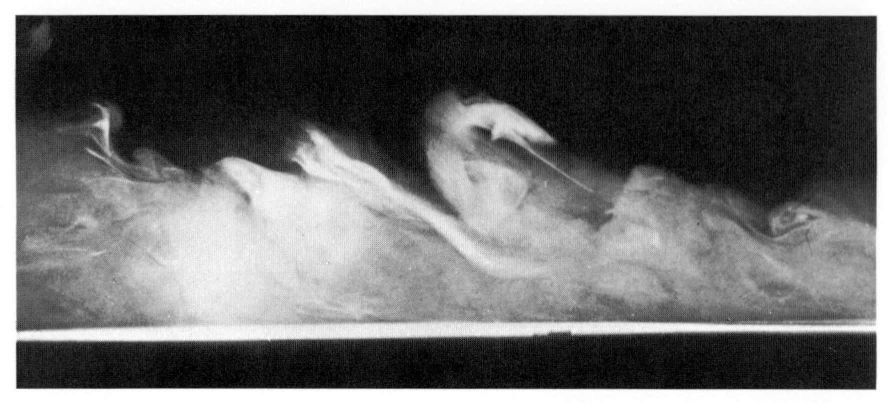

Fig. 1.1. Visualization of a boundary layer by means of smoke. (Photograph courtesy of P. Bradshaw)

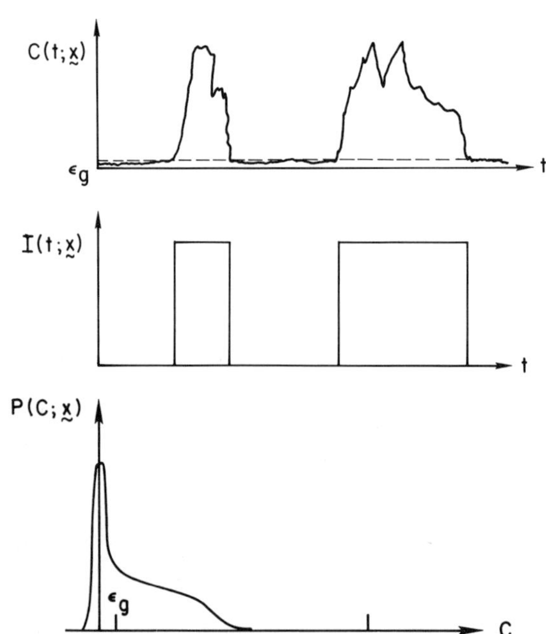

Fig. 1.2. Schematic representation of the turbulent signal from a sensor yielding a scalar and of the intermittency function and probability density function obtained therefrom

two-dimensional jet of a foreign species discharging into a moving airstream. Let the mass fraction of the injected species be denoted by $C(\boldsymbol{x}, t)$ and imagine that we are able to measure, at a fixed spatial point in the flow wherein intermittency occurs, the concentration as a function of time. The history $C(\boldsymbol{x}, t)$ is shown schematically in Fig. 1.2; we see that periods with $C \cong 0$ are interspersed between periods with $C > 0$. The former correspond to intervals when the point is in the external stream whereas the latter correspond to intervals when the point is within the turbulent portion of the flow. Because of inaccuracies in measurement, it is necessary to identify turbulent fluid with $C > \varepsilon_g$ where ε_g is a suitable chosen level or "gate".

With a specified gate the data shown schematically in Fig. 1.2 can be used to establish an intermittency function, defined by

$$I(\boldsymbol{x},t)=\begin{cases}0 & \text{if} \quad C<\varepsilon_g \\ 1 & \text{if} \quad C>\varepsilon_g .\end{cases} \tag{1.34}$$

The mean value of the intermittency function, i.e., $\bar{I}(\boldsymbol{x})$ for stationary flows, is called the intermittency and represents the fraction of the time that the flow is turbulent at the point in question. The intermittency function is important as a basis for conditioned sampling; i.e., the correlation of $I(\boldsymbol{x},t)$ with another variable, e.g., $u_i(\boldsymbol{x},t)$, yields the average of the u_i-velocity component within the turbulent fluid alone. Furthermore, it is possible to use the times when the intermittency function changes value, i.e., from zero to unity and from unity to zero, to identify crossings into and out of turbulence, respectively, and to condition data relative to such times.

The probability-density function $P(C;\boldsymbol{x})$ is also shown in Fig. 1.2. We indicate a spike in the neighborhood of $C=0$ but reflect the reality of the inaccuracies associated with collecting the data by indicating a structure to that spike and in fact negative values of C. In theoretical analyses employing probability-density functions, this observed distribution is idealized; all entries for $C<\varepsilon_g$ are attributed to a delta function at $C=0$ with a strength equal to $1-\bar{I}(\boldsymbol{x})$.

The implication from these considerations is that the multivariable probability-density function used, for example, in (1.33) can, for a point involving intermittency, be decomposed into at least two separate probability-density functions. Suppose that the point in question involves an external, irrotational flow of uniform composition and turbulent fluid with all species present in various amounts. In this case we can write

$$\begin{aligned}P(u_1,u_2,u_3,\varrho,T,Y_1,Y_2,\ldots,Y_N;\boldsymbol{x})&=[1-\bar{I}(\boldsymbol{x})]P_0(u_1,u_2,u_3;\boldsymbol{x})\\&\cdot\delta(\varrho-\varrho_0)\delta(T-T_0)\delta(Y_1-Y_{10})\delta(Y_2-Y_{20})\ldots\delta(Y_N-Y_{N0})\\&+\bar{I}(\boldsymbol{x})P_1(u_1,u_2,u_3,\varrho,T,Y_1,Y_2,\ldots,Y_N;\boldsymbol{x}),\end{aligned} \tag{1.35}$$

where the subscript 0 denotes conditions in the external flow. Each probability-density function in (1.35), the irrotational component P_0 and the rotational component P_1, can be normalized to unity. When (1.35) is introduced into (1.33) it is found that

$$\overline{\varrho u_1'' u_2''}=[1-\bar{I}(\boldsymbol{x})](\overline{\varrho u_1'' u_2''})_0+\bar{I}(\boldsymbol{x})(\overline{\varrho u_1'' u_2''})_1 . \tag{1.36}$$

The correlation on the left is the so-called unconditioned correlation; those on the right are conditioned by the intermittency function.

A further implication of the reality of intermittency is that modeling guided by probability-density considerations should reflect the intermittent nature of the flow. At the present time the strengths of the delta functions must be assumed to be given by experimental data on intermittency, although there exist some early attempts at the calculation of intermittency (cf. [1.28–32]).

1.11 Fourier Decomposition

For statistically stationary flows, treatment of time series as described in Sect. 1.9 does not provide information on multi-time averages, such as the autocorrelation $\overline{u_i'(\mathbf{x}, t_1)u_i'(\mathbf{x}, t_2)}$, for $t_1 \neq t_2$. Information of this type sometimes can be useful in understanding the dynamics of turbulent reacting flows. Since stationarity implies that the stated average depends on t_1 and t_2 only through the time difference $\tau = t_2 - t_1$, a slightly more complicated averaging procedure may provide this quantity from the given time series. Also, through a suitable selection of bins the stated time series can provide, in principle, the one-point, two-time, joint probability-density function $P(u_{i1}, u_{i2})$, with $x_{i1} = x_{i2}$, as defined in Sect. 1.7. Methods of Fourier transforms often have been applied in studying the behavior with τ of such quantities. Some discussion of these Fourier decompositions is desirable here because they have been applied in various forms in theoretical analyses of turbulent reacting flows [1.33].

The Fourier transform of a function $f(\tau)$ is

$$F(\omega) = \int_{-\infty}^{\infty} \mathrm{e}^{\mathrm{i}\omega\tau} f(\tau) d\tau\,, \tag{1.37}$$

and the function $f(\tau)$ can be recovered from $F(\omega)$ by the inverse transform

$$f(\tau) = \frac{1}{2\pi} \int_{-\infty}^{\infty} \mathrm{e}^{-\mathrm{i}\omega\tau} F(\omega) d\omega\,. \tag{1.38}$$

There are mathematical restrictions on $f(\tau)$ for $F(\omega)$ to exist and for (1.38) to apply. In turbulence these restrictions often are circumvented by going to the more general Fourier-Stieltjes transform, but if a sufficiently nonanalytic, delta-function type of behavior for the transforms is allowed, then this is unnecessary. Note that $f(\tau)$ and complex-valued $F(\omega)$ are complementary, in that either one can be generated from the other.

The most fundamental level at which transforms have appeared is in the equation for the probability-density functional [1.20]. In this case the transforms are taken with respect to the velocity components and result in quantities termed characteristic functions, from which probability-density functions can be recovered. At a less fundamental level, transforms often are useful in analyses by perturbation methods (e.g., [1.33]). In general, transforms produce simplification by replacing differentiation with multiplication by the transform variable. For the Fourier decomposition in time to be useful in this manner, however, it is necessary that the process be statistically stationary; otherwise there is an evolution in time typically involving initial conditions that are not readily handled by Fourier methods. Correspondingly, Fourier transforms in space often are introduced, and here homogeneity in the spatial coordinates to be transformed is required.

Fourier decompositions in space have been used extensively in studying homogeneous turbulence [1.5]. In this case, instead of a frequency ω, the transform variable is a wave-number vector $\boldsymbol{k}$ that occurs in a factor of the type $\exp(i\boldsymbol{k}\cdot\boldsymbol{x})$. In transforms of the Navier-Stokes equations, spatial derivatives are replaced by multiplications by $\boldsymbol{k}$. Studies of the transformed equations have resulted in physical interpretations of the dynamics of evolution of turbulence in wave-number space. One such interpretation is the cascade idea; at high Re large eddies are formed which break up into smaller eddies, and this is described by a progression, or cascade, of the turbulence from small to large wave numbers in the transform space. Inertial forces dominate most of the cascade, with viscous forces rapidly taking energy out of the turbulence at a high wave number (small scale) characterized by a length attributed to Kolmogorov. These cascade ideas often underlie approaches to closure of moment equations for describing turbulent reacting flows.

A quantity often considered in studies of homogeneous turbulence is the velocity correlation tensor $R_{ij}(\boldsymbol{x},t)=\overline{u_i'(\boldsymbol{x}_1,t)\,u_j'(\boldsymbol{x}_2,t)}$, where $\boldsymbol{x}=\boldsymbol{x}_2-\boldsymbol{x}_1$. This two-point quantity, analogous to the two-time autocorrelation defined at the beginning of this section, involves more statistical information than one-point quantities, such as the moment $\overline{u_i'u_j'}$, but much less than other quantities, such as the two-point, joint, probability-density function. Study of $R_{ij}(\boldsymbol{x})$ is helpful in understanding turbulent flows [1.3–5]. However, it is more difficult to measure than a one-point, two-time quantity because it requires time series at two different points, for various values of the distance between these two points. It often has been obtained from one-point measurements by introduction of a Taylor hypothesis, which states that the turbulence is convected with negligible change past the measurement station at the average velocity $\bar{\boldsymbol{u}}$, viz., $u_i'(\boldsymbol{x}_1,t_1)=u_i'(\boldsymbol{x}_2,t_2)$, where $\boldsymbol{x}_2=\boldsymbol{x}_1+\bar{\boldsymbol{u}}(t_2-t_1)$. There are restrictions to the use of such a hypothesis, and these restrictions may be especially severe in reacting flows, where the chemistry may cause large changes in turbulence over short distances due to modifications arising from rapid release of heat.

There are quantities analogous to $R_{ij}(\boldsymbol{x})$ for scalar fields such as temperature and chemical composition. Examples are the autocorrelation $\overline{T'(\boldsymbol{x}_1,t)\,T'(\boldsymbol{x}_2,t)}$ and the cross correlation $\overline{T'(\boldsymbol{x}_2,t)\,Y_i'(\boldsymbol{x}_2,t)}$. A few studies have been undertaken of the effects of finite-rate chemistry on these correlations for systems in which the influence of the chemistry on the velocity field is negligible (e.g., [1.34]). The chemistry affects the evolution of these correlations in different ways for different types of reactions.

In homogeneous turbulence, analyses can be facilitated by introducing Fourier decompositions of correlations. These decompositions are termed spectra and are simpler than decompositions of the velocity field per se, for example. An illustration is the spectrum of the velocity correlation tensor,

$$\Phi_{ij}(\boldsymbol{k},t)=\int\limits_{-\infty}^{\infty}\int\limits_{-\infty}^{\infty}\int\limits_{-\infty}^{\infty} e^{i\boldsymbol{k}\cdot\boldsymbol{x}}R_{ij}(\boldsymbol{x},t)dx_1dx_2dx_3, \tag{1.39}$$

where $\boldsymbol{x}=(x_1,x_2,x_3)$ and $\boldsymbol{k}=(k_1,k_2,k_3)$. The spectrum of kinetic energy is the decomposition $\Phi_{kk}(\boldsymbol{k},t)$. In isotropic turbulence, at a given t this function depends on a single variable, the magnitude $|\boldsymbol{k}|=\sqrt{k_1^2+k_2^2+k_3^2}$, since $R_{kk}(\boldsymbol{x},t)$ depends only on $|\boldsymbol{x}|$. Cascade concepts have given the shape of the function $\Phi_{kk}(|\boldsymbol{k}|,t)$, as well as shapes of spectra of correlations of scalar quantities, e.g.,

$$\psi(|\boldsymbol{k}|,t)=\int\limits_{-\infty}^{\infty}\int\limits_{-\infty}^{\infty}\int\limits_{-\infty}^{\infty} e^{i\boldsymbol{k}\cdot\boldsymbol{x}}\,\overline{T'(\boldsymbol{x}_1,t)T'(\boldsymbol{x}_2,t)}\,dx_1dx_2dx_3, \tag{1.40}$$

under conditions such that the scalar is passive, i.e., it exerts no influence on turbulent velocities (e.g., [1.35]). A few studies have been made of the influence of chemistry on spectra of passive scalars.

The inertial and convective terms in the conservation equations are nonlinear. Therefore, when transforms of these equations are taken there arise interactions between components of different wave numbers. The simplest illustration of this is the convolution theorem with time as the independent variable, written in a notation like that of (1.38) and (1.39), viz.,

$$\int\limits_{-\infty}^{\infty} e^{i\omega\tau}f_1(\tau)f_2(\tau)d\tau=2\pi\int\limits_{-\infty}^{\infty} F_1(\xi)F_2(\omega-\xi)d\xi. \tag{1.41}$$

The right-hand side of (1.41) introduces contributions at frequency ω from values of the transforms F_1 and F_2 at all frequencies. This mixing of spectral transfer in frequency or wave-number space gives rise to the cascade process, which occurs at high Re for the energy spectrum $\Phi_{kk}(|\boldsymbol{k}|,t)$ as well as for spectra of scalars. If one reactant experiences a first-order reaction with a negligible temperature dependence, then the chemical production term is linear and does not generate spectral transfer for $\psi(|\boldsymbol{k}|,t)$. Instead, this spectral intensity merely changes exponentially with time, at a rate independent of $|\boldsymbol{k}|$, at each value of $|\boldsymbol{k}|$ as a consequence of the reaction.

However, chemistry usually is not that simple; the production term is often nonlinear in concentrations and usually highly nonlinear in temperature, as discussed in Sect. 1.5. Under these conditions the production term produces spectral transfers of the spectral intensity $\psi(|\boldsymbol{k}|,t)$, thereby modifying cascade aspects of the spectrum. It may be noted that these modifications stem only from the nonlinearity, not from multi-point aspects of the process that arise from terms involving derivatives in the conservation equations. Relatively little study has been given to such influences of chemistry on spectral dynamics. Accordingly, deductions of the behavior of spectra in turbulent reacting flows on the basis of classical spectral theory may be in error.

1.12 Favre-Averaged Conservation Equations

A less esoteric level of description of turbulent reacting flows employs the lowest moments of one-point, one-time probability-density functions. Conservation equations for these moments are derived by taking suitable averages of the equations of Sect. 1.6. Closure is effected by modeling, and the equations then are solved for flows of interest. This is the principal approach found in the literature on turbulent reacting flows. Therefore it is important to present here the equations that are used. Because of the advantages of Favre averaging, discussed in Sect. 1.8, the Favre-averaged forms of the conservation equations in Sects. 1.4, 1.5, and 1.6 are considered. Attention still is restricted to statistically stationary turbulent flows.

The simplest equation is that for overall mass conservation. With Favre averaging, (1.8) becomes

$$\frac{\partial}{\partial x_k}(\bar{\varrho}\tilde{u}_k)=0, \tag{1.42}$$

whereas the equations for momentum conservation (1.9) become

$$\frac{\partial}{\partial x_k}(\bar{\varrho}\tilde{u}_k\tilde{u}_i)=-\frac{\partial\bar{p}}{\partial x_i}+\bar{g}_i+\frac{\partial}{\partial x_k}(\bar{\tau}_{i_k}-\overline{\varrho u_k'' u_i''}). \tag{1.43}$$

It will be sufficient for our purposes and will facilitate exposition to assume simplified transport properties and deal only with Favre-averaged forms of the energy equations given by either (1.15) or (1.16) with $c_{p_i}=c_p$ and $\sigma=1$

$$\frac{\partial}{\partial x_k}(\bar{\varrho}\tilde{u}_k\tilde{h})=\frac{\partial}{\partial x_k}\left(\overline{\mu\frac{\partial h}{dx_k}}-\overline{\varrho u_k'' h''}\right) \tag{1.44}$$

or

$$c_p\frac{\partial}{\partial x_k}(\bar{\varrho}\tilde{u}_k\tilde{T})=\frac{\partial}{\partial x_k}\left[c_p\left(\overline{\mu\frac{\partial T}{\partial x_k}}-\overline{\varrho u_k'' T''}\right)\right]-\sum_{i=1}^{N}\overline{h_i\dot{w}_i}. \tag{1.45}$$

Species conservation (1.17) becomes, according to Favre averaging,

$$\frac{\partial}{\partial x_k}(\bar{\varrho}\tilde{u}_k\tilde{Y}_i)=\frac{\partial}{\partial x_k}\left(\overline{\varrho\mathscr{D}\frac{\partial Y_i}{\partial x_k}}-\overline{\varrho u_k'' Y_i''}\right)+\bar{\dot{w}}_i, \tag{1.46}$$

and element conservation (1.19) becomes

$$\frac{\partial}{\partial x_k}(\bar{\varrho}\tilde{u}_k\tilde{Z}_i)=\frac{\partial}{\partial x_k}\left(\overline{\varrho\mathscr{D}\frac{\partial Z_i}{\partial x_k}}-\overline{\varrho u_k'' Z_i''}\right). \tag{1.47}$$

In (1.43–47) the terms on the left sides represent the mean convection of the various quantities being conserved. Also clearly identifiable are the various molecular terms which are left in primitive form for the time being because Favre averaging does not help to simplify them. Each of these equations also includes a turbulent flux term representing the exchange due to fluctuations; these are the manifestations of the closure problem, and we must deal with them in some fashion to be discussed later and in subsequent chapters. Finally, we see in (1.46) that the influence of chemical reaction on the mean species concentration $\tilde{Y}_i$ appears as the time average of the instantaneous rate of production $\bar{w}_i$.

The equation of state (1.20) upon averaging leads to

$$\bar{p}=R_0\sum_{i=1}^{N}(\bar{\varrho}\tilde{T}\tilde{Y}_i+\overline{\varrho T'' Y_i''})\frac{1}{W_i}. \tag{1.48}$$

When the molecular weights of the participating species do not differ significantly, (1.48) takes on the simpler form

$$\bar{p}=\bar{\varrho}R\tilde{T}. \tag{1.49}$$

1.13 Closure and Additional Transport Equations

Some general considerations regarding closure are of interest in connection with conservation equations for various statistical averages. The simplest solution to the closure problem involves the direct assumption of gradient transport in the equations of the previous section. Such an assumption, usually made in connection with time-averaged quantities, is readily assumed to apply for Favre averaging instead. For the flux of a scalar quantity q, gradient transport assumes that $\overline{\varrho u_i'' q''}=-\bar{\varrho}\nu_{\mathrm{T}}\partial\tilde{q}/\partial x_i$, while for the component of a vector quantity, e.g., u_j, the flux is $\overline{\varrho u_i'' u_j''}=-\bar{\varrho}\nu_{\mathrm{T}}(\partial\tilde{u}_i/\partial x_j+\partial\tilde{u}_j/\partial x_i)$. In both expressions ν_{T} is a turbulent exchange coefficient (not necessarily the same in both cases), an expression for which is to be obtained through modeling in some fashion. More sophisticated closure methods involve development of conservation equations for the various fluxes, which are higher moments of probability density functions. Closure of these so-called second-order equations with models usually is based on gradient transport hypotheses for the various higher-order terms. Thus we see that gradient transport usually enters in one form or another, irrespective of the number of moment equations retained.

The basis of gradient transport approximations resides in small length scales associated with the turbulent fluctuations relative to the length scales of the mean flow (cf., e.g., [1.36]). Unfortunately, this situation seldom prevails in turbulent flows of practical interest, and consequently there is no good basis for the most common closure scheme. Therefore, it is remarkable that theories based on gradient transport predict with reasonable accuracy many quantities of applied interest at least in constant-density flows. The reason for this forgiveness is unclear.

There is some controversy concerning the validity of current second-order closure methods; there arise in these methods a large number of empirical constants which can be adjusted to assure agreement with experiment. Of course, adjustment of the constants from one experiment to another is contrary to the notion of requiring these empirical constants to be "universal". Despite current skepticisms concerning the possibility of achieving universality, the additional physical content of the equations of second-order closures has led to improved methods of describing turbulent flows and can be expected to lead to worthwhile developments in the future.

For turbulent flows with constant density, e.g., for reacting flows with highly dilute reactants in a uniform background gas, the problem of closure involves separately fluid mechanical and chemical aspects of the flow. The fluid mechanical problem is essentially the same as for turbulent flows without chemical reaction and the discussions of *Reynolds* [1.10], *Spalding* [1.14], and others apply.

Essential difficulties arise when these methods are applied to more general turbulent reacting flows. The first trouble encountered is the significant density variation which arises from the heat release in many cases of practical interest. The possible alterations of the modeling used to effect closure, whether simple or sophisticated, must be taken into account, but currently the extent and nature of such alterations are unknown. A healthy skepticism as to the validity of carrying over to reacting flows the models validated for constant density flows is called for although seldom raised in the literature. The second difficulty, which we shall discuss in more detail later, relates to the mean chemical production terms. These involve explicitly the influence of inhomogeneities of the state variables (density, temperature, and composition) on the effective rates of chemical reaction. We shall see that these turbulence effects sometimes hinder and sometimes enhance chemical activity.

It is of interest to exhibit some of the equations involved in second-order closure. Suppose, for example, we seek a transport equation for the mean momentum flux $\overline{\varrho u_i'' u_j''}$ such as appears in (1.43); if (1.9) is multiplied by u_j and (1.9) written for u_j is multiplied by u_i, and if the resulting equations are averaged and added, it is possible by use of (1.8) to establish that

$$\frac{\partial}{\partial x_k}\overline{\varrho u_k u_i u_j} = -\left(\overline{u_j \frac{\partial p}{\partial x_i} + u_i \frac{\partial p}{\partial x_j}}\right) + \overline{u_j g_i} + \overline{u_i g_j} + \left(\overline{u_j \frac{\partial \tau_{ik}}{\partial x_k} + u_i \frac{\partial \tau_{jk}}{\partial x_k}}\right). \tag{1.50}$$

If now the velocity components on the left are decomposed into their mass-averaged and fluctuating components, and if (1.43) for u_i and u_j is taken into account, then the equation

$$\frac{\partial}{\partial x_k}\left(\bar{\varrho}\tilde{u}_k \frac{\overline{\varrho u_i'' u_j''}}{\bar{\varrho}}\right) = -\overline{\varrho u_k'' u_j''}\frac{\partial \tilde{u}_i}{\partial x_k} - \overline{\varrho u_k'' u_i''}\frac{\partial \tilde{u}_j}{\partial x_k} - \frac{\partial}{\partial x_k}\overline{\varrho u_k'' u_i'' u_j''}$$
$$-\left(\overline{u_j''\frac{\partial p}{\partial x_i}} + \overline{u_i''\frac{\partial p}{\partial x_j}}\right) + \overline{u_j'' g_i} + \overline{u_i'' g_j}$$
$$+\overline{u_j''\frac{\partial \tau_{ik}}{\partial x_k}} + \overline{u_i''\frac{\partial \tau_{jk}}{\partial x_k}} \tag{1.51}$$

results. If the indices i and j are equal, then (1.51) is an equation for the mass-averaged intensity of the ith velocity fluctuations. If we let i and j be equal to l and then sum over l, we obtain the transport equation for twice the Favre-averaged turbulent kinetic energy, $\tilde{q} = \frac{1}{2}\overline{\varrho u'' u''}/\bar{\varrho}$; this is an important equation in turbulence phenomenology and thus is written here explicitly:

$$\frac{\partial}{\partial x_k}\bar{\varrho}\tilde{u}_k\tilde{q} = -\overline{\varrho u_k'' u_l''}\frac{\partial \tilde{u}_l}{\partial x_k} - \frac{1}{2}\frac{\partial}{\partial x_k}\overline{\varrho u_k'' u_l'' u_l''}$$
$$-\overline{u_l''\frac{\partial p}{\partial x_l}} + \overline{u_l'' g_l} + \overline{u_l''\frac{\partial \tau_{lk}}{\partial x_k}}. \tag{1.52}$$

It is illuminating to consider the physical content of the various terms on the right side of (1.51); similar remarks pertain to (1.52). The first two terms represent the effect of interaction between the fluctuations as contained in the flux terms and gradients in the mean velocities. The third term describes triple correlations, fluxes of the transported quantity $\varrho u_i'' u_j''$, clearly calling for some modeling. The fourth term represents interaction of velocity fluctuations and the pressure. Body-force effects are contained in the next two terms. The last two terms are special in that they involve molecular effects via the viscosity coefficient μ. Consider the first of these two terms and write

$$\overline{u_j''\frac{\partial \tau_{ik}}{\partial x_k}} = \frac{\partial}{\partial x_k}\overline{u_j''\tau_{ik}} - \overline{\tau_{ik}\frac{\partial u_j''}{\partial x_k}}. \tag{1.53}$$

In turbulent flows at high Re, terms involving the viscosity coefficient can only be significant if large gradients are also involved. Accordingly, the first term in (1.53), containing as it does the gradient of mean quantities, is overwhelmed by the second term; moreover, the second term is dominated by velocity fluctuations with small wave length, i.e., high frequency, as discussed in Sect. 1.11. These considerations indicate the existence of multiple length scales in the transport equations for turbulent flows; these will be discussed later. For the moment it is simply noted that when these considerations are applied to the last term on the right side of (1.52), the result is a negative-definite quantity, termed the velocity dissipation, containing viscous effects which diminish the turbulent kinetic energy.

By the same approach that produced (1.51), transport equations for the flux terms in (1.44–46) can be established; we give one further example. If (1.9) and (1.17) with the index i set to j are considered, then it is found that

$$\frac{\partial}{\partial x_k}\left(\bar{\varrho}\tilde{u}_k \frac{\overline{\varrho u_i'' Y_j''}}{\bar{\varrho}}\right) = -\overline{\varrho u_k'' Y_j''}\frac{\partial \tilde{u}_i}{\partial x_k} - \overline{\varrho u_k'' u_i''}\frac{\partial \tilde{Y}_j}{\partial x_k} - \frac{\partial}{\partial x_k}\overline{\varrho u_i'' u_k'' Y_j''}$$
$$- \overline{Y_j''\frac{\partial p}{\partial x_i}} + \overline{Y_j'' g_i} + \overline{u_i'' \dot{w}_j} + \mu\overline{\frac{\partial u_i}{\partial x_k}\frac{\partial Y_j}{\partial x_k}}. \tag{1.54}$$

The last term on the right side of (1.54) is a dissipation term; we call attention to it because in the phenomenology of constant-density turbulence it is frequently neglected on the basis of the following argument: As indicated earlier, viscous terms such as the dissipation term in question are dominated by high-frequency, short wave-length fluctuations. Local isotropy applies to such fluctuations so that these terms are negligible if they are not invariant to rotation of the coordinate system. But the fluctuations arising in the dissipation term in question involve velocity components u_i, a scalar quantity Y_j, and double differentiation with respect to the coordinates x_k; they are not invariant and are thus frequently neglected. However, it is relatively easy to show that if chemical reaction occurs in laminar flamelets this term is proportional to the chemical source term and not negligible. This consideration illustrates the need for skepticism in carrying over to reacting flows the notions validated and found useful in nonreactive situations.

Consideration of the transport equations such as (1.51, 52, 54) supports the conclusion that the second-order equations contain additional physical content, beyond simple gradient transport considerations, but also require modeling in order to make them useful in predictive methods. The most troublesome terms requiring modeling are those involving the pressure, e.g., $\overline{u_j''(\partial p/\partial x_i)}$ in (1.51). In addition modeling of some terms requires introduction of a turbulence length scale usually associated with the large scale eddies; the choice here is between an algebraically specified length or the calculation of a quantity which combined with other dependent variables can be interpreted as a length. An equation for the velocity dissipation is typically used but involves in itself such severe modeling that its validity, at least for turbulent reacting flows, is dubious.

To a large extent the modeling required to achieve closure for turbulent reacting flows depends on analogy with the more or less successful modeling developed for constant-density flows. Effects associated with chemical reaction can significantly undermine these analogies as we have indicated at several points in this discussion. One effect is the change in density and kinematic viscosity due to the heat release with the consequent possibility of new phenomena not found in constant-density flows. We must also be prepared for effects which might be attributed to direct chemical influences; our discussion of the dissipation term in (1.54) is one example of such an effect. These concerns will be raised often in the chapters that follow.

1.14 Mean Chemical Production

The mean chemical production term that appears in (1.46) requires special consideration. Writing the mean value of a general form for $\dot{w}_i$ as obtained from (1.5) is more confusing than illuminating, and therefore consideration is directed to a series of individual, typical reactions, with forward and backward chemical reactions treated separately. Thus, consider the simple reaction

$$\mathscr{M}_1 + \mathscr{M}_2 \xrightarrow{k_f} \mathscr{M}_3 , \tag{1.55}$$

which will be encountered repeatedly in the following chapters, since it is useful for purposes of exposition. The species $\mathscr{M}_1$ can be considered fuel, $\mathscr{M}_2$ the oxidizer, and $\mathscr{M}_3$ a single product of reaction. Of course, (1.55) represents a highly idealized description of practical chemical systems.

The mean consumption of species $\mathscr{M}_1$ can be written as

$$\begin{aligned} -\overline{\dot{w}}_1 = & \,\overline{\varrho k_f}(\bar{\varrho}\tilde{Y}_1\tilde{Y}_2 + \overline{\varrho Y_1'' Y_2''}) + \overline{\varrho(\varrho k_f)'' Y_2''}\,\tilde{Y}_1 \\ & + \overline{\varrho(\varrho k_f)'' Y_1''}\,\tilde{Y}_2 + \overline{\varrho(\varrho k_f)'' Y_1'' Y_2''} \end{aligned} \tag{1.56}$$

For the time being we postpone detailed consideration of the calculation of the mean and fluctuating components of the reaction rate and discuss the physical significance of the various contributions to the mean chemical and production term.

The first term on the right of (1.56) involves the product of mean quantities. It is the term which would result if various mean quantities are substituted into the time-dependent chemical-production term for the reaction described by (1.55). Thus it represents the mean chemical production with no turbulence effects present; clearly the remaining five terms contain those effects.

The second term on the right of (1.56) contains the influence of inhomogeneities in reactant concentration with an effective reaction-rate constant corresponding to the indicated mean. The remaining terms contain the effect of fluctuations of reactant concentration and reaction-rate constant. It is remarkable that a simple reaction of the sort given by (1.55) leads to such a proliferation of effects, but this situation exemplifies the complexity associated with turbulent reacting flows. Note that two conditions must prevail to reduce significantly the variety of contributions on the right side of (1.56); if the density and the reaction-rate constant both remain constant, then we have

$$-\overline{\dot{w}}_1 = \varrho^2 k_f(\bar{Y}_1\bar{Y}_2 + \overline{Y_1' Y_2'}) . \tag{1.57}$$

In this simplified case, turbulence effects are confined to the influence of inhomogeneities in reactant concentration.

The physics of the turbulence effects, as indicated by either (1.56) or (1.57), are illuminated by consideration of two limiting cases; if the rate of chemical reaction is "slow" relative to the rate of decay of fluctuations in temperature and reactant concentrations, then the fluctuation terms become negligible before much reaction occurs, so that only the first term on the right of the two equations is operative, and there is negligible turbulence effect on mean chemical production. In this "well-mixed" limit, the time-resolved creation term $\dot{w}_1(\boldsymbol{x}, t)$ is constant, being given by the mean values of the reaction-rate constant and reactant concentrations. We shall later be able to be more specific about the definition of "slow".

The other limit corresponds to "fast chemistry", which can be expressed quantitatively by the approximation $Y_1(\boldsymbol{x}, t)\, Y_2(\boldsymbol{x}, t) \simeq 0$, that is, the two reactants do not coexist for otherwise the rapid rate of reaction would have destroyed them. In this case $\overline{\dot{w}}_1(\boldsymbol{x})$ cannot be computed from either (1.56) or (1.57), and $\dot{w}_1(\boldsymbol{x}, t)$ should be considered as a pulse train in which the pulses correspond to the passage by the point in question of oscillating reaction surfaces, where the product is generated by contact of the reactants at the molecular level. We shall discuss the structure of these surfaces subsequently, but for the present we need only observe that in this limit we can consider

$$\overline{\dot{w}}_1 = -\hat{\dot{w}}_1 f_f, \tag{1.58}$$

where $\hat{\dot{w}}_1$ is the mean amount of destruction of species M_1 at each crossing and f_f is the crossing frequency of reaction surfaces at the spatial point $\boldsymbol{x}$. In this case of fast chemistry the correlation coefficient $\overline{Y'_1 Y'_2}/\bar{Y}_1 \bar{Y}_2$ is negative, and in a sense turbulence reduces the mean rate of chemical reaction. Again we shall subsequently be able to sharpen our notions of "fast" versus "slow" chemical behavior.

Evidently these two limiting cases involve quite different manifestations of the effect of turbulence on chemical reactions. It can be anticipated that quite different analyses are appropriate in these two limits and that it is difficult to handle the intermediate situation, in which the chemical reactions are not sufficiently fast or slow relative to the decay of fluctuations in temperature and reactant concentration for either limit to be attained.

The situation is further complicated for reactions of greater complexity; to illustrate, consider the reaction

$$\mathscr{M}_1 + 2\mathscr{M}_2 \xrightarrow{k_f} \mathscr{M}_3 . \tag{1.59}$$

In this case,

$$\begin{aligned} -\overline{\dot{w}}_1 = {} & \widetilde{\varrho^2 k_f}(\bar{\varrho}\tilde{Y}_1 Y_2^2 + \overline{\varrho Y_2''^2}\,\tilde{Y}_1 + 2\overline{\varrho Y_1'' Y_2''}\,\tilde{Y}_2 + \overline{\varrho Y_1'' Y_2''^2}) \\ & + 2\overline{\varrho(\varrho^2 k_f)'' Y_2''}\,\tilde{Y}_1 \tilde{Y}_2 + \overline{\varrho(\varrho^2 k_f)'' Y_2''^2}\,\tilde{Y}_1 + \overline{\varrho(\varrho^2 k_f)'' Y_1''}\,\tilde{Y}_2^2 \\ & + 2\overline{\varrho(\varrho^2 k_f)'' Y_1'' Y_2''}\,\tilde{Y}_2 + \overline{\varrho(\varrho^2 k_f)'' Y_1'' Y_2''^2} . \end{aligned} \tag{1.60}$$

Here proliferation of terms occurs in an exaggerated form. There are a variety of terms describing the alteration of the mean chemical production as given by the product of mean values due to fluctuations in reaction-rate parameters and reactant concentrations.

The extent of proliferation is clearly related to the overall order of the reaction; this suggests that turbulence effects take on their simplest form for a unimolecular reaction, which may be approached from several points of view. One such view is to consider a bimolecular reaction, as in (1.55), in which the concentration of one reactant is much larger than that of the other, causing the rate of chemical production to vary only with the concentration of the deficient reactant since the concentration of the excess reactant remains approximately constant. In this case we can consider as a model

$$\mathcal{M}_1 \xrightarrow{k_\mathrm{f}} \mathcal{M}_2 , \tag{1.61}$$

and we have

$$\overline{\dot{w}}_1 = -(\bar{\varrho}\tilde{k}_\mathrm{f}\tilde{Y}_1 + \overline{\varrho k''_\mathrm{f} Y''_1}) , \tag{1.62}$$

so that the turbulence effect in this case involves a single term; if the reaction-rate constant k_f can be considered constant, then even this term vanishes because of the resulting linearity.

The nonlinearity of the T dependence of the reaction-rate constant k introduces a formidable difficulty. Here we can only indicate the nature of the difficulty and give some indication of methods which have been suggested for its resolution; in subsequent chapters more detailed treatments as appropriate for special cases will be described.

Consider a rate constant of the simplified form

$$k = B\,\mathrm{e}^{-T_\mathrm{a}/T} \tag{1.63}$$

and assume that this applies to the second-order reaction given by (1.55). Generally B and the temperature ratio T_a/T are large, as indicated in Sect. 1.5. The consequent strong temperature dependence of k is the source of the problem.

A straightforward strategy is to replace T in the exponential by its mean and fluctuating components and then to expand the exponential in the form

$$k = B\,\mathrm{e}^{-T_\mathrm{a}/\tilde{T}}\{1 + (T_\mathrm{a}/\tilde{T}^2)T'' + [(T_\mathrm{a}^2/2\tilde{T}^4) - (T_\mathrm{a}/\tilde{T}^3)]T''^2 + \ldots\} . \tag{1.64}$$

The Favre average of this is

$$\tilde{k} = \overline{\varrho k}/\bar{\varrho} = B\,\mathrm{e}^{-T_\mathrm{a}/\tilde{T}}[1 + (T_\mathrm{a}^2/2\tilde{T}^4)\overline{\varrho T''^2}/\bar{\varrho} + \ldots] , \tag{1.65}$$

where $T_a/\tilde{T}^3$ has been neglected in comparison with $T_a^2/2\tilde{T}^4$ on the basis of $T_a/\tilde{T}$ being large. Note, however, that this approach requires $T_a T''/\tilde{T}^2 \ll 1$ if the higher-order terms in the expansion are to be negligible. This inequality places severe limitations on permissible intensities of temperature fluctuations. When the reactants are highly dilute, this approximation may be acceptable, but in many flows of practical interest fluctuations are too intense for (1.65) to be employed. This imposes severe difficulties on attempts to use moment methods, since many higher moments would have to be considered.

Nevertheless, it is interesting to note that (1.65) indicates an increase in the effective reaction rate due to the correction term for temperature fluctuations. This occurs because the nonlinearity of the instantaneous reaction rate is such that for a given magnitude of temperature change there is a greater increase in rate for an increase in temperature than decrease in rate for a decrease in temperature. This enhancement of reaction rate by temperature heterogeneities contrasts with the reduction of reaction rate by concentration heterogeneities under conditions of fast chemistry.

Other strategies are potentially more useful when $T_a/\tilde{T}$ is large. One possibility that has not been investigated is to attempt to treat k rather than T as the dependent variable describing temperature. Another is to make use of existing ideas (e.g., [1.37]) concerning high activation energy to resolve the flow into narrow reaction zones and broader zones in which rates of reaction are negligible. In turbulent flows these zones are transported by the turbulence as shown, for example, by *Clavin* and *Williams* [1.33] and are considered laminar flames or laminar flamelets; they will be discussed in both Chaps. 3 and 4 since they arise with both non-premixed and premixed reactants.

Probability-density functions can provide a convenient means for addressing the problem of calculating average rates of chemical production, one which exposes important physical features and leads to approximate computation schemes in certain cases. Consider again the bimolecular reaction given by (1.55). According to (1.5), for this reaction the production rate depends only on ϱ, T, Y_1, and Y_2. Therefore, in statistically stationary flows $P(\varrho, T, Y_1, Y_2; \boldsymbol{x})$ is the relevant probability-density function. Specifically, the average rate of consumption of reactant $\mathcal{M}_1$ is

$$-\bar{w}_1 = \int_0^\infty \int_0^\infty \int_0^1 \int_0^1 \varrho^2 k Y_1 Y_2 P(\varrho, T, Y_1, Y_2; \boldsymbol{x}) dY_1 dY_2 dT d\varrho \,. \tag{1.66}$$

If the pressure is nearly constant, then $P(\varrho, T, Y_1, Y_2; \boldsymbol{x}) = P(T, Y_1, Y_2; \boldsymbol{x})\,\delta(\varrho - p/RT)$ from (1.21), and there is corresponding simplification in (1.66). It may be observed from this result that knowledge of the joint probability-density function for temperature and reactant concentrations would enable the mean rate of production to be obtained.

There are two alternative approaches to obtaining this probability-density function. One is to derive an equation for its evolution by the methods of Chap.

5; as indicated earlier, this entails modeling to achieve closure. The other is to guess its form from hypotheses concerning the nature of the flow; this approach has been used, for example, in studying nonpremixed flames with "fast" chemistry and will be discussed in Chap. 3; in premixed flames as discussed in Chap. 4 the corresponding calculation involves a single variable $P(c, \boldsymbol{x})$.

When $T_a/\tilde{T}$ is large, it is found that in (1.66) there are significant contributions to the temperature integral only from a narrow range of T. This result is obtained by substituting (1.63) into (1.66). At the lower values of T, the factor $\exp(-T_a/T)$ is very small, and there is negligible contribution to $\bar{w}_1$ in (1.66). If the maximum temperature is sufficiently low, then this situation prevails for all T, and $\bar{w}_1 \approx 0$ from (1.66), thereby reducing the problem to that of nonreactive turbulent mixing. The major contribution to the integral over T in (1.66) thus often comes from a narrow range of temperatures in the vicinity of the maximum temperature, which sometimes is approximately the adiabatic flame temperature in an exothermically reacting system. The decrease in Y_1 and Y_2 in the vicinity of this maximum temperature also influences the integral; this situation has been discussed, for example, by *Williams* [1.37, 38].

Formalisms recently have been developed that aid in accounting for such effects. In these approaches, expansions are made in a large parameter, T_a/T_r, where T_r is a reference temperature characteristic of temperatures existing in the flow at positions where most of the reaction occurs. This reference temperature is determined by generating a dimensionless factor $\hat{B}$ from the quantity B in (1.63). Depending on the problem under consideration the quantities introduced to effect nondimensionalization represent the transient, convective, or diffusive terms in (1.17). There results $\ln \hat{B} = T_a/T_r$ which defines T_r; in some cases a constant which is to be determined in the course of the analysis must be added to the right side of this equation. If for illustrative purposes we assume that this constant is absent in the problem under consideration, then $\hat{B} \exp(-T_a/T_r) = \exp[(T_a/T_r)(1 - T_r/T)]$ which indicates for $T_a/T_r \gg 1$ that the rate constant is exponentially small for $T < T_r$ and exponentially large for $T > T_r$. The parameter T_a/T_r is a nondimensional activation energy, and the formal approach has been termed activation-energy asymptotics. Additional details on the approach may be found in [1.33, 37, 38], for example. At temperatures where k is very large it is necessary that $Y_1 Y_2 \approx 0$, corresponding to the near-equilibrium conditions of fast chemistry. Thus, some physical attributes of the integral in (1.66) can be deduced directly, without resort to predictive approaches.

For nonpremixed systems in the limit of fast chemistry, it has been shown that $\bar{w}_1$ may be obtained more directly by an integral involving a simpler joint probability-density function, namely that for the magnitude and the gradient of a suitably defined conserved scalar of the type Z_i, given in (1.18) (see, for example, [1.38]). Derivation of results of this type will be presented more fully in Chap. 3.

The probability density function $P(T, Y_1, Y_2; \boldsymbol{x})$ can be used not only to evaluate $\bar{w}_1$ but also to obtain other thermochemical quantities; for example, we have

$$\left.\begin{aligned}
\bar{T} &= \int_0^\infty \int_0^1 \int_0^1 TP(T, Y_1, Y_2; \boldsymbol{x})\, dY_1\, dY_2\, dT, \\
\overline{T'^2} &= \int_0^\infty \int_0^1 \int_0^1 T^2 P(T, Y_1, Y_2; \boldsymbol{x})\, dY_1\, dY_2\, dT - \bar{T}^2, \\
\bar{Y}_1 &= \int_0^\infty \int_0^1 \int_0^1 Y_1 P(T, Y_1, Y_2; \boldsymbol{x})\, dY_1\, dY_2\, dT, \\
\overline{T' Y_1'} &= \int_0^\infty \int_0^1 \int_0^1 Y_1 T P(T, Y_1, Y_2; \boldsymbol{x})\, dY_1\, dY_2\, dT - \bar{T}\bar{Y}_1 .
\end{aligned}\right\} \quad (1.67)$$

Other mean values, intensities, and correlations can be calculated in the same fashion. As indicated earlier, conservation equations for these quantities can be established and closed with suitable models for certain terms. If a functional form for $P(T, Y_1, Y_2; \boldsymbol{x})$, involving an appropriate number of parameters $a_i(\boldsymbol{x})$, is adopted as a model for the probability-density function, then these conservation equations become partial differential equations for these parameters. This avenue to flow predictions through approximations of probability-density functions will be discussed more fully in Chap. 4. The accuracy of such a calculation depends on the adopted functional form for P, which must respect for each $\boldsymbol{x}$ the bounds on the variables T, Y_1, and Y_2 and which must be nonnegative. In addition, if intermittency prevails in the flow under consideration, then the integral of the assumed form must account for the intermittency $\bar{I}(\boldsymbol{x})$.

This approach in terms of probability-density functions is related to a problem in statistical theory known as the moment problem which may be stated as follows: Given certain information relative to the statistics of a random variable, determine the probability-density function of that variable. Reference may be made to *Akhiezer* [1.39] for discussion of this topic. Statistical theory also provides bounds on the intensities of the fluctuations of the variables and on the correlations among them, bounds which should be respected by the assumed form for P (cf. [1.40, 41]).

1.15 Coherent Structures

In this introductory survey of turbulent reacting flows some comments must be devoted to the matter of coherent structures which has received considerable attention recently from both the turbulence and the combustion communities. By coherent structures are meant large, rather organized parcels of turbulent fluid with their own dynamic behavior. Their large size implies a relatively long lifetime. Such structures arise in boundary layers, jets, wakes, and mixing layers.

When coherent structures occur, our usual picture of continuous turbulent fluid involving the cascading of large eddies into smaller dissipative scales must be supplemented by the picture of coherent structures which may increase in size in the downstream direction. In some flows this increase is due to pairing of two structures into a single new one, while in other flows the mechanism of the genesis and rate of growth of the parcels results in coherent structures which maintain a more or less constant size relative to the scale of the shear layer, e.g., the turbulent boundary layer thickness, by entraining additional fluid.

One obvious manifestation of coherence is intermittency discussed earlier; in fact the discovery of intermittency by *Corrsin* in 1943 [1.42] may be interpreted as the first evidence of coherent structures. In recent years the methods of conditioned sampling have developed for a variety of basic flows information on the statistical geometry of the surface bounding these coherent structures and on the distribution of fluid-mechanical quantities within these structures. A recent symposium has been devoted to these results (see [1.43]).

The genesis of these large parcels of turbulent fluid is the subject of ongoing research with the two-dimensional turbulent boundary layer and the two-dimensional mixing layer attracting the most attention. In the case of the boundary layer three-dimensional coherent structures originate at the wall but are triggered by older structures which are passing by and which originated further upstream; perhaps the progenitors of all downstream structures are created by the transition process. The two-dimensional mixing layer is special (see [1.44]); under some circumstances the coherent structures in the mixing layer are highly two dimensional, essentially vortices with axes parallel to the splitter plate, and may be due to an instability mechanism. In other circumstances the parcels of turbulent fluid in the two-dimensional mixing layer have the same highly three-dimensional geometry as is found in all other turbulent shear flows.

In terms of the phenomenology of turbulent flows, nonreactive and reactive, the existence of coherent structures implies that unconditioned averaging, whether conventional or Favre in the case of flows with significant density variations, overlooks and disregards an important physical phenomenon. Experiments have shown that in many turbulent flows, large-scale, orderly structures contribute more than half of the total Reynolds stress terms such as $\overline{\varrho u_i'' u_j''}$. At some time in the future it may be possible to combine the present, relatively well-developed statistical approach for the turbulent fluid within the parcels with a more deterministic description of the orderly structures which encapsulate the turbulent fluid. At present little can be said about theoretical descriptions of turbulent flows on the basis of coherent structures. Kovasznay has remarked that there exists no theory of them at present (see [1.43]). Preliminary work on a theory of conditioned turbulence by *Libby* [1.28] and *Dopazo* [1.29] represents an extension of the usual phenomenology to recognize intermittency and thus a start on a new theory.

The widespread acknowledgement of the significance of coherent structures in determining the behavior of turbulent reacting flows has altered our view of turbulence. The behavior of reacting surfaces within reaction zones is now recognized to be dominated by large-scale structures. Thus, experimentalists in combustion now recognize the desirability of applying the methods of conditioned sampling in order to expose important, perhaps relatively rare, events from a bland turbulent background involving only small-scale fluctuations. Some impact of the new view on theories of flows with reactions has begun to occur (see [1.45] and Chap. 6). As yet the resulting theoretical formulations are incomplete and largely untested although they constitute an interesting avenue of research.

1.16 Scales and Similarity Numbers

Dimensionless parameters often occur in the more precise description of turbulent reacting flows. Although it is safe to say that not all such relevant parameters are known today, nevertheless so many have appeared in the literature that it is impractical to list here all that have been used. Some of the more basic parameters will be described in this section. These and others will appear in subsequent chapters; in particular, in Chap. 2 appropriate dimensionless parameters are used to identify the nature of the turbulence in various practical devices.

A suitable starting point is the consideration of fluid mechanical quantities that characterize a particular turbulent reacting system. Most flows have at least one characteristic velocity U and characteristic length scale L of the device in which the flow takes place (see Fig. 2.1). In addition, there is at least one representative density ϱ_0 and representative temperature T_0. Accordingly, a characteristic kinematic viscosity $\nu_0 = \mu_0/\varrho_0$ can be defined, where μ_0 is the coefficient of viscosity at the characteristic temperature T_0. From these quantities, a Reynolds number for the system $\mathrm{Re} = UL/\nu_0$ can be defined. It is worthwhile to note again that since ν is roughly proportional to T^2, large changes in ν can occur in reacting systems involving a significant amount of chemical heat release. A change in T_0 by a factor of four, which is quite modest in combustion, produces a change in Re by more than an order of magnitude. Consequent tendencies for heat release to decrease Re and introduce laminar-like behavior in turbulent flows have been observed experimentally in combustion. This effect complicates efforts to describe turbulent combustion; moreover, changes in Re through heat release must be considered in making order-of-magnitude estimates. Since ν is inversely proportional to the pressure p, and changes in p are usually small, the effects of such changes on ν typically are much less than those of changes of T.

Flow quantities characteristic of the turbulence itself are of more direct relevance to modeling than Re. The turbulent kinetic energy $\tilde{q}$ may be assigned a representative value $\tilde{q}_0$ at a suitable reference point. The relative intensity of the turbulence is then characterized by either $\tilde{q}_0/(\frac{1}{2}U^2)$ or U'/U, where

$U' = \sqrt{2\tilde{q}_0}$ is a representative root-mean-square velocity fluctuation. Weak turbulence corresponds to $U'/U \ll 1$, while intense turbulence has U'/U of order unity.

As discussed in Sect. 1.11, there is a continuous distribution of length scales associated with the turbulent fluctuations of velocity components and of state variables. It is useful to focus on two, widely disparate lengths that determine separate effects in turbulent flows. First, there is a length l_0 characterizing the large eddies, those with low frequencies and long wavelengths. Experimentally l_0 can be defined as a length beyond which various fluid-mechanical quantities become essentially uncorrelated; typically l_0 is less than L but of the same order of magnitude. This length can be used in conjunction with U' to define a turbulence Reynolds number,

$$R_l = U' l_0 / \nu_0 , \tag{1.68}$$

which has more direct bearing on the structures of turbulence in flows than does Re. Large values of R_l can be achieved by intense turbulence, large-scale turbulence, and small values of the kinematic viscosity produced, for example, by low temperatures or high pressures. The cascade view of turbulence dynamics is restricted to large values of R_l. Generally $R_l < \mathrm{Re}$.

The second length scale useful in characterizing the turbulence is that over which molecular effects are significant. This can be introduced in terms of a representative rate of dissipation of velocity fluctuations, $\varepsilon_0 = U'^3/l_0$. This rate estimate corresponds to the idea that the time scale over which velocity fluctuations decay by a factor of $1/e$ is of the order of the turning time of a large eddy. The rate ε_0 increases with increasing turbulent kinetic energy (which is due principally to the large-scale motions) and decreases with increasing size of the large-scale eddies. At the small scales where the molecular dissipation occurs, the relevant parameters are the kinematic viscosity and the rate of dissipation. The only length that can be constructed from these two parameters is the Kolmogorov length, $l_k = (\nu_0^3/\varepsilon_0)^{1/4}$. By substitution,

$$l_k = l_0 / R_l^{3/4} . \tag{1.69}$$

This length is representative of the dimension at which dissipation occurs and defines a cutoff of the spectrum. It is evident from (1.69) that for large R_l there is a large spread of the two extreme lengths characterizing turbulence. This spread is reduced with increasing temperature because of the consequent increase in ν_0.

Considerations analogous to those for velocity apply to scalar fields as well, and lengths analogous to l_k have been introduced for these fields. They differ from l_k by factors involving molecular Prandtl or Schmidt numbers, which differ relatively little from unity for representative gas mixtures. Therefore, in a first approximation for gases l_k may be used for all fields, and the corresponding lengths will not be introduced here.[2]

An additional length, intermediate in size between l_0 and l_k, which often arises in formulations of equations for average quantities in turbulent flows is the Taylor length,

$$\lambda = l_0 / R_1^{1/2}. \tag{1.70}$$

The dissipation term in (1.52) is of order $\mu_0 q_0/\lambda^2$, the strain rate being of order U'/λ.

There are length scales associated with laminar structures in reacting flows. One is the characteristic thickness of a premixed laminar flame, denoted here by δ (see [1.7]). It may be expected that the nature of turbulent reacting flows may differ considerably, depending on whether $\delta < l_k$, $l_k < \delta < \lambda$, $\lambda < \delta < l_0$, or $l_0 < \delta$. Wrinkled laminar flames may occur when $\delta < l_k$ and broadly distributed reactions when $l_0 < \delta$. The nature of turbulent flames implied by these various inequalities has not been completely explored. In theories it is often assumed that $\delta < l_k$.

Characteristic times for chemical conversion can be defined directly from the rate expressions of the form (1.5). To be specific, consider the bimolecular reaction (1.55) with the rate constant given by (1.63). It is necessary to introduce a characteristic temperature T_c, representative of temperatures at which the reaction occurs, as well as representative reactant concentration, Y_{10} and Y_{20}. Then $[Y_{10} Y_{20} \varrho_0 B \exp(-T_a/T_c)]^{-1} = \tau_c$ is a characteristic chemical conversion time for (1.55). Chemical lengths may be constructed from this as $U\tau_c$ or $U'\tau_c$. Comparison of an appropriate chemical length with a fluid-dynamical length provides a nondimensional parameter that has a bearing on the relative rate of the chemical reaction. Nondimensional parameters of this type conventionally are termed Damköhler numbers, or more precisely, Damköhler's first similarity group. An example is

$$D = l_0/(U'\tau_c). \tag{1.71}$$

For large Damköhler numbers, chemistry is fast and reaction sheets of the type discussed in Chaps. 3 and 4 may develop. For small Damköhler numbers, chemistry is slow, and "well-stirred" flows may occur. A difficulty relative to these considerations is that, depending on the choices of Y_{10}, Y_{20}, especially T_c, flow length, and characteristic flow velocity, many different Damköhler numbers can be defined. In fact, in the literature a number of different selections for the definition of Damköhler's first similarity group may be found. In a given turbulent flow the physically most relevant definition usually is not immediately evident. Therefore caution should be exercised in drawing conclusions on the basis of the numerical value of a particular D.

2 The presence of significant amounts of hydrogen alters these considerations somewhat.

The preceding discussion of chemical times has related to a single, bimolecular reaction. In chemical systems of applied interest several, in fact many, reactions may be operative simultaneously. In this case different chemical length scales will arise for each reaction, and each must be subjected to comparison with a suitable flow length. A variety of situations may be encountered, involving some reactions behaving according to the notions of "fast chemistry", others behaving as though "slow chemistry" prevails, and yet others not lending themselves to either category.

There are two other nondimensional numbers relevant to chemistry, which have been introduced by Frank-Kamenetskii and others. These Frank-Kamenetskii numbers are the nondimensional heat release $\alpha = \Delta_p / c_p T_c$, where Δ_p is defined in (1.14), and the nondimensional activation temperature $\beta = T_a / T_c$. Combustion in general, and turbulent combustion in particular, typically are characterized by large values of these numbers. When α is large, the chemistry is likely to exert a large influence on the turbulence. When β is large, the rate of the reaction depends strongly on temperature. It is usually true that the larger β the thinner will be the region in which the principal chemistry occurs. Thus, irrespective of the value of D, reactions tend to be found in thin, convoluted sheets in turbulent flows, for both premixed and nonpremixed systems having large β. In premixed flames it is known that the thickness of the reaction region is of order δ/β (see [1.38]). Different relative sizes of δ/β and fluid-mechanical lengths therefore may introduce additional classes of turbulent reacting flows. The multifaceted character of the subject thus is seen to be impressive.

1.17 Comparison Between Theory and Experiment

Although this volume is limited to the treatment of the theoretical aspects of turbulent reacting flows, the experimental investigation of such flows involving an important but separate area of activity, it is worthwhile to consider briefly the comparison of theoretical predictions with experimental results. Such comparisons always play an important role in achieving advances in the development of theories and of understanding in turbulence, and this is especially true when chemical reactions are involved.

At the outset we note that measurements in turbulent reacting flows involve significant difficulties and many potential sources of error. As a consequence the combustion literature contains important discrepancies among experiments which should apparently be in accord. This places the theoretician in the difficult position of being without reliable guides to some critical aspects of turbulent reacting flows and of making judgements as to which set of data is likely to endure.

Experimental difficulties frequently limit the quantities accessible to comparison. A theoretical analysis may lead to distributions throughout the flow of a variety of fluid-mechanical and thermochemical quantities, only a few of which are measured at a few spatial locations. If acceptable agreement is realized between prediction and the few data available, it is often assumed that the remaining quantities at other locations would also be in agreement if they were provided. On the contrary, if there is disagreement, it is difficult to isolate the source or sources of the error, even under the assumption that the experimental results are suitably accurate. This is the case for several reasons: First, as we mentioned earlier and as will be repeatedly pointed out in the chapters which follow, models must be introduced for various terms in order to achieve a closed set of describing equations. As a general rule these models cannot be directly compared to experimental results. Two examples come immediately to mind; the measurement of static pressure and its correlations with the velocity is difficult even in low-speed isothermal turbulence and is probably hopelessly so in reacting flows. However, such correlations may play an important role in redistributing the several velocity intensities and thus in the behavior of turbulent flows. We have indicated that various dissipation terms arise in the conservation equations employed in methods based on second-order closure. Again in low-speed, isothermal turbulence these terms can be estimated from experimental data on the basis of certain assumptions regarding local isotropy, but it seems likely that some of the dissipation terms of interest in the analysis of turbulent flows will be difficult to measure.

There is another essential difficulty in isolating the source of discrepancy between predictions and experiment. Theoretical analyses frequently involve quantities which facilitate the treatment but which are not directly observable. The element mass fractions Z_i introduced earlier are examples of such quantities; in order to make a comparison between a predicted value for even a mean value of Z_i, e.g., $\tilde{Z}_i$, let alone its intensity $\overline{\varrho Z_i''^2}$ and its correlation with velocity fluctuations, we must obtain information on all the major species contributing to Z_i and to the density. However, in this regard we mention that measurements of a conserved scalar can be interpreted in terms of Z_i. There are other examples of quantities arising in theory and not directly observable.

Finally, we mention that it is difficult in some circumstances to establish whether the source of discrepancy between prediction and experiment lies in the fluid-mechanical or chemical behavior. If the effective rates of heat release are not properly described, the distributions of the statistics of the state variables including the temperature will be in error. But these distributions can also be in error if the diffusive properties of the flow are improperly described. Thus some care is called for in unraveling the various sources of error in turbulent reacting flows.

These remarks will indicate the difficulties in making comparison between theory and experiment in turbulent reacting flows. Despite these difficulties such comparisons are essential if progress is to be achieved. Close coordination between theoreticians and experimentalists is called for; the former must suggest quantities important to measure, the latter must indicate what quantities can be measured.

References

1.1 P.Bradshaw: *An Introduction to Turbulence and Its Measurement* (Pergamon, Oxford 1971)
1.2 P.Bradshaw (ed.): *Turbulence* 2nd, Topics in Applied Physics, Vol. 12 (Springer, Berlin, Heidelberg, New York 1978)
1.3 H.Tennekes, J.L.Lumley: *A First Course in Turbulence* (MIT Press, Cambridge 1972)
1.4 J.O.Hinze: *Turbulence* (McGraw-Hill, New York 1977)
1.5 A.S.Monin, A.M.Yaglom: *Statistical Fluid Mechanics*, 2. Vols. (MIT Press, Cambridge 1971, 1975)
1.6 S.S.Penner: *Chemistry Problems in Jet Propulsion* (Pergamon, New York 1957)
1.7 F.A.Williams: *Combustion Theory* (Addison-Wesley, Reading, Mass. 1965)
1.8 R.A.Strehlow: *Fundamentals of Combustion* (International Textbook, Scranton, Penna. 1968)
1.9 I.Glassman: *Combustion* (Academic, New York 1977)
1.10 W.C.Reynolds: "Computation of Turbulent Flows", *Annual Reviews of Fluid Mechanics*, Vol. 8 (Annual Reviews, Palo Alto 1976)
1.11 J.C.Hill: "Homogeneous Turbulent Mixing with Chemical Reaction", *Annual Reviews of Fluid Mechanics*, Vol. 8 (Annual Reviews, Palo Alto 1976)
1.12 P.A.Libby, F.A.Williams: "Turbulent Flows Involving Chemical Reactions", *Annual Reviews of Fluid Mechanics*, Vol. 8 (Annual Reviews, Palo Alto 1976)
1.13 B.E.Launder, D.B.Spalding: *Mathematical Models of Turbulence* (Academic, London 1972)
1.14 D.B.Spalding: "Turbulence Modelling: Solved and Unsolved Problems", in *Turbulent Mixing in Nonreacting and Reactive Flows*, ed. by S.N.B.Murthy (Plenum, New York 1975) pp. 85–130
1.15 S.Tsugé, K.Sagara: Phys. Fluids **19**, 1478–1485 (1976)
1.16 S.Tsugé, K.Sagara: Combust. Sci. Technol. **18**, 179–189 (1978)
1.17 C.H.Gibson, P.A.Libby: Combust. Sci. Technol. **6**, 29–35 (1972)
1.18 C.H.Gibson: Phys. Fluids **11**, 2316–2327 (1968)
1.19 J.O.Hirschfelder, C.F.Curtiss, R.B.Bird: *Molecular Theory of Gases and Liquids* (Wiley, New York 1954)
1.20 C.Dopazo, E.E.O'Brien: Phys. Fluids **17**, 1968–1975 (1974)
1.21 A.Favre: *Statistical Equations of Turbulent Cases in Problems of Hydrodynamics and Continuum Mechanics* (SIAM, Philadelphia 1969) pp. 231–266
1.22 M.M.Morkovin: "Effects of Compressibility on Turbulent Flows", in *The Mechanics of Turbulence* (Gordon and Breach, New York 1964) pp. 367–380
1.23 R.Bilger: Combust. Sci. Technol. **11**, 215–217 (1975)
1.24 H.H.Fernholz: "External Flows", in Ref. 1.2
1.25 C.van Atta: "Sampling Techniques in Turbulence Measurements", in *Annual Reviews of Fluid Mechanics*, Vol. 6 (Annual Reviews, Palo Alto 1974)
1.26 C.H.Gibson: "Digital Techniques in Turbulence Research", AGARDograph No. 174 (1973)
1.27 L.S.G.Kovasznay, V.Kibens, R.F.Blackwelder: J. Fluid Mech. **41**, 283–325 (1970)
1.28 P.A.Libby: Phys. Fluids **19**, 494–501 (1976)
1.29 C.Dopazo: J. Fluid Mech. **81**, 433–438 (1977)
1.30 C.Dopazo, E.E.O'Brien: "Intermittency in Free Turbulent Shear Flows", Symposium on Turbulent Shear Flows, Pennsylvania State University, 1977, Chap. 1, p. 1
1.31 E.E.O'Brien, C.Dopazo: "Behaviour of Conditioned Variables in Free Turbulent Shear Flows", Symposium on Turbulence, Berlin 1977
1.32 E.E.O'Brien: J. Fluid Mech. **89**, 209–222 (1978)
1.33 P.Clavin, F.A.Williams: J. Fluid Mech. **90**, 589–604 (1979)
1.34 E.E.O'Brien: Phys. Fluids **14**, 1804–1806 (1971)
1.35 C.H.Gibson: Phys. Fluids **11**, 2316–2327 (1968)
1.36 S.Corrsin: "Limitations of Gradient Transport Models in Random Walk and in Turbulence", in *Advances in Geophysics*, Vol. 18A, ed. by F.N.Frenkiel, R.E.Munn (Academic, New York 1974)

1.37 F.A.Williams: "Theory of Combustion in Laminar Flows", *Annual Reviews of Fluid Mechanics*, Vol. 3 (Annual Reviews, Palo Alto 1971) pp. 171–188
1.38 F.A.Williams: "A Review of Some Theoretical Considerations of Turbulent Flame Structure", in *Analytical and Numerical Methods for Investigation of Flow Fields with Chemical Reactions, Especially Related to Combustion*, ed. by M.Barrère, AGARD Conference Proceedings No. 164 (AGARD, Paris 1975) pp. III-1 to III-25
1.39 N.I.Akhiezer: *The Classical Moment Problem* (Hafner, New York 1965)
1.40 S.B.Pope: "Probability Distributions of Scalars in Turbulent Shear Flows", Second Symposium on Turbulent Shear Flows, Imperial College, London (1979)
1.41 A.K.Varma, G.Sandri, C.DuP.Donaldson: "Some Implications of Statistical Constraints for Modeling of Turbulent Reacting Flows", AIAA-80-0138 (1980)
1.42 S.Corrsin: "Investigation of Flow in an Axially Symmetrical Heated Jet of Air", ACR 3L23, NASA Wartime Report W-94 (1943)
1.43 H.Fielder (ed.): *Structure and Mechanisms of Turbulence, I and II*, Lecture Notes in Physics, Vols. 75 and 76 (Springer, Berlin, Heidelberg, New York 1978)
1.44 A.Roshko: AIAA J. **14**, 1349–1357 (1976)
1.45 D.B.Spalding: Combust. Sci. Technol. **13**, 3–25 (1976)

2. Practical Problems in Turbulent Reacting Flows

A. M. Mellor and C. R. Ferguson

With 14 Figures

This chapter discusses the practical problems of turbulent combustion which may be solved, or at least more fully understood, by consideration of the theory of turbulent reacting flows and provides perspective on our current ability to predict the properties of such combustion. The quantities which characterize the turbulence in various practical devices involving turbulent combustion are discussed first. Then cases of quasi-steady flows such as arise in the combustors of gas turbines are considered with emphasis on comparison of the predictions based on current methods of analysis with experimental data. There follows a similar discussion of quasi-periodic flows such as arise in the internal combustion engine.

2.1 Introductory Remarks

The length scales and flow rates associated with all practical devices involving the combustion of fuels in power and propulsion applications result in flows dominated by turbulent processes. Despite this common feature the scales and turbulence characteristics in one device differ greatly from those in another. Thus a wide range of turbulence phenomena is of practical concern.

During the past fifteen years there has been renewed interest in more accurate descriptions of turbulent combustion in order to improve thermal efficiency with minimal noise and air pollution. This renewal has been accompanied by increased capabilities for detailed measurements within turbulent flames both in the laboratory and in industrial devices by means of laser diagnostics. At the same time there have been developed improved predictive methods based on the numerical solution of the describing partial differential equations such as those discussed in Chap. 1 and in the chapters which follow. It is largely the comparison of measurements and predictions which reveals our inadequate understanding of the various interactions between fluid mechanical and chemical effects in turbulent reacting flows. In this chapter we identify selected engineering applications where a more satisfactory ability to model these interactions would yield information of immediate practical significance.

Several reviews of turbulent combustion emphasizing applications are available. A general discussion of modeling of turbulence-chemistry interactions may be found in *Bray* [2.1]. For reciprocating engine combustion, the reader is referred to *Agnew* [2.2, 3], *Blumberg* [2.4], *Henein* [2.5], *Heywood* [2.6, 7], and *Shahed* et al. [2.8]. Gas turbine combustion is reviewed by *Lefebvre* [2.9], *Mellor* [2.10, 11], and *Henderson* and *Blazowski* [2.12], and turbine afterburners by *Zukoski* [2.13]. Furnace and boiler applications are discussed by *Breen* [2.14], *Essenhigh* [2.15], *Beér* [2.16], and *Maček* [2.17], and chemical lasers by *Bronfin* [2.18]. Descriptions of fire and explosion research may be found in *Thomas* [2.19], *Strehlow* [2.20], *Grumer* [2.21], *Williams* [2.22], and *de Ris* [2.23]. These references contain more complete citations of relevant literature.

2.2 Quasi-steady Flames

Our discussion is facilitated if we consider separately quasi-steady and quasi-periodic situations represented, respectively, by gas turbine and furnace combustors and by reciprocating engines. Quasi-steady flames involve well-defined mean values even though at a point in the flow large fluctuations of all flow variables are present. Combustors as discussed in detail in Sect. 2.3 result in such mean values because they operate for extended periods at a fixed power point whereas reciprocating engines as discussed in Sect. 2.4 do not. Fires can frequently be considered quasi-steady since the time scales for global changes are considerably greater than those associated with the turbulence.

In accordance with our earlier observation that turbulent combustion in practical devices can involve a wide variety of length scales and turbulence characteristics it is useful to identify these features in several quasi-steady flows of practical interest. In doing so we shall find some of the notions set forth in Sect. 1.16 significant.

Figure 2.1, taken from *Goulard* et al. [2.24], categorizes turbulent flames in terms of their characteristic velocity U and length L. Shown are the regimes in which the main burners of gas turbines and afterburners operate as well as the regimes corresponding to small scale fires. The flow in furnaces involves characteristic velocities higher than in fires but with roughly the same length scales.

Several other characteristic parameters and similarity groups are represented in the figure. The Froude number U^2/gL, where g is the gravitational acceleration, greater than 10^2, delineates the regime wherein inertial forces overwhelm buoyancy forces. The Reynolds number R_l defined in Sect. 1.16 is evaluated at two pressures: the first value corresponds to a pressure of one atmosphere, the second in parentheses to ten atmospheres. Similarly, the Kolmogorov scale is designated l_k.

Two characteristic times are also sketched in Fig. 2.1; the quantity τ_u is the time for the Kolmogorov scale to be convected past a point in the flow and is taken as l_k/U; for adequate resolution of the turbulent fluctuations in a quasi-

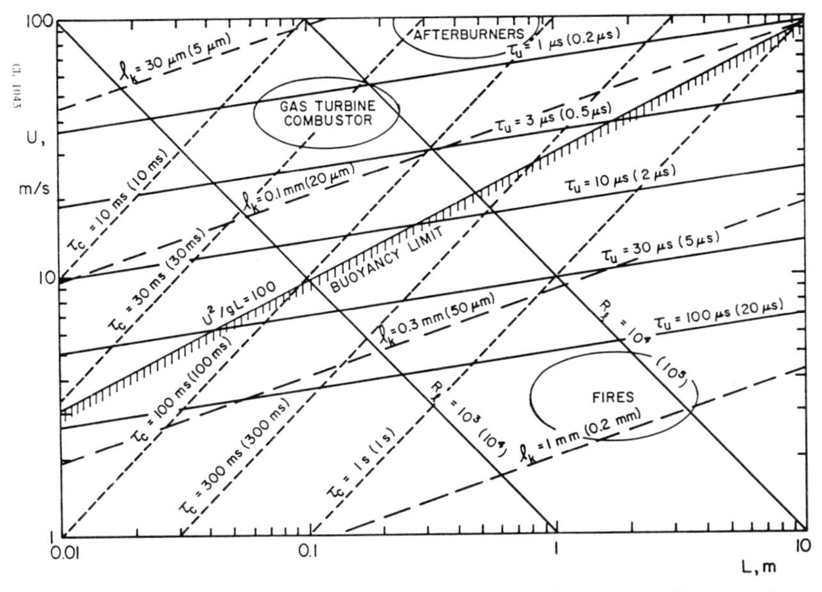

Fig. 2.1. Combustion parameters, in terms of characteristic flow velocity U and combustor size L [2.24]

steady flame *Goulard* et al. [2.24] suggested that measurements provide similar temporal resolution. These authors also suggest that the relevant fluid time to compare with chemical times is the Corrsin time $\tau_c = \lambda^2/6v_0$, where λ is the Taylor microscale given by (1.70). The quantity τ_c refers to the dissipation rate of concentration fluctuations. *Goulard* et al. showed that $\tau_c \sim 10\,L/U$.

Figure 2.1 clearly indicates the ranges of characteristic velocity and length of interest in practical applications: afterburners operate with $L = O(0.5\,\mathrm{m})$ and $U = O(100\,\mathrm{m/s})$ whereas small fires typically involve $L = O(5\,\mathrm{m})$ and $U = O(2\,\mathrm{m/s})$. Fires in forests or resulting from LNG and other fuel spills exhibit even larger length scales.

The majority of the applications cited involve turbulent diffusion flames since the fuel and air are not premixed and furthermore frequently involve liquid and solid fuels. However, in the gas turbine afterburner, liquid fuel is injected upstream of a flameholder, partially vaporizes and mixes with additional air prior to combustion so that the combustion has some of the characteristics of premixed turbulent reactions.

In addition to the regimes of velocity and length characterizing the turbulence discussed here, the combustion of multicomponent hydrocarbon fuels results in a wide range of chemical time scales. One important example of such a range relates to the formation of nitric oxide from air by a reaction which is considerably slower than those associated with the oxidation of the fuel.

2.3 Gas Turbine and Furnace Combustors

As noted in the discussion of Fig. 2.1 furnaces operate with lower characteristic velocities and involve larger length scales than the main burners of gas turbines. Buoyancy effects and lower intensity turbulence are thus encountered in furnaces. Therefore, a more rigorous test of our ability to predict the characteristics of turbulent combustion is presented by the gas turbine combustor, which will accordingly be emphasized here.

Let us consider the geometry of a conventional combustor for a gas turbine. Figure 2.2 represents a cross section through a can combustor or through one fuel injector in an annular combustor. Liquid fuel is injected into the primary zone, which is characterized by a large-scale recirculation, established either by swirling the air as it enters the front of the combustor (the dome), or by impingement of the discrete air jets from large holes in the combustor liner (the skirt). Because of the flow rates of the air and fuel entering separately, there exist: a fuel-rich zone near the injector, where smoke is formed; a stoichiometric contour in the shear layer between the recirculation zone and that portion of the flow moving downstream, important both to flame stabilization and nitric oxide formation; and a fuel-lean region near the wall where air is added through film cooling slots. This latter region is responsible for quenching of the fuel and partially oxidized carbon monoxide reactions.

Generally, the rates of injection of fuel and air into the primary zone are such that the combustion is said to be overall stoichiometric, i.e., if the fuel and air are instantaneously and perfectly mixed, there is sufficient air for complete combustion. However, as noted previously, the primary zone is far from perfectly mixed. More air is added in the secondary zone in an attempt to complete combustion and in the dilution zone to lower the temperatures to values acceptable to the downstream turbomachinery.

It is useful to comment briefly on the implication of the phenomena discussed here relative to methods of prediction of combustor performance. The presence of recirculation in the primary zone results in the describing equations being elliptic so that sophisticated numerical techniques are required. The presence of air penetration holes in the secondary and dilution zones results in the flow being truly three dimensional. Finally, in practice a liquid fuel is injected into the primary zone so that two-phase flows are involved. Note that the complexity associated with the description of such flows is not included in this volume.

A review of our current ability to predict the flow in the combustor of a gas turbine such as described here but with gaseous fuel was provided by *Serag-Eldin* and *Spalding* [2.25]. A three-dimensional flow involving elliptic equations is considered. Time-averaged conservation equations are solved for mass; three components of momentum; turbulent kinetic energy and its dissipation rate; enthalpy; the concentration f defined as $Y_1 - Y_2/s$, where s is the ratio of oxygen (species 2) to fuel (species 1) masses in a stoichiometric mixture; and g, the time-mean square of the fluctuations of f, i.e., (f'^2). Note that f is a

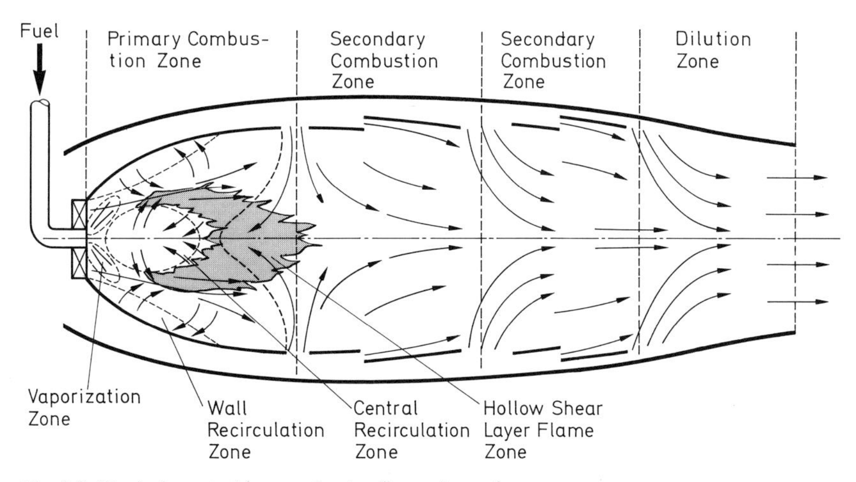

Fig. 2.2. Typical gas turbine combustor flow schematic

conserved scalar which serves the same role as an element mass fraction Z_i of (1.18). Closure is effected by introduction of a local effective eddy transport coefficient involving the turbulent kinetic energy and the dissipation rate. Fast chemistry is assumed to prevail with the probability density function for f assumed to have a battlement shape (cf. Chap. 3).

The analysis is first validated by comparing predictions resulting therefrom with cold flow experiments. Helium is used as a tracer and injected through one of the dilution jets. Helium concentrations are measured at various downstream stations. Figure 2.3 taken from [2.26] compares predicted and experimental radial distributions of helium concentration. Reasonable agreement is shown. Even better agreement is found for the radial profiles of the axial velocity but the swirl velocity and pressure are not as accurately predicted.

These results are generally consistent with those obtained by *Lockwood* et al. [2.27] who used a similar approach but with a simpler formulation for the eddy transport coefficient. Predictions of local mixing without reaction are shown in Fig. 2.4 which presents contours of local mixture fraction denoted by f and proportional to the f defined earlier in connection with the calculation of *Serag-Eldin* and *Spalding* [2.25]. The contours obtained experimentally are shown in the upper portion of the figure, those from prediction in the lower half. The reasonable agreement is seen.

These and similar results demonstrate current capability of calculating with reasonable accuracy turbulent flows with nearly constant density even when recirculation requires the treatment of elliptic equations. In this regard further testing of the mixing model used in [2.25] may be found in *Pope* and *Whitelaw* [2.28].

We now consider the comparison of prediction and experiment from the analysis of *Serag-Eldin* and *Spalding* [2.25] when combustion occurs. In Fig. 2.5 are shown predicted and measured radial profiles of temperature at three

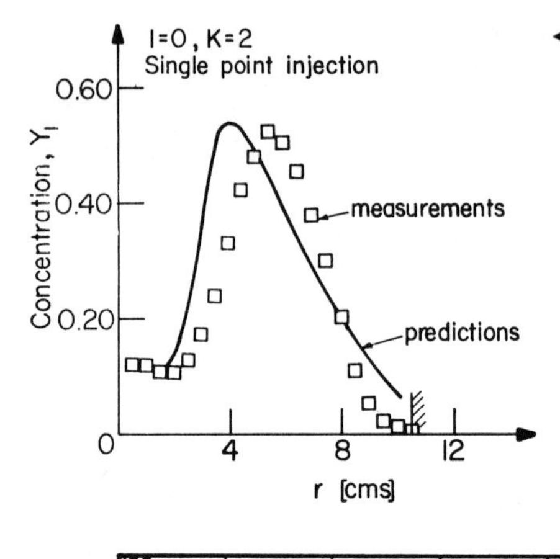

◀ **Fig. 2.3.** Predicted and measured radial profiles of He concentration in cold flow [2.26]

Fig. 2.4. Comparison of predicted and measured mixture fraction contours [2.27]
▼

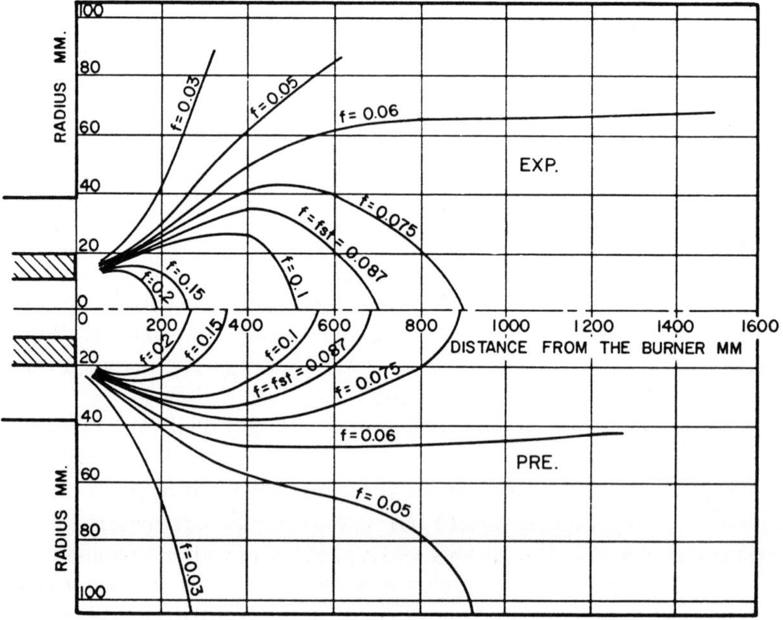

equivalence ratios, i.e., fuel/air flow ratios, at three planes downstream of the dilution holes. We see that although satisfactory agreement is obtained for flame *B*, except near the combustor wall, the discrepancies for flames *A* and *C* can exceed 400 K. Further, predictions are generally too high in flame *C*, but too low in flame *A*; thus the trends are not even qualitatively correct.

The results shown in Fig. 2.5 are typical of those obtained for other reacting flows by other analyses. In terms of many criteria these results are not considered satisfactory and indicate the inadequacy of existing analyses, undoubtedly related to problems of treating variable density turbulence and

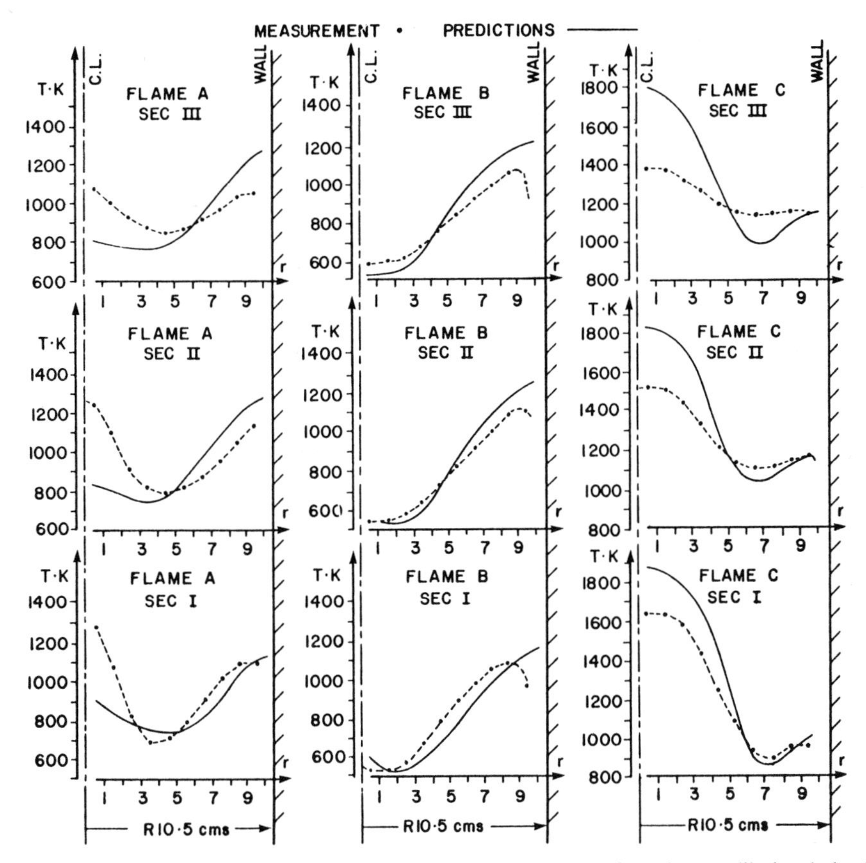

Fig. 2.5. Measured versus predicted temperatures downstream of combustor dilution holes for three flames of differing equivalence ratios [2.25]

the interaction of chemical reaction and turbulence. In view of this inadequacy other performance parameters of practical interest which depend on local heat release, such as nitric oxide emissions, combustion efficiency, smoke emissions, lean flame stabilization, radiant loading on the combustor liner, and ignitability, all matters of interest to the designer, cannot be predicted with sufficient accuracy at the present time. A similar inadequacy prevails in connection with the design of furnaces.

The prediction of some of the performance parameters cited earlier requires consideration of more detailed chemical schemes than that treated in [2.25]. Such consideration requires more species conservation equations [cf. (1.17)] and many more correlations of species concentrations [cf. (1.56)]. Even with suitable modeling to reduce the number of dependent variables, computer requirements increase enormously. A result of this situation is that a variety of predictive methods of an ad hoc nature are employed in attempts to make predictions suitable for design purposes.

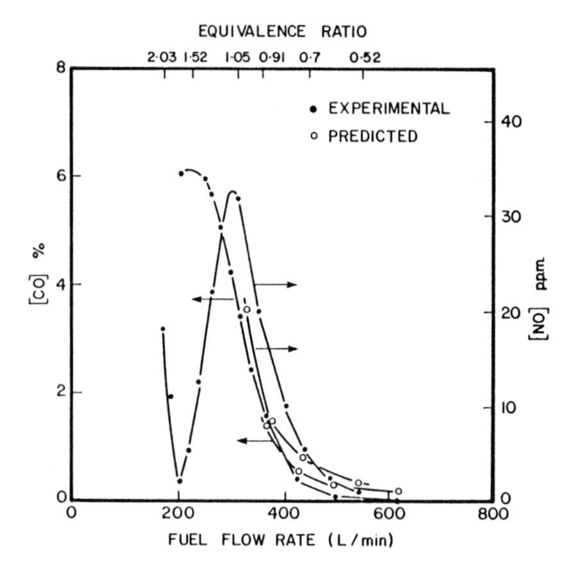

Fig. 2.6. Predicted versus experimental exhaust plane emissions of CO and NO as functions of fuel flow rate [2.29]

In this spirit *Swithenbank* et al. [2.29, 30] used a simplified version of the *Serag-Eldin* and *Spalding* theory to identify intensely stirred and partially stirred regions within the combustor. This information is used to divide the flow into a small number of regions which are treated as interconnected chemical reactors, both well and partially stirred. Detailed kinetics can be included with considerably reduced computational effort, but this description clearly suggests the ad hoc nature of the calculation. In skillful and experienced hands useful results can be obtained.

Although to date the analysis of [2.29, 30] has not been applied to practical combustors, Fig. 2.6 compares prediction and experiment for the emission of NO and CO from the exhaust plane of a simplified, laboratory combustor as functions of equivalence ratio. Unfortunately, these calculations employ an incorrect rate coefficient for a reaction step presumed to be important (see [2.30, 31]) with the consequence that the agreement shown in Fig. 2.6 must be regarded as fortuitous.

Another alternative to use of a detailed model for the entire flow field, such as that of *Serag-Eldin* and *Spalding*, is to identify that subregion of the entire flow responsible for a particular performance feature, and then model this subregion in detail [2.32]. This approach can also be adopted for special parts of the flow, such as the region downstream of the wall film cooling slots, where sufficient spatial resolution from the computational grid cannot be obtained for the complete flow [2.33]. As noted at the beginning of this section, for flame stabilization and nitric oxide formation in a conventional gas turbine combustor, the shear layer surrounding the recirculation zone in the primary is of interest, whereas for unburned fuel and carbon monoxide oxidation, regions adjacent to the combustor wall are more relevant [2.10].

In view of the complex problems which must be solved before accurate performance predictions can be expected from detailed models, semi-empirical scaling procedures in terms of Damköhler numbers have been developed both for pollutant emissions [2.34–38] and for lean flame stabilization and spark ignition [2.13, 39, 40]. As pointed out in Chap. 1, many such numbers can be defined for a given flow field; again it is emphasized that the relevant subregion of the flow must be recognized if successful correlation is expected.

For example, in the original work for a vehicular gas turbine combustor [2.35], the emissions index of nitrogen oxides (NO_xEI, g NO as NO_2/kg fuel) is found to scale linearly with the appropriate Damköhler number for NO formation

$$NO_xEI \sim D_{NO} = \tau_{sl,NO}/\tau_{NO} . \tag{2.1}$$

Here τ_{NO} is the chemical time scale for NO formation, defined as

$$\tau_{NO} = 10^{-12} \exp(66{,}969/T_{\phi=1})\,\mathrm{ms} , \tag{2.2}$$

where $T_a = 67\ (10^3)$ K and the characteristic temperature for NO formation is the stoichiometric adiabatic flame temperature $T_{\phi=1}$. The activation temperature corresponds to the well-known chemistry of NO formation in lean systems, and the characteristic temperature recognizes that NO forms at an appreciable rate only at the highest temperature in the combustor (i.e., around the mean stoichiometric contour in the primary zone).

The numerator of (2.1) is a turbulent mixing time evaluated in the shear layer (sl) where NO forms and is believed proportional to the ratio l_0/U' given in (1.71). By observing the dependence of NO_x exhaust emissions on combustor geometry and inlet conditions, $\tau_{sl,NO}$ is found to be

$$\tau_{sl,NO} = l_{NO}/U_{\phi=1} , \tag{2.3}$$

where

$$l_{NO}^{-1} = l_{sec}^{-1} + d_{comb}^{-1} \tag{2.4}$$

with l_{sec} the axial distance from fuel injector to secondary air penetration jet and d_{comb} combustor diameter at the latter location. Also

$$U_{\phi=1} \equiv \dot{m}_{a\phi=1} R T_{\phi=1}/p(\pi/4) d_{comb}^2 , \tag{2.5}$$

where $\dot{m}_{a\phi=1}$ is the air flow rate through the primary zone, R is the gas constant for air, and p is combustor inlet pressure. The correlation for ten different combustor geometries and three fuel injectors is shown in Fig. 2.7.

Related results have been obtained for CO [2.35] and other combustors [2.36, 38].

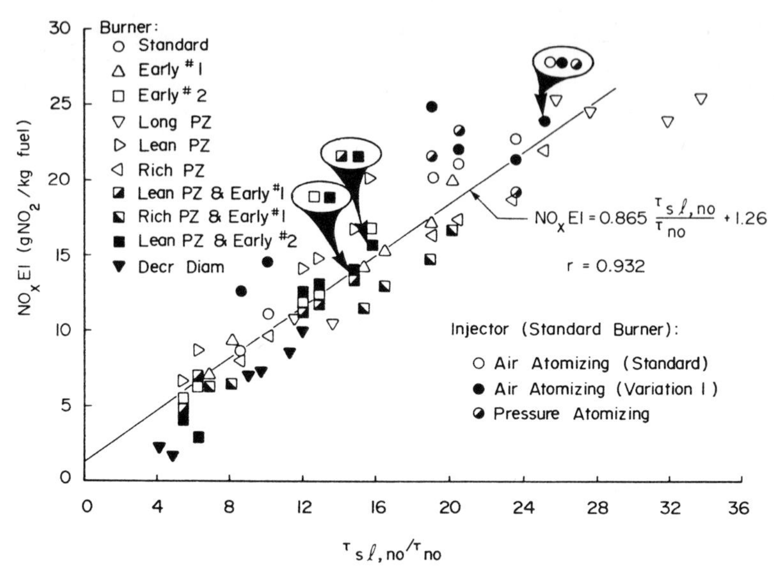

Fig. 2.7. Characteristic time NO_x emissions index correlation for ten vehicular turbine combustors and three differing fuel injectors [2.35]

The predictive ability of the correlation of (2.1) was tested by *Hammond* [2.37] who obtained exhaust emissions for two low NO_x combustors. These are designated Modifications *A* and *B* in Fig. 2.8, and the agreement between prediction and experiment is quite good over most of the engine power range.

Two points about this simple model are surprising. First it is not clear that

$$l_{NO}/U_{\phi=1} \sim l_0/U' \tag{2.6}$$

but resolution of this result will require turbulence measurements inside a combustor with burning in progress.

Second, (2.1) can be derived from first principles [2.34] without consideration of turbulence-chemistry interactions. Perhaps this omission is in part responsible for the scatter around the linear least squares fit of Fig. 2.7, or perhaps the success of the model results from quenching of *large* NO forming eddies by the secondary air penetration jets. Comparison of characteristic time model results with the output of detailed numerical simulations of the combustor flow field may elucidate these questions.

One of the results which might be expected to be obtained from a more adequate theory of turbulent reacting flows for quasi-steady situations is improved explanation of the correlations of the type given by (2.1).

For nonsteady combustion the time subsequent to ignition becomes another important time scale. Perhaps one of the more important practical applications which uses intermittent combustion is the reciprocating engine, the focus of the next section.

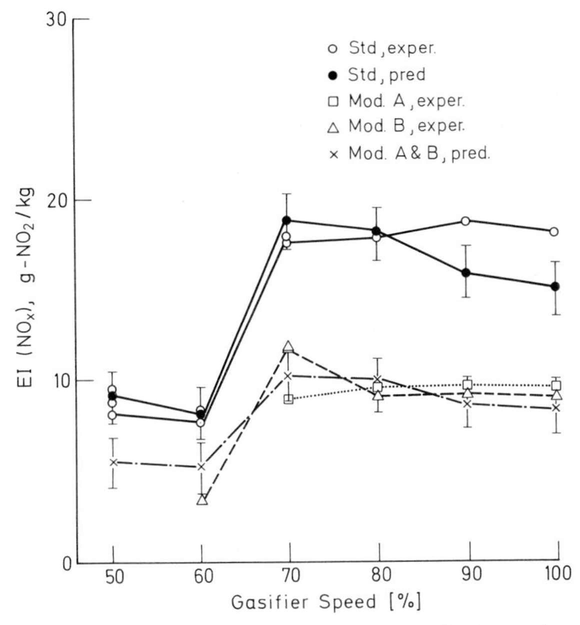

Fig. 2.8. Comparison of predicted and measured exhaust plane NO as functions of engine power [2.37]

2.4 Internal Combustion Engines

In contrast with the flows described so far the processes in an internal combustion engine are quasi-periodic because of the cyclic nature of the engine operation. Hence, the averaging techniques described in Chap. 1 are not applicable. Rather one must average over a number of different experiments performed in the same way, a technique usually referred to as ensemble or sometimes as phase averaging. In terms of the crank angle θ, which is linearly related to time by the engine frequency $\theta = \omega t$, the four-stroke engine is a cyclic device with a period of 4π. It is convenient to average over many cycles of engine operation. For example, *Lancaster* [2.41] has ensemble averaged velocity measurements by the following relation:

$$\bar{u}_i(\boldsymbol{x}, \theta) = \frac{1}{N} \sum_{j=1}^{N} u_i(\boldsymbol{x}, \theta, j), \tag{2.7}$$

where N is the number of cycles averaged and θ varies from 0 to 4π for any cycle. There has not yet been an exposition of the ensemble averaged conservation equations that can be applied to reacting flows in reciprocating engines. Indeed turbulence is difficult to define; *Lancaster* [2.41] pointed out that turbulence measurements of flows in reciprocating engines have trouble distinguishing between cyclically different-structured flows and those fluctuations due to turbulence within any of the structured flows.

To illustrate that point consider the instantaneous velocity of the jth cycle as

$$u_i(\mathbf{x},\theta,j)=\bar{u}_i(\mathbf{x},\theta)+u_i'(\mathbf{x},\theta,j)+u_i''(\mathbf{x},\theta,j)\,, \tag{2.8}$$

where $\bar{u}_i$ is defined by (2.7), u_i' is the turbulent fluctuation, and u_i'' is the cyclic variation of the velocity. In some situations, where u_i'' is slowly varying over some period Θ that is large compared to the turbulence time scale but small compared to a time characteristic of the changing flow, it is possible to separate cyclic variations from the turbulence. During such a period one can write for any cycle j

$$u_i''(\mathbf{x},\theta,j)\approx\frac{1}{\Theta-\theta}\int_{\theta}^{\Theta}u_i(\mathbf{x},\theta,j)d\theta-\bar{u}_i(\mathbf{x},\theta)\,. \tag{2.9}$$

The turbulence at any time in the engine is then determined by ensemble averaging $u_i''^2$ and subtracting from the ensemble averaged mean square fluctuations.

Combustion modeling in reciprocating engines has used (1.8–21) modified by modeling a turbulent viscosity. Presumably then the calculations are applicable to an ensemble of similarly structured flows since $u_i''(\mathbf{x},\theta,j)$ is not considered. The validity of such an approach remains to be demonstrated by comparison between theory and experiment. Such comparisons have recently appeared [2.42].

The mean chemical reaction occurring in reciprocating engines involves both limiting cases of fast chemistry, kinetically controlled reactions, and intermediate cases as well. In diesel engines and stratified charge engines much of the combustion occurs at a rate controlled by mixing; as mentioned in Sect. 1.14 the chemistry is said to be fast. Nitric oxide kinetics in the homogeneous charge spark ignition engine are an example of slow chemistry whereas the chemistry controlling flame propagation is neither fast nor slow. The treatment of all chemistry in modeling combustion engines by the same rate equations one would use for laminar flames, the perfectly mixed case, is probably the most fundamental limitation of current theories.

In addition to the difficulties discussed earlier in connection with predicting the performance of combustors for gas turbines, the predictions for the internal combustion engine involve special difficulties. The cylinder and piston surfaces are generally rough and perhaps covered with oil and/or carbonaceous deposits which are chemically and thermally significant. In addition the surface temperatures are neither spatially nor temporally uniform. Finally, the initial and boundary conditions to be applied in a calculation procedure are uncertain because of the influence of the spark, the fuel injector, the flow into and out of valves, and heat transfer to the surfaces.

However, a great deal is known about the performance and emissions characteristics of most engines from experiments. Computations have been done using the equations described and parameters of turbulence and chemical models fitted so that those characteristics observed to date are calculated correctly. The computations then offer an explanation as to why the performance and emission characteristics are as observed and as such lead to suggestions for improvement.

Calculations were applied to engines involving a homogeneous charge and spark ignition by *Sirignano* [2.43] and *Griffin* et al. [2.44] and to stratified charge engines by *Bellan* and *Sirignano* [2.45], *Westbrook* [2.46], *Boni* et al. [2.47], and *Dwyer* and *Sanders* [2.48]. There are also calculations (more complicated because a fuel spray equation is required) of direct injection stratified charge engines by *Bracco* [2.49] and *Butler* et al. [2.50] but there are no finite difference models of diesel engine combustion. The paper by *Bellan* and *Sirignano* [2.45] is discussed here in detail since it provides a good example of what can be done with some of the difficulties and because it is unique in that a $k-\varepsilon$ model[1] has been applied to the unsteady problem.

The purpose of [2.45] was to investigate the formation of nitric oxide NO during the operation of a stratified charge engine with an open chamber in which the fuel is completely vaporized prior to ignition. Let us first consider their model in the context of some NO emission characteristics reported by *Lavoie* and *Blumberg* [2.51] for such an engine. Figure 2.9 shows the combustion chamber of a Ford programmed combustion (PROCO) engine. Fuel is sprayed into the cylinder during the compression stroke such that substantial vaporization of the fuel and mixing with air occurs prior to ignition.

It is intended that there be a rich fuel/air mixture at the spark plug surrounded by a lean mixture. The rich mixture is spark ignited and a flame spreads rapidly through the chamber because of turbulence generated by the inlet flow and by piston motion in the compression stroke.

Nitric oxide emissions as a function of equivalence ratio for the PROCO engine run with a homogeneous charge, i.e., no stratification, are shown in Fig. 2.10. This figure suggests that to minimize NO emissions this engine should be operated with lean fuel/air mixtures.

There is a problem however with running lean homogeneous mixtures related to the first portion of charge to burn. Due to decreased flame speeds [2.52] lean mixtures require a longer time for the flame front to grow to the integral scale of the turbulence. During this time there are fewer eddies participating in spreading the flame. If more eddies are involved in spreading the flame, then its behavior may be more regular since differences from eddy to eddy may be averaged out. Cycle to cycle variations are reduced by minimizing the time required for the flame to grow to the integral scale; a possible explanation of this observation is that for fixed engine speed, cylinder volume changes less during the time of irregular flame propagation.

1 The $k-\varepsilon$ model is a two-parameter turbulence model where k is the turbulent kinetic energy (which in this book is given the symbol g) and ε is turbulent dissipation rate.

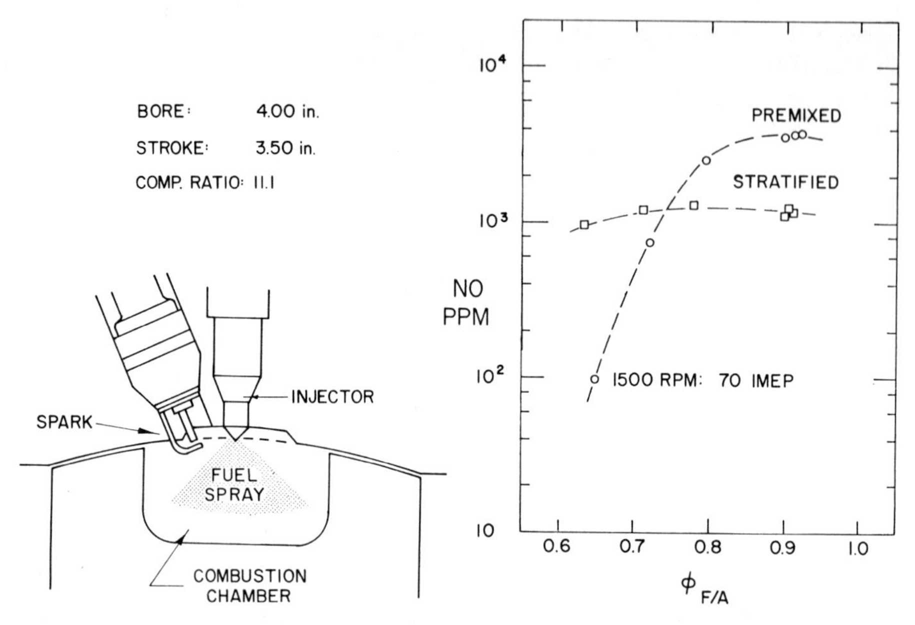

Fig. 2.9. Combustion chamber of a Ford PROCO stratified charge engine [2.51]

Fig. 2.10. Comparison of premixed and stratified nitric oxide emissions of the PROCO engine [2.51]

One motivation for a stratified charge engine is to burn to the integral scale using a rich mixture to minimize this time, which is often referred to as the ignition delay, but to burn the rest of charge so that the engine is running lean overall to maintain lower NO emissions. This can be accomplished as shown by the data in Fig. 2.10 where results are also given for stratified operation of the PROCO engine. Notice that stratification is not beneficial for mixture ratios less than 0.75.

Bellan and *Sirignano* [2.45] did calculations for one-dimensional flame propagation in a cylindrical coordinate system to model such an engine. The calculation starts at a time just after ignition. The distribution of species, temperature, and turbulence diffusivity are specified as initial conditions. The fuel is assumed distributed as a circular core of fuel-rich mixture at the center of the cylinder called Region I. The rest of the cylinder is called Region II and is a fuel-lean mixture.

As mentioned, the turbulence diffusivity is calculated by a $k-\varepsilon$ model. According to this model the diffusivity is $\nu_T = g^2/\varepsilon$ where g is the turbulence kinetic energy and ε is the turbulent dissipation rate. A conservation equation for the turbulent kinetic energy is derived from the momentum equation using ensemble averaging. The derivation results in several diffusionlike terms which must be modeled. The modeled equation solved is

$$\bar{\varrho}\left(\frac{\partial g}{\partial t}+\bar{u}\frac{\partial g}{\partial r}\right)=\frac{1}{r}\frac{\partial}{\partial r}\left(\bar{\varrho}\nu_T\frac{\partial g}{\partial r}\right)+\bar{\varrho}\nu_T\left(\frac{\partial \bar{u}}{\partial r}\right)^2+\bar{\varrho}\nu_T\left(\frac{u_p}{\delta}\right)^2-0.09\bar{\varrho}\varepsilon\,, \tag{2.10}$$

Table 2.1

Parameters case	Overall equivalence ratio	Equivalence ratio in region I	Equivalence ratio in region II
2	0.82	1.26	0.51
10	0.82	1.37	0.50
11	0.78	1.26	0.51

Spark timing: $-10°$.
Compression ratio: 9.7.
Engine speed: 2400 rpm.
Volumetric efficiency: 0.7.
Exhaust gas recirculation: 0%.

where r is the spatial coordinate, δ is the instantaneous cylinder height, and u_p is the piston velocity. Equation (2.10) is the same type of conservation equation as (1.52), but is based on conventional rather than Favre averaging and contains additional modeling assumptions.

The terms on the right-hand side of (2.10) model the diffusion, production by gas motion, production by piston motion, and dissipation of turbulence kinetic energy. The constant 0.09 was chosen to coincide with time-averaged versions of the $k-\varepsilon$ model. Unity turbulent Prandtl and Schmidt numbers have been assumed.

The dissipation rate is also solved for using a modeled equation. In this case

$$\left(\frac{\partial \varepsilon}{\partial t}+\bar{u}\frac{\partial \varepsilon}{\partial r}\right)=\frac{1}{r}\left(\frac{\partial}{\partial r}\frac{\varrho \nu_T r}{1.3}\frac{\partial \varepsilon}{\partial r}\right)+1.45\bar{\varrho} g\left(\frac{\partial \bar{u}}{\partial r}\right)^2-0.18\bar{\varrho}\frac{\varepsilon^2}{g}. \tag{2.11}$$

Again there is a diffusion term, a production term, and a decay term. The constants in the equation were again chosen to coincide with time-averaged versions of the $k-\varepsilon$ model.

Equations (2.10) and (2.11) are used to model the turbulence diffusivity by $\nu_T = g^2/\varepsilon$. These are coupled to conservation equations developed as (1.8–21), averaged and modeled neglecting fluctuation terms in chemical rates, a dubious practice. Initial conditions lead to calculation of the flame propagation.

Table 2.1 shows the parameter variations they used to study the effect of stratification at constant equivalence ratio and the effect of equivalence ratio at constant stratification. They studied 12 different parameter variations and each was identified by a case number. Figures 2.11–14 present some of the results for their case 2. In these figures the coordinate η is a dimensionless distance with respect to cylinder radius that has been scaled with density.

Figure 2.11 presents the variation of the fuel mass fraction Y_F with respect to the distance η for different crank angles (times). The profile at $-10°$ is the initial condition chosen to simulate ignition. The results show that in the first 23.45° of crank angle the fuel is nearly depleted in the rich region; 54% of the fuel is consumed during this period. In the next 14.75° the combustion is 97% complete. The small amount of fuel remaining in the center disappears at 30.25°. Notice that while fuel in the first fuel-rich region is burned, depletion of

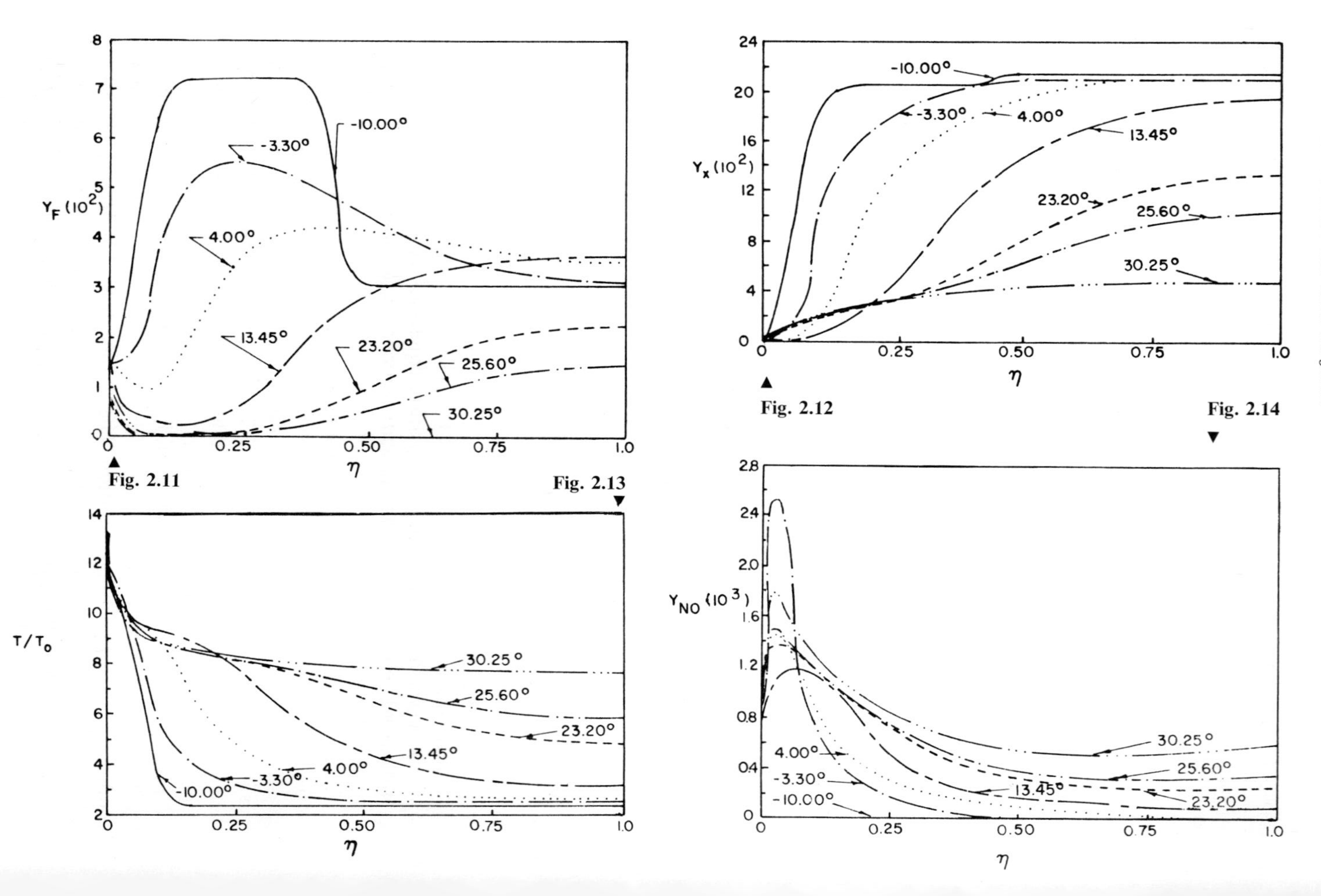

Fig. 2.11

Fig. 2.12

Fig. 2.13

Fig. 2.14

fuel together with a diffusion process ahead of the flame increase the fuel to be burned in the second fuel-lean region.

The spatial variation of the oxygen is given in Fig. 2.12. Comparison with Fig. 2.11 shows that the flame which develops in the first part of the cycle is of the premixed type compared to that existing during the end of combustion. As the flame progresses from the center to the walls of the cylinder, all the oxygen in the first region is consumed. Oxygen from the fuel-lean side then diffuses to meet the fuel which is diffusing from the rich region. Thus at that location the flame is of a diffusion type.

The temperature history is shown in Fig. 2.13. A thin flame propagates initially and thickens with time. When combustion starts, the burning rate is high near the center of the cylinder and negligible towards the walls. As the flame propagates through the lean region temperatures increase most significantly. The hottest spot in the cylinder is at the center where a maximum of 3460 K is reached. At the end of combustion the average temperature in the cylinder is 2125 K.

Finally, the nitric oxide mass fractions are given in Fig. 2.14. Just after ignition NO forms rapidly in the center region of the cylinder because of the high temperatures and the presence of O_2. However, as the O_2 is depleted this NO diffuses from the center and then forms once more as O_2 diffuses to the center. In the lean region O_2 is continuously consumed but never depleted and the temperature always increases for the times shown. Consequently NO here increases monotonically. Eventually temperatures fall due to expansion of the gases by piston motion and NO freezes at concentrations greater than those computed from local equilibrium. To predict the exhaust gas concentration the calculations must be carried out to $+180°$.

Case 10 (see Table 2.1) illustrates the effects of stratification on NO histories. The burning rate is increased by more stratification but the average temperature at the end of combustion is almost the same in both cases. The maximum temperature in case 10 is only 3405 K. There is also less O_2 in the center of the cylinder because the mixture is richer. When combustion is completed there are 606 ppm of NO in the mixture of case 10 but 725 ppm of NO are present in case 2.

In case 11 the stratification of the mixture is conserved but the overall mixture is leaner. Analytically, this is equivalent to narrowing the rich region. There is 4.5% less fuel than in case 2.

◂**Fig. 2.11.** Spatial and temporal distributions of fuel in a stratified charge engine. Case 2 of *Bellan* and *Sirignano* [2.45]

◂**Fig. 2.12.** Spatial and temporal distributions of oxygen in a stratified charge engine. Case 2 of *Bellan* and *Sirignano* [2.45]

◂**Fig. 2.13.** Spatial and temporal distributions of temperature in a stratified charge engine. $T_0 = 258$ K is the temperature at the beginning of the compression stroke. Case 2 of *Bellan* and *Sirignano* [2.45]

◂**Fig. 2.14.** Spatial and temporal distributions of nitric oxide in a stratified charge engine. Case 2 of *Bellan* and *Sirignano* [2.45]

Initially, the burning rates are the same up to $+5.00°$. Combustion then slows down in case 11. The temperature profiles are similar in both cases but a lower average temperature is realized in case 11, 2080 K. When burning is completed 894 ppm of NO are formed in case 11 and 725 ppm in case 2. It is difficult to draw a definite conclusion because at the end of combustion the pressure in case 2 is significantly higher than that of case 11; therefore the rate of formation of NO in case 2 will be higher at the end of combustion. In order to make predictions of exhaust quantities the calculations need to be carried out to $+180°$; they could not be in the present case because of numerical inaccuracies.

It is interesting to note that if, instead of using a spatially varying diffusivity as predicted by the $k-\varepsilon$ model, a spatially invariant diffusivity is used the predicted NO formed is then at all times less for the leaner case and one can safely conclude that there will also be less predicted in the exhaust [2.53].

An interesting question is then raised by the calculations: experimentally it is discovered that as one goes to leaner fuel/air ratios, then NO at the engine exhaust is invariant. When using an algebraic turbulence model the NO is reduced by going leaner. When using a $k-\varepsilon$ model the opposite behavior is observed, at least during the combustion phase.

An important difference between the two turbulence models is that the $k-\varepsilon$ model is coupled to the combustion so that in the less lean case there is a greater generation of turbulence. *Sirignano* [2.54] speculated that less NO is produced in this case because the width of the diffusion flame has been increased by the increased turbulence. This lowers the temperatures and reduces NO formation.

Bellan and *Sirignano* [2.45] indicated that the turbulence production is overpredicted by their $k-\varepsilon$ model. This offers one explanation for the discrepancy between experiment and theory. *Lavoie* and *Blumberg*'s [2.51] zero-dimensional calculations suggest another, that turbulence-chemistry interactions need to be included in the computation.

To summarize, these calculations of turbulent combustion in the reciprocating engine indicate that even trends can be incorrect. A similar result is exhibited for the gas turbine combustor in Fig. 2.5. Further, it has been demonstrated that the source of the inaccuracies within such detailed models is frequently unknown. It is for these reasons that turbulent combustion models to be discussed in the following chapters limit their attention to simple configurations amenable to both theoretical and experimental investigation expected to promise more rapid progress in modeling. Practical combustion systems, although the eventual goal of the modeler, currently involve too many unknowns.

Acknowledgements. The authors' research on engine combustion is supported by the U.S. Army Tank-Automotive Research and Development Command (P. Machala and A. Jaeger, technical monitors), the Army Research Office (J. Murray, technical monitor), the Air Force Office of Scientific Research (B. T. Wolfson, technical monitor), DOT Transportation Systems Center (T. Trella, technical monitor), General Motors Corporation (J. S. Collman and C. Amann, technical monitors), Ford Motor Company (W. R. Wade, technical monitor), and the General Electric Company (C. Wilkes, technical monitor), to all of whom they express their gratitude. The United States Government is authorized to reproduce and distribute reprints for governmental purposes notwithstanding any copyright notation hereon.

References

2.1 K.N.C.Bray: "The Interaction between Turbulence and Combustion", in *Seventeenth Symposium (International) on Combustion* (The Combustion Institute, Pittsburgh 1979) pp. 223–233

2.2 W.G.Agnew: "Automotive Engine Combustion Research Needs", presented at 1977 meeting, Eastern States Section/The Combustion Institute

2.3 W.G.Agnew: Prog. Energy Combust. Sci. **4**, 115–155 (1978)

2.4 P.N.Blumberg: "Requirements, Needs, and Outstanding Combustion Related Problems of the Automotive Community", Workshop on the Numerical Simulation of Combustion for Application to Spark and Compression Ignition Energies, SAI, La Jolla, California (1975) pp. 3-65–3-77

2.5 N.A.Henein: Prog. Energy Combust. Sci. **1**, 165 (1976)

2.6 J.B.Heywood: "Pollutant Formation and Control in Spark Ignition Engines", *Fifteenth Symposium (International) on Combustion* (The Combustion Institute, Pittsburgh 1975) pp. 1191–1211

2.7 J.B.Heywood: Prog. Energy Combust. Sci. **1**, 135 (1976)

2.8 S.M.Shahed, P.E.Flynn, W.T.Lyn: "Diesel Combustion: Review and Prospects", presented at 1978 Meeting, Central States Section/The Combustion Institute

2.9 A.H.Lefebvre: "Pollution Control in Continuous Combustion Engines", Ref. 2.6, pp. 1169–1180

2.10 A.M.Mellor: "Gas Turbine Engine Pollution", *Pollution Formation and Destruction in Flames*, Vol. I of *Progress in Energy and Combustion Science*, ed. by N.A.Chigier (Pergamon, Oxford 1976) pp. 111–133

2.11 A.M.Mellor: "Turbulent-combustion Interaction Models for Practical High Intensity Combustors", Ref. 2.1, pp. 377–387

2.12 R.E.Henderson, W.S.Blazowski: "Turbopropulsion Combustion Technology", in *The Aerothermodynamics of Aircraft Gas Turbine Engines*, AFAPL-TR-78-52 (1978) Chap. 20

2.13 E.Zukoski: "Afterburners", Ref. 2.12, Chap. 21

2.14 B.P.Breen: "Combustion in Large Boilers: Design and Operating Effects on Efficiency and Emissions", *Sixteenth Symposium (International) on Combustion* (The Combustion Institute, Pittsburgh 1977) pp. 19–35

2.15 R.H.Essenhigh: "Combustion and Flame Propagation in Coal Systems: A Review", Ref. 2.14, pp. 353–374

2.16 J.M.Beér: "Fluidized Combustion of Coal", Ref. 2.14, pp. 439–460

2.17 A.Maček: "Coal Combustion in Boilers: a Mature Technology Facing New Constraints", Ref. 2.1, pp. 65–75

2.18 B.R.Bronfin: "Continuous Flow Combustion Lasers", Ref. 2.6, pp. 935–950

2.19 P.H.Thomas: "Behavior of Fires in Enclosures – Some Recent Progress", *Fourteenth Symposium (International) on Combustion* (The Combustion Institute, Pittsburgh 1973) pp. 1007–1020

2.20 R.A.Strehlow: "Unconfined Vapor Cloud Explosions – An Overview", Ref. 2.19, pp. 1189–1200

2.21 J.Grumer: "Recent Research Concerning Extinguishment of Coal Dust Explosions", Ref. 2.6, pp. 103–114

2.22 F.A.Williams: "Mechanisms of Fire Spread", Ref. 2.14, pp. 1281–1294

2.23 J. de Ris: "Fire Radiation – A Review", Ref. 2.1, pp. 1003–1016

2.24 R.Goulard, A.M.Mellor, R.W.Bilger: Combust. Sci. Technol. **14**, 195–219 (1976)
2.25 M.A.Serag-Eldin, D.B.Spalding: "Computation of Three-dimensional Gas Turbine Combustion Chamber Flows", ASME 23rd International Gas Turbine Conference, London (1978)
2.26 M.A.Serag-Eldin: "The Numerical Prediction of the Flow and Combustion Processes in a Three-dimensional Can Combustor", Ph. D. Thesis, London University (1977)
2.27 F.C.Lockwood, F.M.El-Mahallawy, D.B.Spalding: Combust. Flame **23**, 283–293 (1974)
2.28 S.B.Pope, J.H.Whitelaw: J. Fluid Mech. **73**(1), 9–32 (1976)
2.29 P.G.Felton, J.Swithenbank, A.Turan: "Progress in Modelling Combustors", HIC 300, Sheffield University (1977)
2.30 J.Swithenbank, A.Turan, P.G.Felton: "Three-dimensional Two-phase Mathematical Modelling of Gas Turbine Combustors", in *Gas Turbine Combustor Design Problems* (Hemisphere, Washington 1980) pp. 249–314
2.31 R.B.Edelman, P.T.Harsha: "Some Observations on Turbulent Mixing with Chemical Reactions", *Turbulent Combustion* (AIAA, New York 1978) pp. 55–102
2.32 M.M.M.Abou Ellail, A.D.Gosman, F.C.Lockwood, I.E.A.Megahed: "Description and Validation of a Three-dimensional Procedure for Combustion Chamber Flows", Ref. 2.31, pp. 163–190
2.33 H.G.Mongia, K.Smith: "An Empirical/Analytical Design Methodology for Gas Turbine Combustors", AIAA Paper No. 78–998 (1978)
2.34 J.H.Tuttle, M.B.Colket, R.W.Bilger, A.M.Mellor: "Characteristic Times for Combustion and Pollutant Formation in Spray Combustion", Ref. 2.14, pp. 209–219
2.35 A.M.Mellor: AIAA J. Energy **1**, 244–249 (1977)
2.36 A.M.Mellor: AIAA J. Energy **1**, 257–262 (1977)
2.37 D.C.Hammond, Jr.: AIAA J. Energy **1**, 250–256 (1977)
2.38 A.M.Mellor, R.M.Washam: AIAA J. Energy **3**, 250–253 (1979)
2.39 S.L.Plee, A.M.Mellor: Combust. Flame **35**, 61–80 (1979)
2.40 J.E.Peters, A.M.Mellor: Combust. Flame **38**, 65–74 (1980)
2.41 D.R.Lancaster: SAE Trans. **85**, 651–670, paper 760159 (1976)
2.42 H.C.Gupta, R.L.Steinberger, F.V.Bracco: Combust. Sci. Technol. **22**, 63–82 (1980)
2.43 W.A.Sirignano: Combust. Sci. Technol. **1**, 99–108 (1973)
2.44 M.D.Griffin, R.Diwaker, J.D.Anderson: "Computational Fluid Dynamics Applied to Flows in an Internal Combustion Engine", AIAA Paper 78–57 (1978)
2.45 J.R.Bellan, W.A.Sirignano: Combust. Sci. Technol. **12**, 75–104 (1976)
2.46 C.K.Westbrook: Acta Astron. **5**, 1185–1189 (1978)
2.47 A.A.Boni, M.Chapman, J.L.Cook, G.P.Schneyer: "Computer Simulation of Combustion in a Stratified Charge Engine", Ref. 2.14, pp. 1527–1542
2.48 H.A.Dwyer, B.R.Sanders: "Unsteady Flow and Flame Propagation in a Prechamber of a Stratified Charge Engine", The Institute of Mechanical Engineers, Conference on Stratified Charge Engine, London, England (1976)
2.49 F.V.Bracco: Combust. Sci. Technol. **8**, 69–84 (1973)
2.50 T.D.Butler, L.D.Cloutman, J.K.Dukowicz, J.D.Ramshaw, R.B.Krieger: "Toward a Comprehensive Model for Combustion in a Direct-injection Stratified-charge Engine", in *Combustion Modeling in Reciprocating Engines* (Plenum, New York 1980) pp. 231–264
2.51 G.A.Lavoie, P.N.Blumberg: Combust. Sci. Technol. **8**, 25–38 (1973)
2.52 R.J.Tabaczynski, C.R.Ferguson, K.Radhakrishnan: SAE Trans. **87**, 2414–2433, paper 770647 (1977)
2.53 J.R.Bellan, W.A.Sirignano: Combust. Sci. Technol. **8**, 51–69 (1973)
2.54 W.A.Sirignano: Private communication (1978)

3. Turbulent Flows with Nonpremixed Reactants

R. W. Bilger

With 6 Figures

An essential feature of many of the practical problems described in Chap. 2 is that the rate of chemical reaction is limited, to some extent at least, by the rate of mixing of the reactants. In many of these problems the reactants enter the field of interest in two streams or feeds of distinct but constant chemical form. In these cases the assumption of equal diffusivities for all species leads to the result that the state of mixing is uniquely determined by one conserved scalar variable and this leads to a great simplification of the analysis and of the conceptual description of the flow.

Problems of particular interest in the modeling of these flows lie in the handling of the mean chemical production term (see Sect. 1.14) and the effects of the heat release on the turbulence structure and hence on the turbulence flux closure models (see Sect. 1.13). These are the most pertinent questions of the so-called chemistry/turbulence interactions. For the two-feed system in the limit of fast chemistry the first of these problems is largely obviated by the fact that the molecular species are instantaneously related to the conserved scalar and the statistics of all thermodynamic variables are obtainable from sufficient knowledge of the statistics of that scalar (Sect. 3.1).

For finite rate chemistry a two-variable modification of this approach is possible with one variable being the conserved scalar and the other being a molecular species or progress variable for which the chemical production term must be modeled. A second variable of some promise is that giving the perturbation from the fast chemistry solution (Sect. 3.2). An alternative approach is to attempt a second-order closure of the mean chemical production term for all species. Section 3.3 discusses the methods that have been tried and their probable limitation to flows with only moderate kinetic rates. A set of further alternatives derives from models arising from fast mixing assumptions such as the theory of the perfectly stirred reactor. These are considered in Sect. 3.4.

Spectral considerations and the appropriate nondimensional numbers to determine the relative regime are dealt with in Sect. 3.5. The modeling of the turbulent flux terms and questions of flame generated turbulence are considered in Sect. 3.6.

Questions of flame structure have long fascinated workers in this field: is there essentially a highly contorted thin flame sheet or is the reaction distributed over a broad zone even on an instantaneous basis? Even if the chemistry may be considered fast for the main heat releasing reactions, what about the formation of nitric oxide, nitrogen dioxide, and soot which have slow kinetics? And the burn-out of soot and carbon monoxide? How does one handle flame radiation? These and other questions of great relevance in energy and environmental applications lie behind the more rigorous treatment of the subject now being developed.

3.1 The Conserved Scalar Approach

In many problems of practical interest the chemical reaction rates are fast so that the reaction is completed as soon as the reactants are mixed. The classical approach to the solution of these problems is to describe the mixing by obtaining the solution for a conserved scalar. The fast chemistry assumption implies that the instantaneous molecular species concentrations and temperature are functions only of the conserved scalar concentration at that instant. The functions are nonlinear, however, and the central problem becomes that of linking the means and higher moments of the species and temperature to those of the conserved scalar. This is usually accomplished by considering the complete probability density function (pdf) of the conserved scalar.

Figure 3.1 shows a simple mixing flow with its mixing pattern described in terms of a conserved scalar ξ. Such mixing patterns are measured experimentally by tagging one of the flows with an inert species or with refractory particles and obtaining the species or particle concentration using laser Raman or Mie scattering, respectively. Alternatively, element concentrations can be

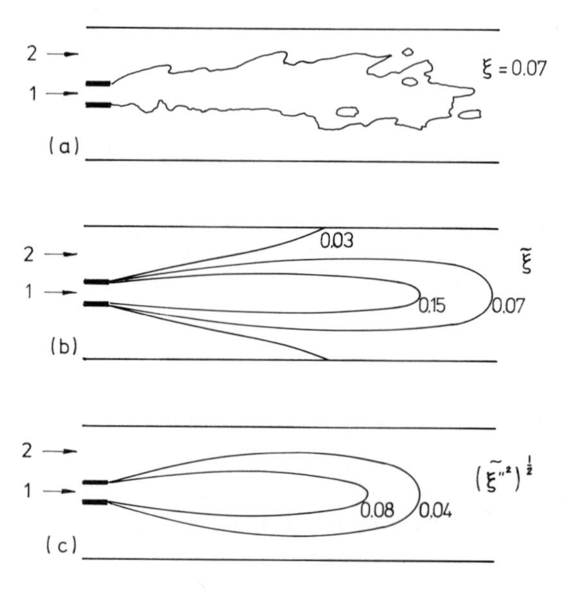

Fig. 3.1a–c. Mixing patterns in a typical turbulent reacting flow with nonpremixed reactants entering with feeds 1 and 2. (**a**) Instantaneous isopleth for a conserved scalar ξ; (**b**) isopleths for the Favre average of ξ; (**c**) contours of root mean square fluctuation of ξ

obtained by laser spark spectroscopy in which a very small volume of the gas is pulsed with a high-intensity laser converting it into plasma, the spectroscopic analysis of the light emission showing characteristic atomic and ionic lines with intensities related to atomic abundances. The conserved scalar approach can be used for systems very much more complex than that shown in Fig. 3.1 but it may help the physical interpretation of what follows to think of a simple system such as this.

3.1.1 Conserved Scalars

There are several scalar variables which are conserved under chemical reaction and which can be used as a basis for describing the mixing in a nonpremixed reactant flow. The mass fraction Z_i of element i is such a variable and the absence of a chemical source term in its conservation equation (1.19) has already been remarked upon. There are $M-1$ such variables where M is the number of elements and $\sum_{i=1}^{M} Z_i=1$. Solution of the $M-1$ equations for $\tilde{Z}_i$ (1.47) yields the mean elemental composition throughout the field and this may be looked upon as a description of the mixing of the fluid. Fast chemistry assumptions can yield the molecular species composition from this elemental composition.

Where the chemistry can be reduced to a one-step reaction such as

$$F+r\cdot 0 \rightleftharpoons (r+1)\,P\,, \tag{3.1}$$

simpler conserved scalars may be used. This reaction is written here in mass quantities rather than the more conventional molar form and may be read as "1 kg of fuel plus r kg of oxidant yields $(r+1)$ kg of product". We have

$$\dot{w}_F=\dot{w}_O/r=-\dot{w}_P/(r+1) \tag{3.2}$$

so that

$$\beta_{FO}\equiv Y_F-Y_O/r\,, \tag{3.3a}$$

$$\beta_{FP}\equiv Y_F+Y_P/(r+1)\,, \tag{3.3b}$$

$$\beta_{OP}\equiv Y_O+rY_P/(r+1)\,, \tag{3.3c}$$

are conserved scalars. These are examples of the so-called Shvab-Zeldovich coupling functions [3.1] and others may be formed using the sensible enthalpy.

There has been a great variation in the choice of conserved scalar in the literature [3.2–31] and some confusion has arisen with respect to the merits of various choices. In many circumstances, the conserved scalars are all linearly related so that solution for one yields all of the others; the choice is then arbitrary and a matter of taste. Under other circumstances there may be factors which favor the choice of one or more relative to the others. Such factors include the turbulence Reynolds number (nonequal diffusivity effects), the number and uniformity of the reactant feeds, and the complexity of the chemical mechanism.

Although the chemical production term is eliminated by the above conserved scalars, the balance equation will only take the form of (1.19) if the diffusivities of the molecular species are equal. If not, a source term results from differential diffusion effects. In turbulent flows of moderate to high Reynolds numbers such differential diffusion effects are usually neglected under an assumption analogous to the Reynolds number similarity hypothesis [3.32]. Reynolds number similarity, or the insensitivity of all but the finest scales of the turbulence structure to the flow Reynolds number, is a common feature of many turbulent flows. This independence of viscosity implies that turbulent fluxes dominate molecular fluxes and this can be expected to carry over to other molecular effects such as diffusion. Thus in (1.47) the molecular diffusion terms can be neglected in comparison with the turbulent diffusion $\overline{\varrho u_k'' Z_i''}$ and this will apply for any differential diffusion source terms too. Differential diffusion will only affect the high wave-number structure of the turbulent species composition. For flows with relatively low turbulence Reynolds number these differential effects may be significant for species such as H_2 and H which have very high diffusivities relative to other species. For such flows it may be prudent to form conserved scalars only from species of similar diffusivities and molecular weights. It will probably be necessary then to solve the nonconserved molecular species equation (1.17, 46)[1] for at least one molecular species to obtain closure.

Linear relationships among all the conserved scalars exist only when there are two uniform reactant feeds. This restriction on the reactant feeds only extends to elemental composition; each feed may be in several streams each of which may have any state of chemical aggregation, e.g., it may be partially reacted or pyrolyzed. If enthalpies are to be included in the conserved scalar set, then the enthalpies of the two feeds must be uniform. Uniformity here implies spatial and temporal constancy. When there are three feeds there are two independent conserved scalars and so on.

Conserved scalars of the Shvab-Zeldovich (3.3) type are only useful when a single one-step reaction such as (3.1) is involved. The analysis has been extended to two-step reactions [3.33]. In many chemical systems, particularly those involving high heat release, many multistep and parallel reactions occur and it is better to use conserved scalars based on elements.

For many problems the two-feed and equal diffusivity assumptions are good approximations and warrant the great simplification that results. All the

1 $\mathscr{D}_i$ may be used in place of $\mathscr{D}$ in these equations if the species i is relatively dilute; $\mathscr{D}_i$ is the binary diffusion coefficient of species i in the diluent.

conserved scalars are then linearly related and any one may be preferred. Here we prefer the mixture fraction. *Becker* [3.34] has studied three-feed mixing problems but we shall not discuss these further here.

3.1.2 The Mixture Fraction and Mixing Patterns

With the assumption of equal diffusivities the balance equation for a chemical element (1.19) has no source term and it is exactly similar for all elements. In a two-feed problem[2] the element mass fractions can be normalized in the form $(Z_i - Z_{i_2})/(Z_{i_1} - Z_{i_2})$ and then the boundary conditions become identical for all i. It can be shown [3.22] that such normalized variables are equal and furthermore that all conserved scalars are linearly related so that we may define

$$\xi \equiv \frac{Z_i - Z_{i_2}}{Z_{i_1} - Z_{i_2}} = \frac{\beta - \beta_2}{\beta_1 - \beta_2}, \tag{3.4}$$

which is called the mixture fraction. The subscripts 1 and 2 refer to the (uniform) composition in the two feeds and β is a Shvab-Zeldovich function of the type in (3.3). It is seen that $\xi = 1$ in feed 1 and 0 in feed 2, and that ξ may be physically interpreted as the mass fraction of material in the mixture which originated in feed 1, with $1 - \xi$ originating in feed 2. Thus the result of turbulent interdiffusion on the composition at a point is the same as if we took a mass ξ of feed 1 and mixed it with a mass $1 - \xi$ of feed 2 and let reaction occur.

Equation (1.19) becomes

$$\frac{\partial}{\partial t}(\varrho\xi) + \frac{\partial}{\partial x_k}(\varrho u_k \xi) = \frac{\partial}{\partial x_k}\left(\varrho \mathscr{D} \frac{\partial \xi}{\partial x_k}\right) \tag{3.5}$$

and the behavior of ξ is that of a conserved scalar in the flow. We can expect that its statistics will be like that of other conserved scalars such as inert species, and, to the first order, like that of temperature or concentration in a nonreacting slightly heated or mixing flow of similar geometry. For stationary turbulent flow of moderate to high Reynolds number (1.47) becomes

$$\frac{\partial}{\partial x_k}(\bar{\varrho}\tilde{u}_k\tilde{\xi}) = -\frac{\partial}{\partial x_k}(\overline{\varrho u_k'' \xi''}). \tag{3.6}$$

We shall discuss later (Sect. 3.6) the ways in which this equation can be modeled and solved. We may also write an equation for the Favre variance $\widetilde{\xi''^2} \equiv \overline{\varrho \xi''^2}/\bar{\varrho}$,

$$\bar{\varrho}\tilde{u}_k \frac{\partial \widetilde{\xi''^2}}{\partial x_k} = -2\overline{\varrho u_k'' \xi''}\frac{\partial \tilde{\xi}}{\partial x_k} - \frac{\partial}{\partial x_k}(\overline{\varrho u_k'' \xi''^2}) - 2\overline{\varrho \mathscr{D} \frac{\partial \xi''}{\partial x_k}\frac{\partial \xi''}{\partial x_k}}. \tag{3.7}$$

---

2 A no-flux condition applies on all bounding surfaces.

The terms on the RHS represent, respectively, the production by the mean gradient, diffusion by the velocity fluctuations, and dissipation by molecular diffusion. These can be modeled and the equation solved to give a measure of the "concentration fluctuations" that occur in the turbulent field [3.35], the fluctuations being in the conserved scalar.

Solutions of (3.5–7) will yield the mixing pattern for the flow with a typical example shown in Fig. 3.1. The instantaneous pattern of Fig. 3.1a will not be obtained with current solution methods and the mixing is described by solutions of (3.6) and (3.7) which result in patterns such as Fig. 3.1b and 3.1c. Typically $(\widetilde{\xi''^2})^{1/2}/\tilde{\xi} \sim 0.4$[3] so that the effect of concentration fluctuations can be very significant in determining the local instantaneous mixture strength.

In the two-feed problems discussed here there will always be a value ξ_s of the mixture fraction at which the feed materials are in exact stoichiometric proportion. For the one-step reaction (3.1) with no oxidant in the fuel feed and no fuel in the oxidant feed but perhaps an inert species in both,

$$\xi_s = \frac{Y_{O_2}}{rY_{F_1} + Y_{O_2}}. \tag{3.8}$$

This stoichiometric value of the conserved scalar corresponds to β_{FO} of (3.3a) equal to zero, and has a particular significance. For a one-step irreversible reaction with infinitely fast chemistry $\xi = \xi_s$ corresponds to the instantaneous location of the flame sheet and the point $\tilde{\xi} = \xi_s$ to its mean location. For reversible, multiple reaction and slower chemistry these locations are the centers of the instantaneous and mean reaction zones. For a system in which $\xi_s = 0.07$, Fig. 3.1 gives an indication of the instantaneous and time average reaction zone structure.

The assumption of unity Lewis number ($\mathscr{D} = k/\varrho C_p$, k being thermal conductivity) and neglect of the pressure fluctuation term $\partial p/\partial t$ bring the enthalpy conservation equation (1.15) into the same form as (1.19) and (3.5). If there are two feeds each with uniform enthalpy and if the system is adiabatic (for both radiative transfer and convective and conductive exchanges through wall boundaries), the enthalpy solutions will be fully similar to those for the other conserved scalars yielding

$$\frac{h - h_2}{h_1 - h_2} = \xi. \tag{3.9}$$

This is a valuable simplifying assumption in flows with significant heat release. The neglect of $\partial p/\partial t$ is justified in most systems involving low Mach numbers except where the spatially averaged pressure is varying with time as in an internal combustion engine.

---

3 In intermittent parts of the flow values very much higher than this are obtained.

3.1.3 Fast Chemistry Assumptions

For fast chemistry and the one-step irreversible reaction [reaction (3.1) in the forward direction only] there will be no oxidant present for mixtures richer than stoichiometric ($\xi > \xi_s$) and no fuel present when the mixture is weaker than stoichiometric ($\xi < \xi_s$). Both will be zero when the mixture is stoichiometric. This yields the functional relationships shown in Fig. 3.2a,

$$\beta_{FO} \leqq 0, \xi \leqq \xi_s : Y_F = 0\,; Y_O = r\beta_{FO} = r Y_B(\xi_s - \xi)\,; \quad Y_p = (r+1)\, Y_B \xi (1-\xi_s)\,, \tag{3.10a}$$

$$\beta_{FO} \geqq 0, \xi \geqq \xi_s : Y_F = Y_B(\xi - \xi_s) = \beta_{FO}\,; Y_O = 0\,; \quad Y_p = (r+1)\, Y_B \xi_s (1-\xi)\,, \tag{3.10b}$$

where

$$Y_B \equiv Y_{F_1} + \frac{1}{r} Y_{O_2} = Y_{F_1}/(1-\xi_s)\,. \tag{3.10c}$$

Thus at a given point in the flow the value of ξ at any particular instant defines the complete composition of the mixture at that instant through these functions.

If the reaction is reversible [back reaction in (3.1) finite], there will be both fuel and oxidant present near the stoichiometric value ξ_s due to the dissociative back reaction. The fast chemistry assumption implies that the forward and back reaction rates are fast compared with the turbulent mixing processes. Thus the composition of the mixture at a given point at any particular instant will be the same as that if the mixture were isolated and allowed to come to chemical equilibrium. The species concentrations are thus unique functions of ξ as illustrated in Fig. 3.2b. Mathematically, closure is obtained by the use of the equilibrium constant for the reaction. Where there is significant temperature variation this can be obtained from (3.9) and the equilibrium constant will be a function of ξ alone.

Even for many species and multiple reactions the fast chemistry assumption implies that these will be effectively equilibrated. There will be sufficient equilibrium constants to enable the molecular species to be calculated from the elemental composition and enthalpy given by (3.4) and (3.9). Computer codes exist for doing these calculations for even the most complex systems [3.36]. Results are shown in Figs. 3.2c and 3.2d for hydrogen/air mixtures.

The results above may be summarized by

$$Y_i = Y_i^e(\xi)\,. \tag{3.11}$$

The superscript e may be thought of as denoting equilibrium. In some situations it will denote equilibrium of only some of the species and some of the

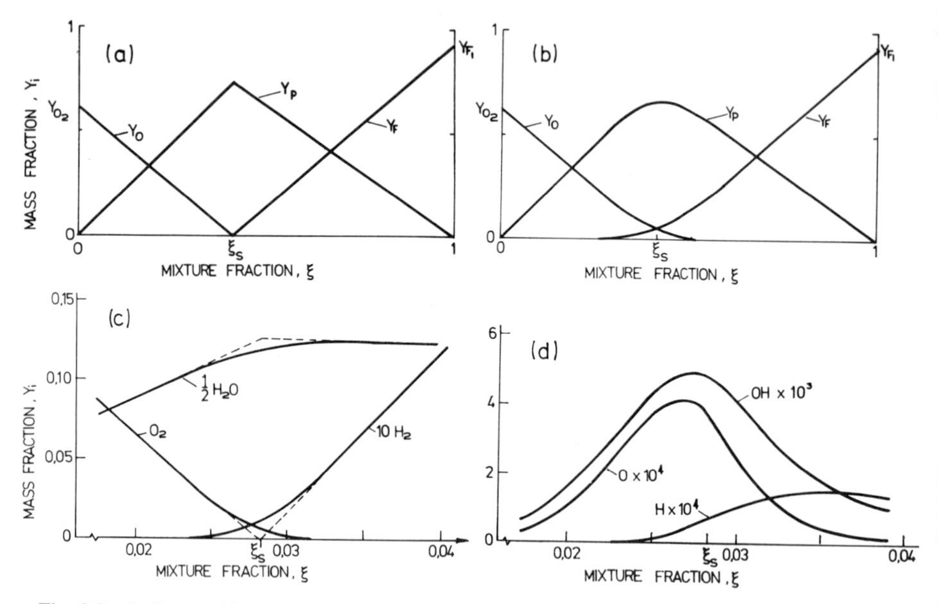

Fig. 3.2a–d. Composition structure functions for fast chemistry. (**a**) One step-irreversible reaction, (**b**) one-step reversible reaction, (**c**) equilibrium hydrogen/air, main species, (**d**) equilibrium hydrogen/air, minor species. Note that the effects of turbulence on (**a**) produce a mean composition structure like that shown in (**b**)

possible reactions. Other reactions may be effectively frozen. [For hydrocarbon diffusion flames the relations (3.11) that are used may be entirely empirical. It has been found [3.37] that in laminar diffusion flames of hydrocarbon fuels $Y_i = Y_i(\xi)$ quite closely independent of the flow condition and these relations may form a suitable basis for turbulent flame computations.] For systems in which (3.9) applies we will also have

$$T = T^e(\xi), \tag{3.12}$$

$$\varrho = \varrho^e(\xi). \tag{3.13}$$

3.1.4 Mean Properties

At a given point the relationships (3.11–13) hold at any particular instant of time. As a consequence means and higher moments of these thermochemical variables at this point may be obtained from the pdf of ξ. The averaging processes described for stationary flows in Sect. 1.9 may be modified in the circumstances where the variable Q is a single-valued function of ξ alone: $Q = Q(\xi)$. We have

$$P(Q;\mathbf{x})dQ = P(\xi;\mathbf{x})d\xi,$$

it being noted that the functions P are different depending on the argument. [Often this is emphasized by subscripting P as $P_Q(Q;\boldsymbol{x})$. This notation is too cumbersome to use here.]

Relations of the form of (1.27, 28) become

$$\bar{Q}(\boldsymbol{x})=\int_0^1 Q(\xi)\,P(\xi;\boldsymbol{x})d\xi\,, \tag{3.14}$$

$$\overline{Q^2(\boldsymbol{x})}=\int_0^1 Q^2(\xi)\,P(\xi;\boldsymbol{x})\,d\xi\,. \tag{3.15}$$

We may define the Favre pdf [3.38]

$$\tilde{P}(\xi;\boldsymbol{x})\equiv\frac{1}{\bar{\varrho}}\int_0^\infty \varrho P(\varrho,\xi;\boldsymbol{x})\,d\varrho\,,$$

which with $\varrho=\varrho^{\mathrm{e}}(\xi)$ becomes

$$\tilde{P}(\xi;\boldsymbol{x})=\frac{\varrho^{\mathrm{e}}(\xi)}{\bar{\varrho}(\boldsymbol{x})}\,P(\xi;\boldsymbol{x})\,. \tag{3.16}$$

The mean density may be obtained from

$$\bar{\varrho}(\boldsymbol{x})=\int_0^1 \varrho^{\mathrm{e}}(\xi)\,P(\xi;\boldsymbol{x})\,d\xi \tag{3.17a}$$

$$=\left\{\int_0^1 [\varrho^{\mathrm{e}}(\xi)]^{-1}\tilde{P}(\xi;\boldsymbol{x})d\xi\right\}^{-1}. \tag{3.17b}$$

The determination of $\bar{\varrho}$ is important because it is the direct means by which the thermochemical aspects of the flow are coupled to the fluid-mechanical aspects.

The conventional averages for the thermodynamic variables may be formed in the manner of (3.14) and (3.15) using the fast chemistry relations (3.11–13). For Favre averaging we have

$$\tilde{Y}_i(\boldsymbol{x})=\int_0^1 Y_i^{\mathrm{e}}(\xi)\,\tilde{P}(\xi;\boldsymbol{x})\,d\xi\,, \tag{3.18}$$

$$\widetilde{Y_i''^2}(\boldsymbol{x})=\int_0^1 [Y_i^{\mathrm{e}}(\xi)-\tilde{Y}_i(\boldsymbol{x})]^2\,\tilde{P}(\xi;\boldsymbol{x})\,d\xi\,, \tag{3.19}$$

$$\tilde{T}(\boldsymbol{x})=\int_0^1 T^{\mathrm{e}}(\xi)\,\tilde{P}(\xi;\boldsymbol{x})\,d\xi\,, \tag{3.20}$$

$$\widetilde{T''^2}(\boldsymbol{x})=\int_0^1 [T^{\mathrm{e}}(\xi)-\tilde{T}(\boldsymbol{x})]^2\,\tilde{P}(\xi;\boldsymbol{x})\,d\xi\,. \tag{3.21}$$

Through the use of (3.16) it is possible to form both conventional and Favre averages when either the conventional or Favre pdf is known.

For the one-step irreversible reaction of (3.10) and Fig. 3.2 we have

$$\tilde{Y}_F = Y_B \int_{\xi_s}^{1} (\xi - \xi_s)\, \tilde{P}(\xi; \boldsymbol{x})\, d\xi\,, \tag{3.22a}$$

$$\tilde{Y}_O = r Y_B \int_{0}^{\xi_s} (\xi_s - \xi)\, \tilde{P}(\xi; \boldsymbol{x})\, d\xi\,, \tag{3.22b}$$

$$\tilde{Y}_P = (r+1) Y_B \left[(1-\xi_s) \int_{0}^{\xi_s} \xi \tilde{P}(\xi; \boldsymbol{x})\, d\xi + \xi_s \int_{\xi_s}^{1} (1-\xi)\, \tilde{P}(\xi; \boldsymbol{x})\, d\xi \right]. \tag{3.22c}$$

All of the integrals on the right-hand side of (3.22) are related so that these results may be expressed

$$\tilde{Y}_F = Y_F^{\mathrm{e}}(\tilde{\xi}) + \alpha Y_B \breve{\xi} J_1[(\xi_s - \tilde{\xi})/\breve{\xi}]\,, \tag{3.23a}$$

$$\tilde{Y}_O = Y_O^{\mathrm{e}}(\tilde{\xi}) + \alpha r Y_B \breve{\xi} J_1[(\xi_s - \tilde{\xi})/\breve{\xi}]\,, \tag{3.23b}$$

$$\tilde{Y}_P = Y_P^{\mathrm{e}}(\tilde{\xi}) - \alpha (r+1) Y_B \breve{\xi} J_1[(\xi_s - \tilde{\xi})/\breve{\xi}]\,, \tag{3.23c}$$

where

$$J_1[(\xi_s - \tilde{\xi})/\breve{\xi}] = \int_{0}^{\xi_s} [(\xi_s - \xi)/\breve{\xi}]\, \tilde{P}(\xi; \boldsymbol{x})\, d\xi - H(\xi_s - \tilde{\xi}) \cdot (\xi_s - \tilde{\xi})/\breve{\xi}\,. \tag{3.24}$$

For the one-step irreversible reaction $\alpha = 1$. For reversible and multistep reactions α is a correction factor which is defined by (3.23). $H(\xi_s - \tilde{\xi})$ is the Heaviside unit step function and

$$\breve{\xi} = (\widetilde{\xi''^2})^{1/2}\,.$$

A similar result can be obtained for conventional averaging. This integral is a fundamental property of the pdf and may be called the unmixedness integral. Unmixedness is the term coined by *Hawthorne* et al. [3.2] to describe the effect of the concentration fluctuations on the mean composition. As J_1 is non-negative it is seen from (3.23) that the effect of the turbulence is to increase the mean concentration of reactants above the "laminar" value at the mean mixture fraction. The mean product concentration is correspondingly reduced. The $\tilde{Y} \sim \tilde{\xi}$ relationships which result are schematically similar to that shown in Fig. 3.2b. Figure 3.3 shows values of J_1 calculated for different pdf types. It is seen that this integral is not too sensitive to the form of the pdf for values of $\tilde{\xi}$ close to ξ_s. For $\tilde{\xi}$ more distant from ξ_s the value of the integral is quite sensitive to the form of the pdf but there the turbulence unmixedness "correction" is

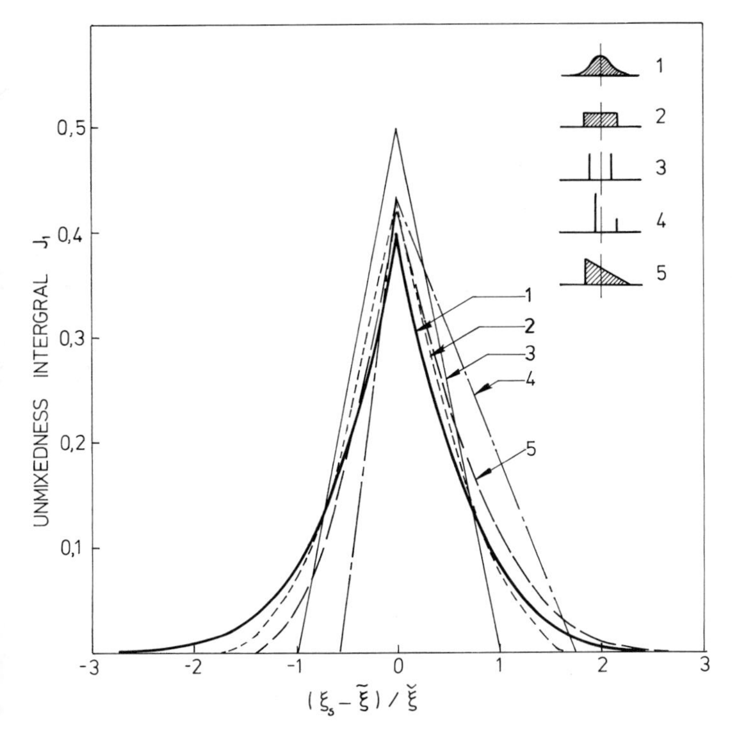

Fig. 3.3. Unmixedness integral, (3.24), for different pdf forms. (*1*) Gaussian, (*2*) conditionally uniform, (*3*) symmetrical bivariate, (*4*) unsymmetrical bivariate, and (*5*) conditionally triangular

quite small. It may be quite important, however, in the determination of visible flame length and combustion efficiency.

When the structure functions (3.11) show a more gentle transition at ξ_s as for reversible reactions shown in Figs. 3.2b and 3.2c, the unmixedness correction can be expected to be less than for the irreversible case, that is, $\alpha < 1$. If (3.11–13) are completely linear there will be no correction, $\alpha = 0$; such linear relations occur when there is no chemical reaction. The structure functions for $1/\varrho$ and T are usually not piecewise linear. Unmixedness corrections for these will be, by analogy,

$$\frac{1}{\bar{\varrho}} - \frac{1}{\varrho(\tilde{\xi})} = -\alpha_\varrho \frac{\{v(\xi_s) - [(1-\xi_s)v(0) + \xi_s v(1)]\}}{\xi_s(1-\xi_s)} \check{\xi} J_1 , \tag{3.25}$$

$$\tilde{T} - T(\tilde{\xi}) = -\alpha_T \frac{\{T(\xi_s) - [(1-\xi_s)\,T(0) + \xi_s T(1)]\}}{\xi_s(1-\xi_s)} \check{\xi} J_1 , \tag{3.26}$$

where $v \equiv 1/\varrho$ is the specific volume. α_ϱ and α_T can be expected to be quite close to but somewhat less than 1 for one-step irreversible reactions. $v(0)$, $T(0)$ are values at $\xi=0$ and $v(1)$ and $T(1)$ at $\xi=1$. The factors involving these quantities are found by analogy with (3.23), the departure of the "knee" at ξ_s from the straight line between $v(0)$ and $v(1)$ or $T(0)$ and $T(1)$ being the parameter that describes the nonlinearity around ξ_s.

Other moments and correlations of thermochemical properties may be obtained in a similar fashion. Indeed the whole pdf for any property may be obtained by suitable calculation from the pdf of the conserved scalar [3.14, 23, 26]. For the irreversible one-step reaction with fast chemistry the oxidant pdf is a weighted transform of that part of $\tilde{P}(\xi)$ between 0 and ξ_s and the fuel pdf similarly between ξ_s and 1.

3.1.5 Pdf's of Conserved Scalars

Solution for the mean and other moments of the thermochemical properties by the conserved scalar approach requires knowledge of the pdf of that scalar. At first this may seem to be a severe limitation to the method as the form of the pdf will depend on the flow conditions and will be coupled with the chemical heat release. It appears however that the constraints imposed by the mean and variance computed from (3.6) and (3.7) limit the errors that can be made using even quite arbitrary forms for the detailed shape of the pdf. Computations have been made using a wide range of assumed forms for the pdf including beta functions [3.13, 39], Gaussians, and clipped Gaussians [3.10, 18, 27] with quite good agreement being obtained with experimental data for the conserved scalar field and for the mean composition. This result is consistent with (3.23, 25, 26) and Fig. 3.3 where it is shown that the turbulent corrections to the "laminar" properties are insensitive to the form of the pdf particularly where $\tilde{\xi}$ is near ξ_s. However, there will be cases when more accurate forms of the pdf are required. In Chap. 5 methods for the direct calculation of the pdf are described. In this section we discuss semi-empirical approaches to modeling the pdf.

The pdf of the mixture fraction can be obtained from continuous or discrete measurements of the concentration of an inert species or of fine particles in the flow. Not many measurements have been made in reacting flows. Where there is not any substantial heat release the pdf can be expected to be similar to that of any other conserved scalar in a fluid mechanically similar flow. Pdfs of temperature fluctuations have been made in a wide range of nonreacting flows including mixing layers [3.40–42], jets [3.43–47], wakes [3.48], boundary layers [3.49, 50], and the free atmosphere [3.51]. Figure 3.4 summarizes in schematic form the types of pdf found. The broad arrow at $\xi=0$ or 1 signifies the delta function associated with the nonturbulent unmixed fluid (see Sect. 1.10). In practice measured pdfs show these delta functions smeared into a Gaussian-like peak [3.52] by electronic noise and residual temperature fluctuations in the free stream. In combustion situations there is unlikely to be any contamination of one feed stream by reactants from the other feed. Vorticity fluctuations and foreign species can only enter the free stream by molecular

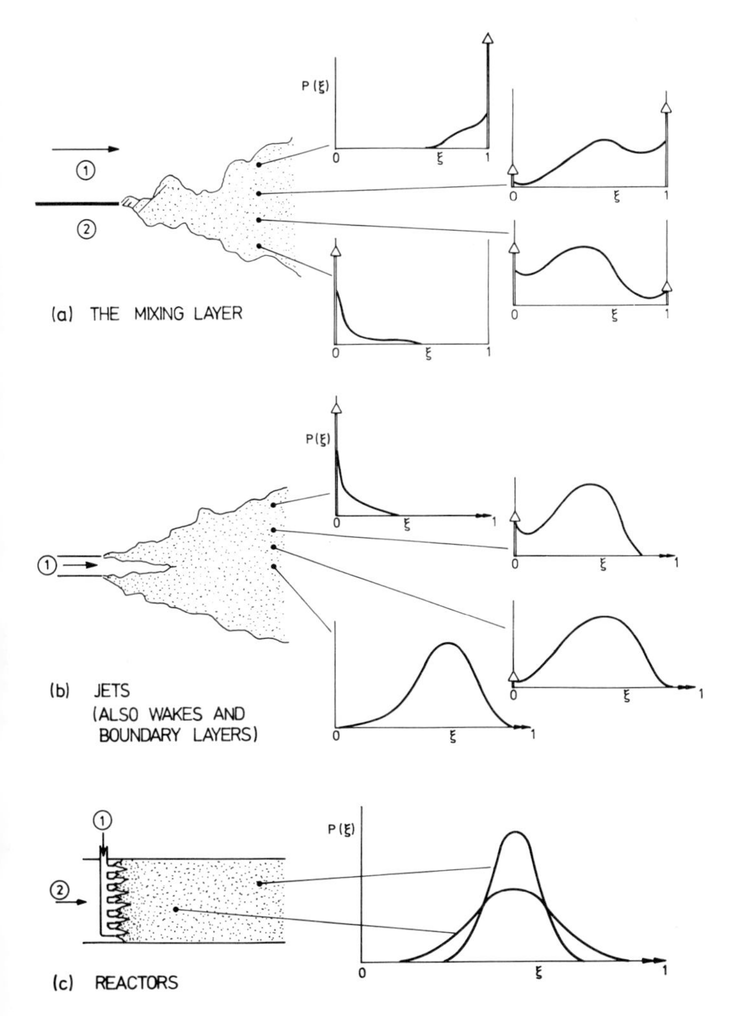

Fig. 3.4a–c. Probability density function forms for a conserved scalar in various types of flow (schematic only)

processes and the turbulence and concentration interfaces are very close to one another. For jets away from the potential core, wakes, and wall boundary layers there is intermittency only with respect to the free stream and we have, following (1.35),

$$P(\xi)=[1-\bar{I}(\boldsymbol{x})]\,\delta(\xi)+\bar{I}(\boldsymbol{x})\,P_1(\xi)\,. \tag{3.27}$$

$P_1(\xi)$ is the pdf of the turbulent fluid. Where $\bar{I}(\boldsymbol{x})$ is greater than about 0.6 it is pseudo-Gaussian as shown in Fig. 3.4b. For low values of the intermittency it appears to be dominated by fluid in the neighborhood of the interface which

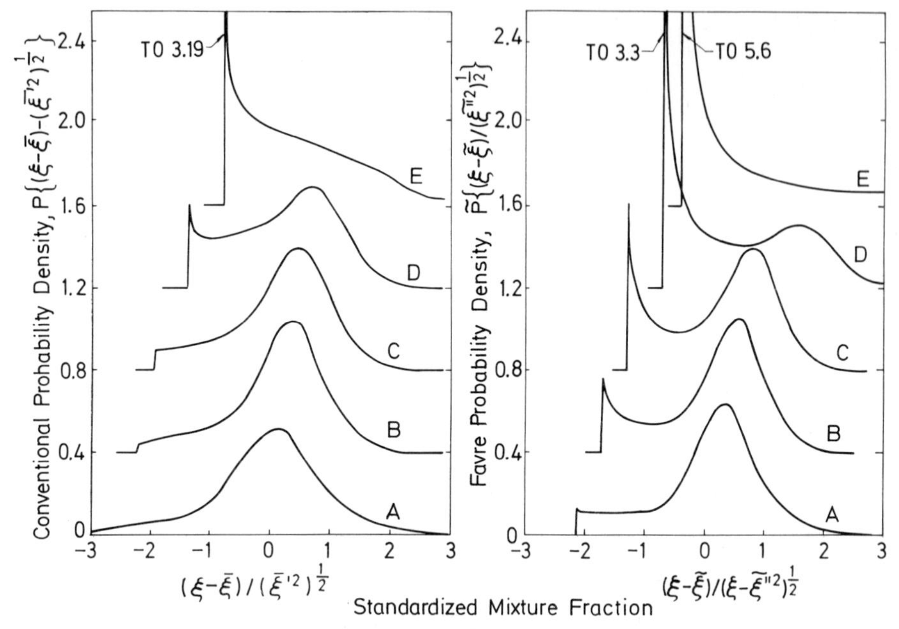

Fig. 3.5. Conventional and Favre pdfs measured in a hydrogen/air diffusion flame at 40 jet diameters from the nozzle by *Kennedy* and *Kent* [3.55]. Radial distances in nozzle radii: (*A*) 0.0, (*B*) 1.16, (*C*) 2.33, (*D*) 4.65, (*E*) 9.31

Corrsin has called the viscous superlayer [3.53]; it has a more triangular form. In reactors where there is no free stream the pdf can be expected to be close to Gaussian as shown in Fig. 3.4c.

Within the turbulent fluid, that is $0<\xi<1$, the distribution is smooth and continuous and this is of considerable physical significance. There is no preferred mixed concentration indicating that as products are formed at or near $\xi=\xi_s$ they are concurrently being diffused and mixed into the surrounding material; there are no "pure product" ($\xi=\xi_s$) eddies leading to large values of the pdf at $\xi=\xi_s$. Pure reactants exist only in the intermittent nonturbulent parts of the flow and so that here there are no "pure fuel" ($\xi=1$) or "pure oxidant" ($\xi=0$) eddies either.

For flows with significant heat release there are as yet few data available on the conserved scalar pdfs. Measurements have been made by *Ebrahimi* and *Kleine* [3.54] and *Kennedy* and *Kent* [3.55] in turbulent jet diffusion flames using the light scatter technique with the fuel stream seeded with fine particles. Only the data of *Kennedy* and *Kent* [3.55] have been processed to give pdfs of ξ from the measurements of $\varrho\xi$. This was done by making the fast chemistry assumption (3.13) for the hydrogen/air system and computer processing the data. They report both conventional and Favre pdfs. Figure 3.5 shows these on a radial traverse across a turbulent diffusion flame of hydrogen jetting into still air at an axial distance of $x/D=40$. It can be seen that the Favre pdfs emphasize the intermittency spike.

For the Favre pdf (3.27) becomes

$$\tilde{P}(\xi)=(1-\tilde{I})\,\delta(\xi)+\tilde{I}\tilde{P}_1(\xi)\,, \tag{3.28}$$

where the Favre intermittency is related to the conventional $\bar{I}$ by

$$\bar{\varrho}(1-\tilde{I})=\varrho_0(1-\bar{I})\,, \tag{3.29a}$$

$$\bar{\varrho}\tilde{I}=\bar{\varrho}_t\bar{I} \tag{3.29b}$$

with $\bar{\varrho}_t$ the mean density in the turbulent fluid. A question of considerable interest is whether the Favre pdfs or the conventional pdfs of ξ resemble more the pdfs of conserved scalars in uniform density flows. A close resemblance would indicate an insensitivity of this type of pdf to heat release and variable density effects. As a consequence it could be expected that the generality of the pdf modeling would be improved by the use of that type of pdf. Both *Bilger* [3.56] and *Kennedy* and *Kent* [3.55] addressed this question. Although the latter paper favors the Favre pdf as being more invariant, this cannot as yet be regarded as conclusive. Both types of pdf are quite similar in form to those found in uniform density and nonreacting flows. This gives encouragement for the use of such data where no reacting flow data exist.

The pdfs in Fig. 3.5 show no special structure near $\xi=\xi_s$ and like the nonreacting pdfs have a continuous distribution in the turbulent fluid. There will thus be no "pure product" eddies. For a given value of $\tilde{I}$ or $\bar{I}$ the pdfs of *Kennedy* and *Kent* [3.55] and the pdfs in nonreacting jet flows show a general similarity of form for all distances from the fuel nozzle. There is a strong suggestion in the data, which needs further consideration, that the three parameters $\tilde{I}$, $\tilde{\xi}$, and $\breve{\xi}\equiv(\widetilde{\xi''^2})^{1/2}$ give a universal characterization of the pdf. Furthermore there appears to be a strong correlation between $\breve{\xi}/\tilde{\xi}$ and $\tilde{I}$. *Kent* and *Bilger* [3.25] showed such a relationship for a wide range of nonreacting jet and wake data. Thus

$$\tilde{P}_1(\xi\,;\boldsymbol{x})d\xi=P_1(\xi/\breve{\xi}\,;\tilde{I})\,d(\xi/\breve{\xi}) \tag{3.30}$$

may be a good empirical model for the turbulent part of the pdf. Since only $\breve{\xi}$ and $\tilde{\xi}$ are calculated in the usual models the first estimate of $\tilde{I}$ that can be obtained from $\breve{\xi}/\tilde{\xi}$ may need improving. This could be done by assuming the error function behavior with respect to the cross stream distance which is usually found in these flows [3.53]. Alternatively a balance equation for $\tilde{I}$ can be formulated and solved as has been done by *Libby* [3.57]. If only mean quantities are required for one-step irreversible reactions an analysis of the experimental data will probably yield an almost universal set of $J_1[(\xi_s-\tilde{\xi})/\breve{\xi}\,;\breve{\xi}/\tilde{\xi}]$ functions. All of the above remarks apply for singly bounded pdfs such as found in jets, wakes, and boundary layers. Suitable modifications to this approach can be made to obtain universal empirical models for doubly bounded pdfs such as occur in mixing layers and the essentially unbounded pdfs which are found in reactors remote from the injectors.

As an intermediate approach that adopted by *Kent* and *Bilger* [3.25] is probably most suitable. The Favre pdf is split as in (3.28),

$$\tilde{P}(\xi)=(1-\tilde{I})\,\delta(\xi)+\tilde{I}\tilde{P}_1(\xi)\,, \tag{3.31}$$

with $\tilde{P}_1$ being a Gaussian with mean $\tilde{\xi}_t$ and variance $\widetilde{\xi_t''^2}$ so that

$$\tilde{\xi}=\tilde{I}\tilde{\xi}_t \tag{3.32}$$

and

$$\widetilde{\xi''^2}=\tilde{I}[\widetilde{\xi_t''^2}+(\tilde{\xi}_t-\tilde{\xi})^2]+(1-\tilde{I})\tilde{\xi}^2\,. \tag{3.33}$$

$\tilde{I}$ is estimated from the empirical correlation

$$\tilde{I}=(K+1)/(\widetilde{\xi''^2}/\tilde{\xi}^2+1)\,, \tag{3.34a}$$

which implies

$$\widetilde{\xi_t''^2}=K\tilde{\xi}_t^2\,, \tag{3.34b}$$

and $K\approx 0.25$. A slight clipping of the Gaussian on the low side at $\xi=0$ results and this can be accommodated without significant error in the above formulae by a slight adjustment to the constant so that

$$\tilde{P}_1(\xi)=\frac{1.023}{\sqrt{2\pi}}\exp[-\tfrac{1}{2}(\xi-\tilde{\xi}_t)^2/\widetilde{\xi_t''^2}]\,. \tag{3.35}$$

3.1.6 The Reaction Rate

The essence of the conserved scalar approach is the use of (3.11–13) to obviate the complex problem of handling the chemical production term (see Sect. 1.14). As a consequence of this approach a simple formula for the instantaneous reaction rate results [3.22]. Substitution of (3.11) into (1.17) yields

$$\begin{aligned}\dot{w}_i=&\frac{dY_i^e}{d\xi}\left[\frac{\partial}{\partial t}(\varrho\xi)+\frac{\partial}{\partial x_k}(\varrho u_k\xi)-\frac{\partial}{\partial x_k}\left(\varrho\mathscr{D}\frac{\partial\xi}{\partial x_k}\right)\right]\\&-\varrho\mathscr{D}\frac{\partial\xi}{\partial x_k}\frac{\partial\xi}{\partial x_k}\frac{d^2Y_i^e}{d\xi^2}\\=&-\varrho\mathscr{D}\frac{\partial\xi}{\partial x_k}\frac{\partial\xi}{\partial x_k}\frac{d^2Y_i^e}{d\xi^2}\,,\end{aligned} \tag{3.36}$$

where (3.5) has been used. This is a specialized case of a more general result [3.58] discussed in Sect. 3.2.

Taking the mean of (3.36) yields

$$\bar{w}_i = -\frac{1}{2}\bar{\varrho}\int_0^1\int_0^\infty \chi \frac{d^2 Y_i^e}{d\xi^2}\tilde{P}(\chi,\xi)\,d\chi d\xi\,, \tag{3.37}$$

where

$$\chi \equiv 2D\frac{\partial\xi''}{\partial x_k}\frac{\partial\xi''}{\partial x_k} \approx 2D\frac{\partial\xi}{\partial x_k}\frac{\partial\xi}{\partial x_k} \tag{3.38}$$

in moderate to high Reynolds number turbulence, and $\tilde{P}(\chi,\xi)$ is the joint Favre pdf of χ and ξ. The mean of χ is the scalar dissipation in (3.7). For the one-step irreversible reaction with fast chemistry of (3.10), $d^2Y_F^e/d\xi^2$ has properties like those of a delta function at $\xi=\xi_s$ and

$$\bar{w}_F = -\tfrac{1}{2}\bar{\varrho}Y_B\tilde{\chi}(\xi_s)\tilde{P}(\xi_s)\,, \tag{3.39}$$

where $\tilde{\chi}(\xi_s)$ is the conditional Favre expectation of χ for $\xi=\xi_s$.

The physical interpretation of these results will be developed more fully later. Reference to the last term of (3.7) indicates that the reaction rate is proportional to the scalar dissipation, while reference to Fig. 3.2 indicates that the reaction is associated with the nonlinearities in the relationships (3.11). *Williams* [1.33] gave intuitive arguments for a somewhat similar dependence of the reaction rate on the scalar dissipation. The dependence of reaction rates on scalar dissipation has also been found for premixed systems and this is discussed in Chap. 4.

3.1.7 Application in Homogeneous Turbulence

Homogeneous turbulence is such that the properties do not vary with position in the field; that is, they are translationally invariant. It is somewhat less restrictive than isotropic turbulence which also requires rotational invariance. Results for homogeneous turbulence are often considered to be of limited practical value but they do represent well the phenomena in certain types of chemical reactors. Of particular interest are reactors of the plug flow type in which the reactants are introduced in a grid of injectors. Variations in the cross stream direction are negligible and are sufficiently small in the streamwise direction for the homogeneous turbulence assumptions to be approximately valid in a Lagrangian frame. Application of the conserved scalar approach to such problems is largely due to *Toor* [3.5, 7] and his followers. An excellent review of this work was given by *Hill* [3.59]. The theory results in the ability to design and scale up chemical reactors with confidence, a development of considerable economic consequence. More general forms of the results are presented here and there may be further important results yet to be explored.

Two of the most important consequences of the homogeneous turbulence assumptions are that $\tilde{\xi}$ is constant throughout the field and (3.7) may be reduced to

$$\frac{d\widetilde{\xi''^2}}{dt}=2\breve{\xi}\frac{d\breve{\xi}}{dt}=-\tilde{\chi}, \tag{3.40}$$

where time is in a Lagrangian reference frame travelling with the flow. An implicit assumption frequently made is that the pdf of $(\xi-\tilde{\xi})/\breve{\xi}$ is invariant. In fact it is usually assumed that the pdf is Gaussian which automatically assures this invariance. The invariance assumption may be stated:

$$\tilde{P}[(\xi-\tilde{\xi})/\breve{\xi};t]=\tilde{P}[(\xi-\tilde{\xi})/\breve{\xi}]. \tag{3.41}$$

For a stoichiometric ($\tilde{\xi}=\xi_s$) system with a one-step irreversible reaction (3.23) become

$$\tilde{Y}_F(t)=Y_B J_1(0)\,\breve{\xi}, \tag{3.42a}$$

$$\tilde{Y}_O(t)=r\,Y_B J_1(0)\,\breve{\xi}, \tag{3.42b}$$

$$\tilde{Y}_P(t)=Y_P^e(\xi_s)-(r+1)\,Y_B J_1(0)\,\breve{\xi}, \tag{3.42c}$$

which correspond to the results obtained by *Toor* [3.5]. Equations (3.23) are far more general due to their applicability in off-stoichiometric systems and in systems with significant density fluctuations. The value of Favre averaging is seen in the fact that $\bar{\varrho}\tilde{Y}_i V$ is precisely the mass of i in the system of volume V. For off-stoichiometric systems it is seen that the reactants and product are expressed in terms of the proximity to their ultimate concentrations,

$$\tilde{Y}_F(t)=Y_F(\infty)+\alpha Y_B\breve{\xi}J_1(z_s), \tag{3.43a}$$

$$\tilde{Y}_O(t)=Y_O(\infty)+\alpha r Y_B\breve{\xi}J_1(z_s), \tag{3.43b}$$

$$\tilde{Y}_P(t)=Y_P(\infty)-\alpha(r+1)\,Y_B\breve{\xi}J_1(z_s), \tag{3.43c}$$

where $z_s\equiv(\xi_s-\tilde{\xi})/\breve{\xi}$. For a Gaussian pdf

$$J_1(z_s)=\frac{1}{\sqrt{2\pi}}\mathrm{e}^{-z_s^2/2}-\frac{|z_s|}{2}[1-\mathrm{erf}(|z_s|/2)]. \tag{3.44}$$

This is plotted in Fig. 3.3. It is seen that z_s will increase with time and $J_1(z_s)$ will fall due to the decay of $\breve{\xi}$. (It should not be concluded however that the reaction rate is faster in off-stoichiometric systems – see further on.)

The generalization to chemical systems with reverse reactions $\alpha \neq 1$ has not been extensively explored. In general α will vary with time in homogeneous flows. When the concentration fluctuations $\breve{\xi}$ are relatively small a Taylor expansion for Y_i about $\tilde{\xi}$ yields

$$Y_i = Y_i^e(\tilde{\xi}) + (dY_i^e/d\xi)_{\tilde{\xi}}(\xi - \tilde{\xi}) + \frac{1}{2}\left(\frac{dY_i^e}{d\xi^2}\right)_{\tilde{\xi}}(\xi - \tilde{\xi})^2 + \dots$$

and convolution of this with the pdf of ξ yields

$$\tilde{Y}_i(t) \approx Y_i(\infty) + \tfrac{1}{2}(d^2 Y_i^e/d\xi^2)_{\tilde{\xi}}\breve{\xi}^2, \tag{3.45}$$

where $Y_i(\infty) = Y_i^e(\tilde{\xi})$. This should be contrasted with the one-step irreversible result for stoichiometric reactants (3.42) where the decay is with the first power of $\breve{\xi}$. Obviously the higher derivatives of $Y_i^e(\xi)$ must not be large if (3.45) is to hold.

The reaction rate may be obtained by differentiating (3.43) in the irreversible one-step case ($\alpha = 1$)

$$\bar{\dot{w}}_F = \bar{\varrho}\frac{d\tilde{Y}_F}{dt} = \bar{\varrho} Y_B \int_0^{\xi_s} (\xi_s - \xi)\frac{\partial \tilde{P}}{\partial t}\, d\xi \tag{3.46a}$$

$$= -\bar{\varrho} Y_B \frac{1}{\breve{\xi}}\frac{d\breve{\xi}}{dt}\int_0^{\xi_s} (\xi - \tilde{\xi})\, \tilde{P}(\xi, t)\, d\xi\,, \tag{3.46b}$$

where use has been made of (3.41) in deriving (3.46b). Similar expressions can be obtained for $\bar{\dot{w}}_O$ and $\bar{\dot{w}}_P$. The earlier reaction rate expression (3.39) may be simpler to use in many cases.

Equating (3.46) and (3.39) may lead to some useful results. If $\tilde{P}(\xi, t)$ is Gaussian (3.46) becomes

$$\begin{aligned} \bar{\dot{w}}_F &= \bar{\varrho} Y_B \tilde{P}(z_s)\frac{d\breve{\xi}}{dt} \\ &= -\tfrac{1}{2}\bar{\varrho} Y_B \tilde{\chi} \tilde{P}(\xi_s)\,, \end{aligned} \tag{3.47}$$

where use has been made of $\tilde{P}(z_s) = \breve{\xi}\tilde{P}(\xi_s)$ and (3.40). Comparison with (3.39) yields

$$\tilde{\chi}(\xi_s) = \tilde{\chi}\,. \tag{3.48}$$

Thus for this Gaussian case the conditioned expectation of χ is equal to its unconditioned expectation. This is a necessary but not sufficient condition for the statistical independence of χ and ξ which would result in

$$\tilde{P}(\chi, \xi) = \tilde{P}(\chi)\, \tilde{P}(\xi)\,. \tag{3.49}$$

It appears that the assumption of (3.49) may be a more realistic one than was at first thought [3.22].

Returning to (3.46) we see that off-stoichiometric reaction rates will be slower than for stoichiometric systems assuming of course that the mode and mean of the pdf coincide. The usefulness of this result in chemical reactor design is yet to be explored. The result is further generalized to systems with moderately fast chemistry in Sect. 3.2.

Notice that differentiating (3.45) will yield a result comparable with (3.37) if (3.49) is applied to the latter. A detailed physical interpretation of these results is left to Sect. 3.2.

3.1.8 Application in Shear Flows

To crystallize some of the issues involved we consider an axisymmetric flow such as in Fig. 3.1 and write the balance equations (1.42, 43; 3.6, 7) in their modeled axisymmetric boundary layer form, namely

$$\frac{\partial}{\partial x}(\bar{\varrho}\tilde{u})+\frac{1}{r}\frac{\partial}{\partial r}(r\bar{\varrho}\tilde{v})=0\,, \tag{3.50}$$

$$\bar{\varrho}\tilde{u}\frac{\partial \tilde{u}}{\partial x}+\bar{\varrho}\tilde{v}\frac{\partial \tilde{u}}{\partial r}=\frac{1}{r}\frac{\partial}{\partial r}\left(r\mu_{\mathrm{t}}\frac{\partial \tilde{u}}{\partial r}\right)-\frac{d\bar{p}}{dx}+\bar{\varrho}g_x\,, \tag{3.51}$$

$$\bar{\varrho}\tilde{u}\frac{\partial \tilde{\xi}}{\partial x}+\bar{\varrho}\tilde{v}\frac{\partial \tilde{\xi}}{\partial r}=\frac{1}{r}\frac{\partial}{\partial r}\left[r(\mu_{\mathrm{t}}/\sigma_{\xi})\frac{\partial \tilde{\xi}}{\partial r}\right], \tag{3.52}$$

$$\bar{\varrho}\tilde{u}\frac{\partial \widetilde{\xi''^2}}{\partial x}+\bar{\varrho}\tilde{v}\frac{\partial \widetilde{\xi''^2}}{\partial r}=C_{g_1}\mu_{\mathrm{t}}\left(\frac{\partial \tilde{\xi}}{\partial r}\right)^2+\frac{1}{r}\frac{\partial}{\partial r}\left[r(\mu_{\mathrm{t}}/\sigma_g)\frac{\partial \widetilde{\xi''^2}}{\partial r}\right]$$
$$-C_{g_2}\bar{\varrho}\varepsilon\widetilde{\xi''^2}/\tilde{q}\,, \tag{3.53}$$

where μ_{t} is the turbulent diffusivity, ε the turbulence dissipation, $\tilde{q}$ the Favre turbulence kinetic energy, g_x the component of the gravitational acceleration in the x direction, $\bar{p}$ the mean pressure, and σ_{ξ}, σ_g, C_{g_1}, and C_{g_2} model constants. Details of the modeling involved in these equations and other possible models will be considered in Sect. 3.6. Separate balance equations are usually used to determine ε and $\tilde{q}$. There are thus six partial differential equations (pde's) in the six dependent variables $\tilde{u}$, $\tilde{v}$, $\tilde{\xi}$, $\widetilde{\xi''^2}$, $\tilde{q}$, and ε. The axial pressure gradient is either specified or computed from the free stream velocity or from integrals of (3.50, 51) over the duct area. The turbulent diffusivity μ_{t} is usually modeled as proportional to $\tilde{q}^2/\varepsilon$ and g_x, σ_{ξ}, σ_g, C_{g_1}, and C_{g_2} taken as constants independent of the rate of chemical reaction. Only $\bar{\varrho}$ needs to be determined in terms of the other quantities to close the set of equations. Our concern here is with the importance of this closure in $\bar{\varrho}$. It may be noted in passing that the solution procedure for these parabolic equations is usually numerical and involves a marching in x [3.8–13, 25, 28, 30, 31] although for some flows a similarity solution will exist [3.20].

The solution for $\tilde{u}$, $\tilde{v}$, and $\tilde{\xi}$ defines the main features of the flow field and the mixing pattern. Velocities and the mean location of the stoichiometric ($\tilde{\xi}=\xi_s$) contour are determined. This latter defines the middle of the reaction zone and so determines the mean flame size and shape. It is important to note that (3.50–53) and their solutions are independent of considerations of fast or slow chemistry, the number of steps and irreversibility or otherwise of the chemical reactions, or of the form of pdfs for concentration fluctuations, except insofar as they enter through $\bar{\varrho}$. A question of significance then is the relative importance of correctly taking into account the effects of chemistry and of concentration fluctuations on $\bar{\varrho}$. And how significant are errors in $\bar{\varrho}$ in the determination of the basic $\tilde{u}$, $\tilde{v}$, $\tilde{\xi}$ field? These questions have not been adequately addressed in the literature, most authors making arbitrary assumptions about the effects of either the chemistry or the fluctuations or both on $\bar{\varrho}$.

The effects of concentration fluctuations on the mean density is inextricably linked with the Favre averaging employed in (3.52, 53). Many authors use these equations but for the conventional averages $\bar{\xi}$ and $\overline{\xi'^2}$ rather than $\tilde{\xi}$ and $\widetilde{\xi''^2}$. They neglect, often implicitly, the difference $\tilde{\xi}-\bar{\xi}=\overline{\varrho'\xi'}/\bar{\varrho}$ which can be very significant. The measurements of *Kennedy* and *Kent* [3.55] indicate that the Favre average $\tilde{\xi}$ can be half the conventional average $\bar{\xi}$. The effect of concentration fluctuations on the density can be viewed in the form

$$\bar{\varrho}=\varrho^{e}(\langle\xi\rangle)+f\{\check{\xi}\},$$

where $\langle\xi\rangle$ is a mean (Favre or conventional) and $f\{\check{\xi}\}$ expresses the effect of fluctuations. Differences arising from the choice of $\langle\xi\rangle$ as $\tilde{\xi}$ or $\bar{\xi}$ can outweigh the contribution of $f\{\check{\xi}\}$. Such problems lie behind early attempts [3.12, 60] to incorporate the effects of fluctuations into the mean density. For hydrogen/air diffusion flames *Kent* and *Bilger* [3.25] found that the use of the "laminar" density $\varrho^{e}(\tilde{\xi})$ gave little difference from the correct formula (3.17b) but that the use of (3.17a) gave large errors. There may be more difference between the use of "laminar" and turbulent mean density for hydrocarbon/air systems where the relationships between specific volume and ξ are more nonlinear. Alternative approaches are to use (1.48) or (1.49) but since these involve calculation of $\tilde{T}$ by the pdf method (3.20) or modeling the chemical production term in (1.45) there appears to be no advantage over direct calculation through the use of (3.17b). It appears that many of the imprecise treatments of this problem arise from a lack of appreciation of the concepts rather than operational difficulties. Studies by *Kent* and *Bilger* [3.25] indicated that $\bar{\varrho}$ and the flow field are insensitive to the precise form of the pdf used in (3.17b). What is important is that the correct form of (3.17) be used and that the pdf has the correct moments $\tilde{\xi}$ and $\widetilde{\xi''^2}$ and an adequate treatment of intermittency where present.

The effects of nonequilibrium chemistry and radiant heat losses may have more effect on the mean density than the details of the form of the pdf. The nonequilibrium calculations of *Janicka* and *Kollman* [3.31] indicate that $\tilde{T}$ may be lowered by as much as 100 K due to nonequilibrium effects, while *Stårner* [3.61] indicated that heat losses due to radiation may be as high as 10%, both for the hydrogen/air flame of *Kent* and *Bilger* [3.11, 62]. For hydrocarbon/air flames these effects could be expected to be considerably greater. *Tamanini* [3.63] described two methods by which radiation losses can be accounted for in a manner consistent with the conserved scalar approach.

3.1.9 Reaction Zone Structure

In the limit of fast chemistry the reaction will be confined to the surface given by $\xi = \xi_s$ for systems with one-step irreversible reactions. This is evident from the structure of (3.10) and the reaction rate formula (3.36). This surface will be highly contorted and possibly multiply connected as shown for the contour $\xi = 0.07$ in Fig. 3.1. *Libby* [3.26] has studied the statistics of such surfaces and suggests that they are somewhat similar to those of the interface which divides turbulent from nonturbulent fluid in free turbulent flows. The "intermittency" of when $\xi > \xi_s$ and its crossing frequency are normally distributed while the lengths of $\xi > \xi_s$ have a Poisson-like distribution.

For finite rate chemistry the reaction zone will be broadened around the $\xi = \xi_s$ contour. *Gibson* and *Libby* [3.64] estimated this thickness to be $L_R \sim L_B D_{1T}^{1/3}$ where L_B is the Batchelor length scale $(\nu \mathscr{D}^2/\varepsilon)^{1/4}$ and D_{1T} is a turbulent Damköhler number $(\nu k^2/\varepsilon)^{1/2}$ with k the chemical rate constant. *Bilger* [3.65] pointed out that in fully turbulent flows a better estimate may be $L_R \sim \mathscr{D}^{1/2} k^{1/3} \chi^{1/6}$. An analysis based on the reaction rate formula (3.36) appears to be warranted but as yet has not been carried out.

In flames and other reacting flows of practical concern the reaction mechanism is complex and involves significant reverse reactions. In these cases the $Y_i \sim \xi$ relationship becomes as in Fig. 3.2b with a finite second derivative $d^2 Y_i^e / d\xi^2$ over a range of ξ near ξ_s. The reaction rate equation (3.36) indicates that chemical reaction occurs over this range in ξ with the reaction zone becoming "equilibrium-broadened" around ξ_s. This broadening is due to the back reactions shifting the equilibrium products on either side of ξ_s. For hydrogen/air mixtures the second derivative is significant over the range $\xi_s \pm 0.3 \xi_s$ and this range of ξ is comparable with $\tilde{\xi} \pm \check{\xi}$. This indicates that the reaction will be occurring over a broad zone around the $\tilde{\xi} = \xi_s$ contour at any instant in time. In fact the breadth of the reaction zone at any instant will be almost as wide as the time-averaged reaction zone. It will be of the same order of magnitude as the size of the largest eddies in the turbulence, so that smaller scale eddies will be contained within the reaction zone. If the fast chemistry assumption holds, the equilibrium relations (3.11, 12) will still apply instantaneously. The reactions that are occurring are just those required to shift the equilibrium and there will be no significant departure of the concentrations from equilibrium.

Williams [3.66] conceived turbulent flames structure under some circumstances to be an ensemble of stretched laminar diffusion flamelets. The stretching is due to the turbulent strain in the flow producing steep concentration gradients and increasing the "loading" on the reaction zone. By "loading" is meant the rate at which reactants diffuse toward the reaction zone and this can become so great that the chemical kinetics cannot keep pace, leading to flamelet extinction. For laminar hydrocarbon/air diffusion flames *Bilger* [3.37] showed that there appears to be a quasi-equilibrium structure $Y_i = Y_i^e(\xi)$ (where the superscript now signifies quasi-equilibrium rather than true equilibrium) independent of the flame loading over a wide range of flame loading. A test of whether the ensemble model is valid should thus be that mean concentrations will be given by (3.18). *Moseley* [3.67] has carried out experiments on LPG (liquefied petroleum gas) flames and found that his measurements in turbulent flames are consistent with the laminar flame structure using (3.18).

3.1.10 Pollutant Formation

One of the most interesting and important applications of the theory of turbulent reacting flow centers on the formation of nitric oxide which is produced by the relatively slow Zeldovich chain reactions

$$O + N_2 \xrightarrow{k_4} NO + N\,, \tag{3.54a}$$

$$N + O_2 \xrightarrow{k_5} NO + O\,. \tag{3.54b}$$

Bowman [3.68] gave a good review of the kinetics of these reactions. They are several orders of magnitude slower than those of the main heat release reactions and the concentrations involved are such that neither the temperature or main species compositions are affected. The formation of nitric oxide is thus kinetically limited.

The first of these reactions is rate limiting and this yields [3.68] under conditions in which reverse reactions can be neglected

$$w_{NO} = 2\varrho^2 k_4 Y_O Y_{N_2}\,. \tag{3.55}$$

The rate constant k_4 for the forward rate of reaction (3.54a) is purely a function of temperature. If the O atom concentration is assumed to be in equilibrium as was done in many early flame studies [3.68], Y_O, k_4, ϱ, and Y_{N_2} are functions only of the mixture fraction ξ. We then have [3.10, 11]

$$\dot{w}_{NO}/\varrho = R_{NO}(\xi)$$

and

$$\overline{\dot{w}_{\mathrm{NO}}}=\bar{\varrho}\int_0^1 R_{\mathrm{NO}}(\xi)\,\tilde{P}(\xi)\,d\xi\,. \tag{3.56}$$

This mean chemical production term can be used in (1.46) to obtain concentrations of nitric oxide. This approach is consistent with the fast chemistry/conserved scalar method outlined in this section. Since $\tilde{P}(\xi)$ is considered to be essentially independent of Reynolds number, a consequence of this model is that the nitric oxide concentration will be proportional to L/V where L is a characteristic length scale of the flow and V a characteristic velocity (see Sect. 1.16). However, *Bilger* and *Beck* [3.69] found for a series of fluid-mechanically similar flows that concentrations are proportional to $(L/V)\,\mathrm{Re}^{-1/2}$. The modeling in (3.56) also tends to overestimate nitric oxide concentrations and does not give the "rich shift" found experimentally [3.11, 69]. It is found that the peak NO concentrations and NO formation rate are shifted toward the fuel-rich side of the flame compared with that implied by (3.56).

The principal reasons for the failure of the conserved scalar/fast chemistry approach to the prediction of NO formation are thought to lie in the assumption that O atom concentrations and the temperature are those given by equilibrium. It is well known in combustion [3.68] that while the conversion of the main reactants may be fast the approach of free radicals such as O, OH, and H to equilibrium is limited by the relatively slow three-body recombination reactions such as $H+OH+M\rightarrow H_2O+M$. Typically free radical concentrations can be almost an order of magnitude higher than their equilibrium values even though the main reactants are essentially completely converted. There is also some temperature nonequilibrium due to the energy associated with these free radicals. The prediction of these effects in a fundamentally sound way is one of the challenges of turbulent reacting flow theory. It will be discussed in Sect. 3.2.

In the same manner the formation and burn-out of soot and carbon monoxide are thought to be kinetically limited but have not as yet been addressed by the conserved scalar approch. Nitrogen dioxide is formed in gas turbine combustion chambers possibly by the reaction of NO with H_2O in the quench regions surrounding air injection ports. This is again a kinetically controlled process which has not been treated by the conserved scalar approach.

It is seen that the conserved scalar approach, as it is currently formulated, needs to be extended to enable it to be used for calculation of the kinetically limited processes of pollutant formation and burn-out. It is not sufficient to consider these processes as independent of the main reactions since it is the main reaction rates which determine the level of the free-radical concentrations required in the kinetic calculations. This extension of the theory is addressed in the next section.

3.2 Two-Variable Approaches

We consider analyses in which the conserved scalar approach is extended to cases of finite rate chemistry by the consideration of a further variable which is affected by the chemical kinetics. Problems are formulated such that all the other variables of interest, such as the molecular species concentrations, density, and temperature can be expressed in terms of these two variables. Unlike the conserved scalar, this other variable must have a chemical production term and the closure problem for this term (Sect. 1.14) must be addressed.

3.2.1 Second Variables

In many problems if one molecular species is determined in addition to the conserved scalar then all the other species and the temperature can be determined. This is easily seen for one-step reversible or irreversible reactions (3.1). The Shvab-Zeldovich relations (3.3) can be determined from the conserved scalar and specification of any one of Y_F, Y_O or Y_p determines the other two. The temperature is then determined from (3.9) and (1.13).

For more complicated multistep chemical mechanisms it would be possible to use third, fourth, etc., variables. However, a simpler approach is possible using the notion of a progress variable for the chemical reaction. For premixed flames, which of course have a fixed mixture fraction ξ, it has been found useful ([3.70] and Chap. 4) to express all of the concentrations and temperature in terms of a reaction progress variable which is zero when the system is fully unreacted and unity when it is fully reacted. Concentrations are linearly related to the progress variable so that the specification of any one specifies the others. The instantaneous reaction rate is then also fully determined by the progress variable (or one species concentration) and in our case the mixture fraction.

In some systems with complex multistep mechanisms, other, more fundamentally based, simplifications are possible. Such is the case for the hydrogen/air system studied by *Janicka* and *Kollman* [3.31]. Here the species of interest in the main reaction are H_2, O_2, H_2O, N_2, O, OH, and H: seven in all. Three equations are provided by the specification of the mixture fraction which determines the element mass fractions. Another three equations are provided by assumption of equilibrium for the fast two-body shuffle reactions

$$H+O_2 \rightleftharpoons OH+O, \tag{3.57a}$$

$$O+H_2 \rightleftharpoons OH+H, \tag{3.57b}$$

$$H_2+OH \rightleftharpoons H_2O+H, \tag{3.57c}$$

$$2OH \rightleftharpoons H_2O+O. \tag{3.57d}$$

The composition is thus determined if any one of H_2, O_2, H_2O, O, H, or OH can be specified or calculated. A difficulty in calculating one of these species is that its reaction rate contains many terms including the net rates of the reactions (3.57). This difficulty has long been recognized in chemical kinetics and it is obviated by a technique drawing from the work of *Kaskan* and *Schott* [3.71] and *Dixon-Lewis* (see, e.g., [3.72]). The technique is to use a combined variable such as $Y^*_{H_2}$, defined by

$$Y^*_{H_2} \equiv Y_{H_2} + \tfrac{1}{2}(W_{H_2}/W_{OH})\,Y_{OH} + (W_{H_2}/W_O)\,Y_O + \tfrac{3}{2}(W_{H_2}/W_H)\,Y_H\,, \tag{3.58}$$

which eliminates consideration of the net rates of reactions (3.57). It has a production term solely dependent on the net rates of the three body recombination reactions

$$H + OH + M \rightleftharpoons H_2O + M\,, \tag{3.59a}$$

$$H + O + M \rightleftharpoons OH + M\,, \tag{3.59b}$$

$$H + H + M \rightleftharpoons H_2 + M\,. \tag{3.59c}$$

It may be thought of as a conserved scalar of the type (3.3) except that it is only conserved for the reactions (3.57) and not conserved for the reactions (3.58). The two variables ξ and $Y^*_{H_2}$ thus determine all the species concentrations and through (3.9) and (1.13) the temperature. As a consequence the rates of the reactions (3.59) are only dependent on and $Y^*_{H_2}$ and so also is the rate of chemical production of $Y^*_{H_2}$, $\dot{w}^*_{H_2}$:

$$\dot{w}^*_{H_2} = \dot{w}^*_{H_2}(\xi, Y^*_{H_2})\,. \tag{3.60}$$

Note also that $\varrho = \varrho(\xi, Y^*_{H_2})$ so that either Favre or conventional averaging may be used.

3.2.2 Chemical Production Term Closure for the Second Variable

All of the problems associated with this closure for a single species as discussed in Sect. 1.14 apply to the closure for the chemical production term of the second variable. Some authors choose to use merely a global kinetic rate based on the mean quantities

$$\overline{\dot{w}_F} = -\bar{k}\bar{\varrho}^2\,\bar{Y}_F\,\bar{Y}_O\exp(-T_a/\bar{T})\,. \tag{3.61}$$

This is unsatisfactory since it neglects the effects of fluctuations and the important role of mixing in determining the reaction rates. Others such as

Hutchinson et al. [3.73] followed *Borghi* [3.74] and applied a correction factor $(1+F)$ to (3.61) where

$$F=F(\bar{T},\overline{T'^2},\bar{Y}_O,\bar{Y}_F,\overline{Y'_O T'},\overline{Y'_F Y'_O})$$

is obtained by a truncation of the form of the expansions (1.56) and (1.64). The correlations $\overline{Y'_O T'}$ and $\overline{Y'_F Y'_O}$ have to be obtained by second-order closure of their balance equations and the method seems to gain little advantage from employing the conserved scalar. Methods of this type are discussed further in Sect. 3.3.

Spalding [3.21] attempted to combine the effects of mixing and kinetics through the use of the eddy break-up rate R_{EBU} and the kinetic rate of reaction R_{L} in a laminar premixed flame

$$\frac{1}{\bar{\dot{w}}_p}=\left(\frac{\mathrm{Re}}{R_{\mathrm{EBU}}}+\frac{\mathrm{Re}+A}{R_{\mathrm{L}}}\right)/(\mathrm{Re}+A)\,, \tag{3.62}$$

where Re is the local turbulence Reynolds number, A a constant. Various formulations have been proposed for R_{EBU} [3.9, 75], the most satisfying of which appear to be of the form

$$R_{\mathrm{EBU}}=\mathrm{const}\,\breve{\xi}\varepsilon/\tilde{q}\,. \tag{3.63}$$

This approach is semi-empirical and not soundly based in theory. It does however have similarities to the perturbation approach presented in Sect. 3.2.3.

A more satisfactory way of approaching the chemical closure problem posed here is that used by *Janicka* and *Kollman*, namely, modeling the joint pdf of ξ and (in their case) $Y^*_{\mathrm{H}_2}$. Several constraints can be placed on this pdf. It must comply with the moments $\widetilde{Y^*_{\mathrm{H}_2}}$, $\widetilde{Y^{*\prime\prime 2}_{\mathrm{H}_2}}$, $\overline{\varrho\xi'' Y^{*\prime\prime}_{\mathrm{H}_2}}/\bar{\varrho}$ $\tilde{\xi}$, and $\widetilde{\xi''^2}$ and its range must lie between 0 and 1 on ξ and between the unreacted and fully reacted (equilibrium) states for $Y^*_{\mathrm{H}_2}$ at any given value of ξ. The quantities $\widetilde{Y^{*\prime\prime 2}_{\mathrm{H}_2}}$ and $\overline{\varrho\xi'' Y''_{\mathrm{H}_2}}/\bar{\varrho}$ are obtainable from modeled balance equations with the chemical production terms $2\,\overline{\dot{w}^*_{\mathrm{H}_2} Y^{*\prime\prime}_{\mathrm{H}_2}}$ and $\overline{\dot{w}^*_{\mathrm{H}_2}\xi''}$ readily obtained from the joint pdf. The interesting and very plausible results obtained by *Janicka* and *Kollman* could probably be improved by incorporation of the ideas on pdf form and intermittency expressed in Sect. 3.1.5 and employing the correlation $\overline{\varrho\xi'' Y^{*\prime\prime}_{\mathrm{H}_2}}/\bar{\varrho}$. The non-equilibrium O atom concentrations are obtained directly from $Y^*_{\mathrm{H}_2}$ and ξ so that the instantaneous nitric oxide production rate $\dot{w}_{\mathrm{NO}}=\dot{w}_{\mathrm{NO}}(\xi, Y^*_{\mathrm{H}_2})$. The NO concentrations can be calculated using the joint pdf to obtain $\overline{\dot{w}_{\mathrm{NO}}}$. This approach appears to have promise.

A basic problem with all of these two-variable approaches is that they are not well suited to systems in which the chemistry is reasonably fast and the composition is close to equilibrium. Small errors in the estimation of the second variable can lead to large errors in its difference or departure from equilibrium.

In many systems of practical interest, such as flames with nonequilibrium radical contributions to nitric oxide formation (see Sect. 3.1.10), it is the departure from the equilibrium solution which is of interest. Accordingly, a perturbation approach is suggested.

3.2.3 Perturbation Variables

An alternative approach [3.58] is to use a perturbation of the fast chemistry relations (3.11); let

$$Y_i = Y_i^e(\xi) + y_i, \tag{3.64}$$

where y_i is the departure of species i from its equilibrium or quasi-equilibrium relationship $Y_i^e(\xi)$, and is not necessarily small. Substitution into (1.17) yields with the use of (3.5) [see (3.36)]

$$\frac{\partial}{\partial t}(\varrho y_i) + \frac{\partial}{\partial x_k}(\varrho u_k y_i) - \frac{\partial}{\partial x_k}\left(\varrho \mathscr{D} \frac{\partial y_i}{\partial x_k}\right) = \dot{w}_i + \varrho \mathscr{D} \frac{\partial \xi}{\partial x_k} \frac{\partial \xi}{\partial x_k} \frac{d^2 Y_i^e}{d\xi^2}. \tag{3.65}$$

For stationary flows this may be averaged to yield

$$\bar{\varrho} \tilde{u}_k \frac{\partial \tilde{y}_i}{\partial x_k} = -\frac{\partial}{\partial x_k}(\overline{\varrho u_k'' y_i''}) + \overline{\dot{w}}_i + \overline{\varrho \mathscr{D} \frac{\partial \xi}{\partial x_k} \frac{\partial \xi}{\partial x_k} \frac{d^2 Y_i^e}{d\xi^2}}. \tag{3.66}$$

These equations provide a clearer picture of the reaction rate expression given earlier in (3.36, 37). The term $\overline{\varrho \mathscr{D} \frac{\partial \xi}{\partial x_k} \frac{\partial \xi}{\partial x_k} d^2 Y_i^e / d\xi^2}$ is the rate at which out-of-equilibrium material is produced by fine scale turbulent mixing; we shall call it the microscale mixing source term. (In [3.76] this physical interpretation is supported by a more detailed mechanistic argument.) It is the source term for $\tilde{y}_i$, that is, for material which is capable of being reacted. The term $\overline{\dot{w}_i}$ is the chemical sink term for this material (it being negative if i is a reactant). In the limit of fast chemistry $\tilde{y}_i \rightarrow 0$ and the chemical sink term becomes everywhere equal to the microscale mixing source term, yielding (3.36, 37). This behavior is confirmed by an asymptotic analysis [3.76] which is discussed in the following subsection.

At first sight (3.66) does not appear to have much of an advantage over direct closure methods. We are still left with the closure problem for $\overline{\dot{w}_i}$ and furthermore we have to deal with the microscale mixing source term. In the following subsection we examine the interesting possibility that expression of $\dot{w}_i$ in terms of y_i is better conditioned than the usual problem with the result that a low-order closure for $\overline{\dot{w}_i}$ is sufficient.

The treatment of the microscale mixing source term has been discussed for homogeneous turbulence in Sect. 3.1.7. In turbulent shear flows satisfactory measurements of the joint pdf of ξ and χ have yet to be made. A preliminary analysis has been made of the tape recorded data for experiments in laboratory [3.49] and atmospheric boundary layers [3.51] and although the results are not of sufficient quality to obtain satisfactory joint pdfs they do indicate that the correlation is quite small, a correlation coefficient of the order of -0.1 being obtained in the results processed so far. Further work is needed in this area. However, in view of these results and the result (3.48) in Gaussian homogeneous turbulence it appears that the assumption of statistical independence (3.49) may be a good first approximation. With $d^2 Y_i^e/d\xi^2$ obtained directly from chemical equilibrium the term becomes

$$\overline{\varrho \mathscr{D} \frac{\partial \xi}{\partial x_k} \frac{\partial \xi}{\partial x_k} \frac{d^2 Y_i^e}{d\xi^2}} = \tfrac{1}{2} \bar{\varrho} \tilde{\chi} \int_0^1 \frac{d^2 Y_i^e}{d\xi^2} \tilde{P}(\xi)\, d\xi \tag{3.67}$$

$$\doteqdot \tfrac{1}{2} \bar{\varrho} \tilde{\chi} \tilde{P}(\xi_s/\check{\xi}) \Delta_s/\check{\xi}\,. \tag{3.68}$$

The latter approximation applies when $d^2 Y_i^e/d\xi^2$ is only significant over a range in ξ small compared with $\check{\xi}$, Δ_s being the change in $dY_i^e/d\xi$ over this region.

Some remarks on terminology and generality are in order here. The term "perturbation" has been used here in the sense of a departure from the equilibrium or fast chemistry solution. Equations (3.65, 66) do not depend on whether this perturbation is small or large and incorporate none of the usual assumptions of small perturbation theory. It should also be noted that the microscale mixing source term is theoretically derived and not an intuitive or semi-empirical formulation as has been used by other authors [1.33, 3.9]. In fact the whole of (3.65, 66) are theoretically derived and only involve the equal diffusivity and two-feed assumptions. They can obviously be extended to three or more feeds.

3.2.4 Perturbation Closure of $\overline{\dot{w}_i}$

We now turn to the $\overline{\dot{w}_i}$ term in (3.66) when a two-variable formulation, as discussed in Sect. 3.2.1, is possible. We have then that only one y_i quantity, denoted by y, is needed and the entire composition and temperature are determined by ξ and y. Moreover

$$\dot{w}_y = \dot{w}_y(\xi, y)\,. \tag{3.69}$$

For the *Janicka* and *Kollman* treatment of the hydrogen/air system we define

$$y \equiv (Y_{H_2}^*)^e - Y_{H_2}^*\,, \tag{3.70}$$

where $(Y^*_{H_2})^e$ is the equilibrium value. Thermochemical calculations for $\dot{w}_y$ in this case at atmospheric pressure are well correlated by [3.76]

$$\dot{w}_y/\varrho = -1.6\times 10^6 y^2\,\mathrm{s}^{-1}, 0.02 \leqq \xi \leqq 0.05\,. \tag{3.71}$$

The result that the reaction rate is independent of ξ over the range of interest near stoichiometric ($\xi_s = 0.028$) is particularly fortuitous. It is thought to derive from the zero activation energy associated with the recombination direction of the reactions (3.59).

A simple form for the reaction rate in terms of the perturbation variable is obtained also for some one-step reversible reactions [3.58]. A particular example is that for the photochemical smog system of dilute NO_2, NO, and O_3 in air under strong sunlight. It is shown in [3.76] that the reaction rate is of the form

$$\begin{aligned}\dot{w}_y/\varrho &= -f_1(\xi)y + f_2(\xi)y^2\\ &\approx -f_1(\xi)y\,.\end{aligned} \tag{3.72}$$

The linearization is possible over much of the range of interest and furthermore $f_1(\xi)$ is found to be slowly varying over a wide range of ξ.

By no means all reaction rates will be expressible as such simple functions of ξ and y. However it is probable that they will exhibit a minimum of a first-order dependence on y so that $\dot{w}_y \to 0$ as $y \to 0$ as required for equilibrium. It is this direct dependence of the reaction rate on y, the departure from equilibrium, which indicates that the perturbation approach may be better than the direct approach at least for systems of moderately fast chemistry. In the direct approach the departure from equilibrium is arrived at from calculations of $\tilde{Y}_i$ and $\tilde{Y}^e_i(\tilde{\xi}, \breve{\xi})$ and other moments of these quantities. If the chemistry is moderately fast, the departure from equilibrium is small and y is obtained as a small difference between estimates of relatively large quantities. The perturbation approach of (3.66) is thus likely to yield a much better estimate of $\tilde{y}$.

With this formulation of the reaction rate one approach to the closure problem is to neglect the effect of fluctuations and take

$$\overline{\dot{w}_y} \approx \dot{w}_y(\tilde{\xi}, \tilde{y})\,. \tag{3.73}$$

For the hydrogen/air system this would become from (3.71)

$$\overline{\dot{w}_y}/\bar{\varrho} \approx -1.6\times 10^6 \tilde{y}^2\,\mathrm{s}^{-1} \tag{3.74}$$

and for the one-step reversible photochemical smog system of (3.72)

$$\overline{\dot{w}_y} \approx -\bar{\varrho} f_1(\bar{\xi})\bar{y}\,. \tag{3.75}$$

The size of the truncation errors in such approximation has yet to be explored. However it can be expected that these first-order approximations will give first-order estimates for $\tilde{y}$ when used in the balance equation (3.66).

Closure by means of estimating the joint pdf of ξ and y has not been investigated. The moments $\tilde{\xi}$, $\widetilde{\xi''^2}$, and $\tilde{y}$ will be available from (3.6, 7, 66). Moment equations for $\overline{y''^2}$ and $\overline{\varrho\xi'' Y''}$ may be written but they will involve the terms $\overline{\varrho y''\chi d^2 Y^e/d\xi^2}$ and $\overline{\varrho\xi''\chi d^2 Y^e/d\xi^2}$ which will need to be modeled. The second of these terms should not present a significant problem, particularly as more information on the joint pdf of χ and ξ becomes available. In the first term y'' can be expected to have strong positive peaks near $\xi=\xi_s$ where $d^2 Y^e/d\xi^2$ is a maximum. This behavior should also be reflected in the joint pdf with the structure for $y>0$ centered on $\xi=\xi_s$ rather than $\xi=\tilde{\xi}$.

The expression of the mean reaction rate in the simple forms of (3.74, 75), namely

$$\overline{w_y} = -\bar{\varrho}k\tilde{y}^n \tag{3.76}$$

allows [3.76] an asymptotic solution for $\tilde{y}$ to be obtained in terms of a reciprocal Damköhler number ε,

$$\varepsilon \equiv U/(LkY_B^{n-1}).$$

Here U is a typical mean flow velocity, L a mean flow length scale, and Y_B a typical concentration of the species of which y is a perturbation. For flows in which the microscale mixing term of (3.66) is the only source of y, the solution is of the form

$$y = \varepsilon^{1/n} y^{(1)} + \varepsilon^{2/n} y^{(2)} + O\{\varepsilon^{2/n}\}. \tag{3.77}$$

Here $y^{(1)}$ involves a normalized ratio of the microscale mixing term in (3.66) to the chemical sink term and $y^{(2)}$ the normalized ratio of the transport terms $\bar{\varrho}\tilde{u}_k \dfrac{\partial \tilde{y}}{\partial x_k} + \dfrac{\partial}{\partial x_k}(\overline{u_k'' y''})$ to the chemical sink term.

It is seen from (3.77) that the lowest order term does not involve the transport terms and so that in the limit of fast chemistry $\varepsilon \to 0$, $y \to \varepsilon^{1/n} y^{(1)}$ which is the solution that is obtained if (3.37) is used rather than (3.66). This solution is used in [3.76] to make estimates of the departure from equilibrium in the turbulent diffusion flame of hydrogen in a coflowing stream of air as studied in [3.11, 69]. It is found that in the reaction zone that $\tilde{y} \sim 3 \times 10^{-4}$ with y defined by (3.70). This corresponds to a temperature depression below equilibrium of about 20 K which is much less than that found by *Janicka* and *Kollman* [3.31]. For reasons already stated it is believed that the perturbation estimate will be more accurate as well as being simpler to obtain.

3.3 Direct Closure Approaches

Although the conserved scalar approach is well founded, well established, and in general to be preferred, there are some problems in which it is either impossible or inappropriate to apply. Such problems probably include those in which the turbulence Reynolds number is low and differential diffusion effects are important and those in which the reactant feeds are not uniform and constant in composition and enthalpy. In these cases it is either not possible to define an appropriate conserved scalar or there are not unique relationships among the conserved scalars. In other cases there are complex chemical reactions far from equilibrium and thus little advantage lies in the two-variable approach. In such cases it is necessary to attempt direct closure of the chemical production term by either the moment or the pdf methods discussed briefly in Sect. 1.14. So far these methods are developed for relatively simple chemistry and for equal molecular diffusivity and to some extent have treated the same problems that are most appropriately tackled by the conserved scalar approach. We shall describe them here and indicate how more difficult problems may be treated.

3.3.1 Moment Closure Methods

For simplicity we follow the development of *Borghi* [3.74] for a simple one-step irreversible reaction (1.55) using the notation of (3.1) and conventional averaging. A similar development can be followed for Favre averaging. The reaction rate is given by

$$-\dot{w}_F = k\varrho^2 Y_F Y_O \tag{3.78}$$

with

$$k = BT^N \exp(-T_a/T) . \tag{3.79}$$

Using (1.21), assuming negligible variations in molecular weight, and allowing for conventional fluctuations around conventional means yields

$$-\dot{w}_F = \frac{BP^2}{R^2}(\bar{T}+T')^{N-2}\exp\left(-\frac{T_a}{\bar{T}+T'}\right)(\bar{Y}_F+Y'_F)(\bar{Y}_O+Y'_O) . \tag{3.80}$$

The exponential may be written

$$\exp\left(-\frac{T_a}{\bar{T}+T'}\right) = \exp\left(-\frac{T_a}{\bar{T}}\right)\exp\left(\frac{T_a}{\bar{T}}\,\frac{T'/\bar{T}}{1+T'/\bar{T}}\right),$$

which may be expanded to yield

$$-\overline{\dot{w}_F} = \frac{BP^2}{R^2}\,\bar{T}^{N-2}\exp\left(-\frac{T_a}{\bar{T}}\right)\bar{Y}_F\bar{Y}_O(1+F) \tag{3.81}$$

with

$$F \equiv \frac{\overline{Y'_F Y'_O}}{\bar{Y}_F \bar{Y}_O} + (P_2 + Q_2 + P_1 Q_1)\frac{\overline{T'^2}}{\bar{T}^2}$$

$$+ (P_1 + Q_1)\left[\frac{\overline{T'Y'_F}}{\bar{T}\bar{Y}_F} + \frac{\overline{T'Y'_O}}{\bar{T}\bar{Y}_O}\right] + P_1 \frac{\overline{T'Y'_F Y'_O}}{\bar{T}\bar{Y}_F \bar{Y}_O}$$

$$+ P_2\left[\frac{\overline{T'^2 Y'_F}}{\bar{T}^2 \bar{Y}_F} + \frac{\overline{T'^2 Y'_O}}{\bar{T}^2 \bar{Y}_O}\right] + (P_3 + Q_3)\frac{\overline{T'^3}}{\bar{T}^3} + \dots, \tag{3.82}$$

where

$$P_n \equiv \sum_{k=1}^{n} (-1)^{n-k} \frac{(n-1)!}{(n-k)!\,((k-1)!)^2 k} \left(\frac{T_a}{\bar{T}}\right)^k, \tag{3.83}$$

$$Q_n \equiv \frac{(N-2)(N-1)\dots(N+1+n)}{n!}. \tag{3.84}$$

The series (3.82) is only convergent if T_a is not large compared with $\bar{T}$ and the fluctuation levels are small. These are severe restrictions, from the practical point of view, since generally $T_a/\bar{T} \gg 1$. This should be kept in mind as we proceed.

Balance equations may be written for the second-order correlations $\overline{Y'_F Y'_O}$, $\overline{Y'_F T'}$, $\overline{Y'_O T'}$, $\overline{T'^2}$. For example multiplying (1.17) by Y''_j and the equivalent equation in Y_j by Y''_i adding and averaging yields, in the present notation,

$$\bar{\varrho}\tilde{u}_k \frac{\partial}{\partial x_k}(\overline{\varrho Y''_F Y''_O}/\bar{\varrho}) = -\overline{\varrho u''_k Y''_F}\frac{\partial \tilde{Y}_O}{\partial x_k} - \overline{\varrho u''_k Y''_O}\frac{\partial \tilde{Y}_F}{\partial x_k}$$

$$-\frac{\partial}{\partial x_k}(\overline{\varrho u''_k Y''_F Y''_O})$$

$$-\overline{2\varrho D \frac{\partial Y''_F}{\partial x_k}\frac{\partial Y''_O}{\partial x_k}}$$

$$+\overline{Y''_F \dot{w}_O} + \overline{Y''_O \dot{w}_F}. \tag{3.85}$$

Favre averaging is adopted here but conventional averaging can also be used. The first three terms on the right-hand side are usually modeled in terms of gradient transport and the fourth "dissipation" term modeled in the form [3.77] (see also Chap. 1)

$$-\overline{2\varrho D \frac{\partial Y''_F}{\partial x_k}\frac{\partial Y''_O}{\partial x_k}} \approx \text{const}\frac{\varepsilon}{\tilde{q}}\overline{\varrho Y''_F Y''_O}\left(1 + \frac{a_3}{R_T}\right), \tag{3.86}$$

where a_3 is a constant and R_T the turbulence Reynolds number. The covariances $\overline{Y''_F \dot{w}_O}\,\overline{Y''_O \dot{w}_F}$ can be expanded in a form similar to (3.81, 82). Correlations involving T can be obtained from (1.16, 17) in a similar manner. *Borghi* [3.74] avoids the complex chemical production terms that result by considering a conserved scalar of the form $[T + \Delta_p(r+1)\,Y_F/C_p]$ where Δ_p is the heat of combustion as in (1.14).

Closure is obtained by appropriate treatment of the third and higher order correlations which appear in the chemical production terms. *Borghi* [3.74] adopted the common procedure of setting these equal to zero. This procedure can lead to erroneous results since these zero values may be inconsistent with the inequalities that exist between moments of various orders. *Lin* and *O'Brien* [3.78] have proposed closure approximations which preserve these inequalities. It should also be noted here that solution of equations of the type of (3.85) can lead to second-order correlations which are inconsistent with such moment relationships. *Akhiezer* [1.39] treated some of the mathematics that lies behind these moment inequalities. Further work is required to generalize the problem to covariances between two or more variables. Some progress has been made by *Varma* et al. [3.79].

An alternative approach to closure is to model the higher moments in terms of the second-order and lower moments. Most of the present methods of doing this employ notions of the multivariable pdf either explicitly or through such concepts as the "typical eddy". These will be discussed in the next subsection.

Application of the methods of moment closure appears to be best suited to isothermal or near-isothermal problems or reactions with low activation temperature. *Borghi* [3.74] obtained plausible results for the burn-out of carbon monoxide in a jet-engine plume and *Donaldson* and *Hilst* [3.80] have applied the method to atmospheric reactions forming photochemical oxidants. The number of pde's that need to be solved simultaneously is large and the limited validity of the expansion of the exponential function cannot be too highly stressed.

3.3.2 Pdf Closure Methods

The pdf closure of (1.66) can be generalized

$$\dot{w}_i = \dot{w}_i(\varrho, T, Y_1, Y_2, \ldots, Y_N) \tag{3.87}$$

$$\overline{\dot{w}_i} = \int_0^1 \ldots \int_0^1 \int_0^1 \int_0^\infty \int_0^\infty \dot{w}_i P(\varrho, T, Y_1, Y_2, \ldots, Y_N; \boldsymbol{x})\, d\varrho\, dT\, dY_1\, dY_2 \ldots dY_N \,. \tag{3.88}$$

Specification of the multivariable pdf $P(\varrho, T, Y_1, Y_2, \ldots, Y_N; \boldsymbol{x})$ can be constrained by its moments $\bar{\varrho}$, $\tilde{T}$, $\tilde{Y}_i$, $\widetilde{T''^2}$, $\overline{\varrho Y''_1 Y''_2}$, etc., as defined by equations of the form (1.67) and determined as a function of $\boldsymbol{x}$ by solution of balance equations of the form of (1.48, 45, 46; 3.85). The second-order correlation equations involve terms of the type $\overline{\dot{w}_i Y_j''}$ and $\overline{\dot{w}_i T''}$. These are determined directly as

$$\overline{\dot{w}_i Y_j''} = \int_0^1 \dots \int_0^1 \int_0^1 \int_0^\infty \int_0^\infty (Y_j - \tilde{Y}_i)\dot{w}_i P(\varrho, T, Y_1, Y_2, \dots, Y_N; \boldsymbol{x}) \, d\varrho dT dY_1 dY_2 \dots dY_N . \tag{3.89}$$

There is no experimental evidence available to guide the structuring of this multivariable pdf subject to these moment constraints. *Bonniot* et al. [3.81] discussed the constraints on the pdf due to considerations of pure mixing or pure reaction. Under certain conditions the pdf can be collapsed into a bivariate pdf equivalent to that for a conserved scalar mixing variable and a reaction progress variable as discussed in Sect. 3.2.

A manifestation of this pdf modeling is the "typical eddy" proposed by *Donaldson* et al. [3.82–84]. The pdf is represented by seven Dirac delta functions corresponding to the pure components (two reactants and one product), half and half mixtures of reactants and products, two at a time, and the seventh an equally proportioned mixture of each reactant and the product. A particular, but unspecified temperature is assumed to be associated with each of these compositional "events". The computed moments and other constraints (such as the sum of the concentrations must be unity) are sufficient to determine the strength of the delta functions and the associated temperatures. Some problems are experienced [3.84–86] with delta functions of negative strengths but these can be obviated by allowing the relative proportions of reactants and products to vary in the "events" having such values. The name "typical eddy" arises from a physical realization of this pdf as a series of cells or boxes of these compositions passing the field point each having a duration proportional to the strength of the delta function. The need for series expansions for the reaction rate as in (3.81, 82) and for the density is obviated as each cell or event has a unique value of reaction rate and density associated with it. This is also true of the reaction rate composition correlation or reaction rate enthalpy correlation needed to obtain closure of the second-order correlation equations such as (3.85).

A criticism that can be levelled at the "typical eddy" approach is that the composition of the cells is purely arbitrary and that any allowable value of a given third-order correlation could be obtained by adjusting the proportion of reactants and products in each of these cells. Furthermore the reaction rate is nonzero in only two of the cells, that containing all three species and that containing the two reactants only. With heat release present, and noting the strong dependence of reaction rate on temperature, it seems that the reaction rate will be quite fortuitously determined by the particular distribution of temperature achieved among the cells. The fact that reasonable results are obtained suggests that the closure is not particularly sensitive to the modeling of terms like the third-order correlations.

3.4 Other Approaches

In chemical reactor theory the concept of a particle "age" within the reactor volume is used to describe the concentration time history experienced by a particle travelling through the system. This concept is found to be particularly useful in reactors of the stirred tank type where there is continual back-mixing of reacted and partially reacted material with the fresh reactants. This age is a fluid mechanical property of the reactor which can be determined experimentally by such methods as the injection of pulses of a tracer. Another fluid mechanical property of interest is the degree of cross-stream mixing as evidenced by the concentration fluctuations of the mixture fraction or other conserved scalar. *Pratt* [3.24] gave a comprehensive review of how these fluid mechanical properties are defined and how they may be used to determine the mean reaction rate. The coalescence dispersion equation of *Curl* [3.87] was used to describe the concentration distribution under the action of cross-stream mixing, back-mixing, and chemical reaction. This may be put in discrete form as done by *Flagan* and *Appleton* [3.88] for combustion in a plug-flow reactor. Monte Carlo methods may be used to simplify the computation. The formulation of the problem that results is quite similar to the pdf method discussed in Chap. 5. While the approach shows some promise it has yet to be completely formulated or used in a comprehensive way to calculate mean reaction rates in a system with nonpremixed reactants. The reader is referred to *Pratt* [3.24] for a full development.

Another approach currently receiving some attention is that of making detailed computations on a hypothetical element of the turbulent reaction zone. This is seen as a multilayered sandwich of alternately oxidant-rich and fuel-rich layers undergoing a shear or stretching strain. The behavior of the sandwich is calculated as an unsteady laminar flow problem so that the full multicomponent diffusion effects and complex free radical chemical mechanisms can be incorporated. A simple early version of this approach is that given by *Mao* and *Toor* [3.89]. They considered simple one-step irreversible chemistry and alternate slabs of initially unmixed oxidizer and fuel with thickness δ. The model is purely diffusional and there is no stretching or shear strain on the slabs. Results are presented in terms of a Damköhler number $\alpha \equiv k\delta^2 Y_{O_0}/D$, a stoichiometry parameter β, and dimensionless time Dt/δ^2. By an appropriate choice of δ they were able to fit the data of *Vassilatos* and *Toor* [3.90] over a wide range of α and β. It is not clear how their choice of δ was actually made and the values used, $\delta \sim 10^{-3}$ cm, seem to bear little relationship to actual turbulent scales in the flow. *Hill* [3.59] pointed out that the claim of insensitivity to the details of the mixing process is probably unfounded. The most recent application of these ideas is the "ESCIMO" theory of *Spalding* [3.91]. The acronym stands for Engulfment, Stretching, Coherence, Interdiffusion, Moving Observer and incorporates both Lagrangian and Eulerian approaches. Chemical reactions of any complexity can be solved by numerical solution of the parabolic differential equations. There are however problems of incorporating these results for the stretched laminar sandwiches into a description of the turbulence structure. A complete formulation and application to a problem of practical interest involving moderately fast chemistry is yet to appear.

3.5 Spectra and Nondimensional Numbers

The methods of approach discussed above have advantages relative to one another depending on how fast the chemistry is compared to the turbulent mixing. However, there is a problem in defining an appropriate time scale ratio to enable classification of the various regimes. Turbulence has several time scales, some associated with the large-scale structure, others with the fine-scale structure and the ratio of these depends on the turbulence Reynolds number $R_1 = U'l_0/\nu_0$ (1.68). Classification of the regimes will depend on the information being sought. Chemistry which is fast relative to the time scale of the large-scale motions may be slow compared to the time scale of the fine-scale motions. These considerations have not received adequate treatment in the literature.

3.5.1 Spectra

Hill [3.59] reviewed work that has been done on the spectral characteristics of reacting systems involving one species only. There is apparently no work on premixed systems with two species, let alone on nonpremixed systems.

A starting point for our discussion is the spectrum of the conserved scalar. There is a wealth of material in the literature on the nature of spectra of scalars in nonreacting flows. In reacting flows it can be expected that the spectra of conserved scalars will have similar properties. Even in flows where significant heat release accompanies chemical reaction, spectra are very like those in nonreacting flows [3.54, 92]; no unusual features are present that can be associated with the heat release or chemical reaction. In these laboratory combustion flows the turbulence Reynolds number is relatively low due to the high kinematic viscosity at high temperatures and there is not a pronounced inertial subrange as in other low Reynolds number turbulence. An inertial subrange can be expected in many practical combustion systems where either the high turbulent velocities or large length scale will make for a high turbulence Reynolds number.

Our discussion of spectra is once again perhaps best approached by considering the fast chemistry limit in which (3.11) applies. For systems of reversible reactions and small fluctuations we may define a linearity parameter

$$C_\xi \equiv \check{\xi}[(d^2 Y_i^e/d\xi^2)/(dY_i^e/d\xi)]_{\xi=\bar{\xi}}. \tag{3.90}$$

When $C_\xi \ll 1$ we have

$$Y_i^{e'} \equiv Y_i^e - \overline{Y_i^e} = (dY_i/d\xi)_{\xi=\bar{\xi}}\xi' \tag{3.91}$$

so that the fluctuations are linearly related and the normalized spectra of reactants and products are the same as that for ξ. This may be expressed (see Sect. 1.11 for notation)

$$\psi_y(k) = (dY_i/d\xi)^2_{\xi=\bar{\xi}}\psi_\xi(k), \tag{3.92}$$

where $k = |\boldsymbol{k}|$. A function of some interest is the co-spectrum $\psi_{a,b}$ of two reactants A and B with fluctuations Y_a' and Y_b'. This co-spectrum, or cross-spectral density function, is the Fourier transform of the cross-correlation function. It may be viewed as giving the spectral structure of the correlation $\overline{Y_a' Y_b'}$ and in general has real and imaginary components. It may be split into an amplitude and phase with the normalized amplitude $\gamma^2_{a,b}$ being defined [3.93]

$$\gamma^2_{a,b}(k) = \frac{|\psi_{a,b}(k)|^2}{\psi_a(k)\psi_b(k)} \tag{3.93}$$

and the phase

$$\theta_{a,b}(k) = \tan^{-1}[\mathrm{Im}\{\Psi_{a,b}(k)\}/\mathrm{Re}\{\Psi_{a,b}(k)\}]. \tag{3.94}$$

For initially unmixed reactants A and B following (3.91) $\gamma^2_{a,b}(k) = 1$ and $\theta_{a,b}(k) = -180°$, indicating that Y_a and Y_b are coherent and exactly out of phase over all wave numbers. Departure from this behavior indicates a breakdown of the fast chemistry assumption.

For the one-step irreversible reaction of (3.1) with fast chemistry (3.10), Y_F and Y_O have fluctuations as shown in Fig. 3.6 with

$$Y_F'(t) = Y_B[\xi(t) - \xi_s]\,I_s(t) - \bar{Y}_F, \tag{3.95a}$$

$$Y_O'(t) = rY_B[\xi_s - \xi(t)]\,[1 - I_s(t)] - \bar{Y}_O, \tag{3.95b}$$

and their product

$$\begin{aligned} Y_F' Y_O'(t) = {} & \bar{Y}_F \bar{Y}_O - r\bar{Y}_F Y_B[\xi_s - \xi(t)]\,[1 - I_s(t)] \\ & - \bar{Y}_O Y_B[\xi(t) - \xi_s]\,I_s(t). \end{aligned} \tag{3.96}$$

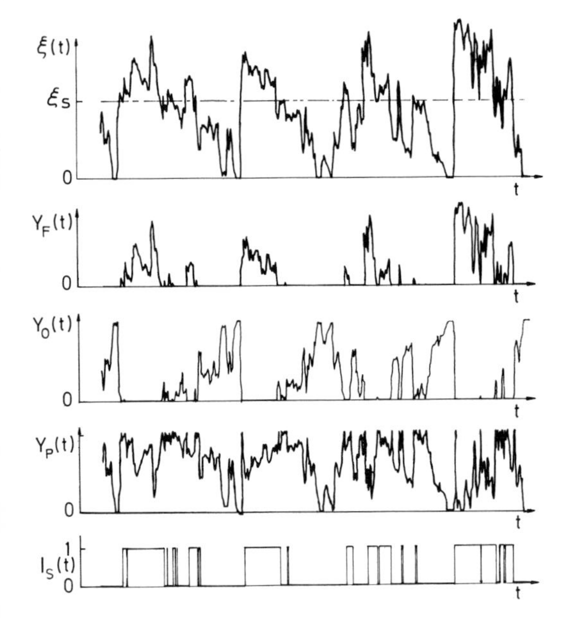

Fig. 3.6. Time series for the conserved scalar, reactant products and the switching function $I_s(t)$ where $I_s(t)=1$ for $\xi_s > \xi_s$, $I_s(t)=0$ for $\xi < \xi_s$. The conserved scalar has an intermittency of 0.90. A one-step irreversible reaction with fast chemistry is assumed

Here $I_s(t)=1$ when $\xi \geqq \xi_s$ and $=0$ when $\xi < \xi_s$ so that the product $I_s(t)[1-I_s(t)]=0$. It can be seen from these equations that Y'_F and Y'_O are signals similar to $\xi(t)$ but made intermittent by the conditioning function $I_s(t)$. The product signal is also of this character. For turbulent flows in which the intermittency is due to the turbulent/nonturbulent interface (see Sect. 1.10) the presence of the intermittency has little effect on the normalized spectrum of a scalar. If this is true here, and it may well not be, the normalized spectra of Y'_F and Y'_O can be expected to be similar to that of ξ'. The integral over the co-spectrum yields

$$\int_0^\infty \psi_{F,O}(k)dk = \overline{Y'_F Y'_O} = -\bar{Y}_F \bar{Y}_O , \tag{3.97}$$

and since most of the energy in the signals is at low wave number we must expect the coherence to be near unity with a 180° phase angle at low wave numbers. This can be seen from (3.95) where the correlation between Y'_F and Y'_O is seen to correspond to the ξ signal with the negative of the fluctuation in the switching function, $-I'_s(t)$. At low wave numbers this correlation is high and negative. At high wave numbers the correlations will be zero as there is no fine-scale structure in the switching function. We can thus expect the coherence to be high at low wave numbers dropping to zero at high wave numbers, with a phase going from 180° to indeterminate. The drop can be expected to occur near the wave number corresponding to the mean crossing frequency of $\xi(t)=\xi_s$. This will be of the order of the wave number corresponding to the large-scale structure in the turbulence.

The effects of departure from the fast chemistry limit are less easy to determine. The perturbation approach of (3.64) can be used and consideration given to the spectrum of y_i and its co-spectrum with ξ. The balance equation is rather like that considered by *Corrsin* [3.94, 95] for single species reacting flows particularly when it is recognized that the reaction rate has a linear or higher order dependence on y_i as developed in Sect. 3.2.3. However, there is some dependence also on ξ. Furthermore the source term $1/2\varrho\chi d^2 Y_i^e/d\xi^2$ appearing in (3.65) needs to be considered as well. It is not clear at this stage what the behavior of this term will be. What is clear is that for a pair of reactants the y_i contributions have the same sign so that they will tend to produce a positive correlation between the reactants. This should reduce the coherence and phase of the reactant co-spectrum from the equilibrium values discussed earlier.

Kewley and *Bilger* [3.96] have measured the co-spectrum of ozone and nitric oxide in photochemical smog in the atmosphere. In this reaction the NO_2 is decomposed by the action of ultraviolet light and the reaction is effectively reversible

$$O_3 + NO \underset{k_1}{\overset{k_3}{\rightleftharpoons}} NO_2 + O_2 . \tag{3.98}$$

Near a source of fresh emissions of NO a two-feed mixing and reaction model is appropriate and such an analysis is carried out in [3.97]. In this system $C_\xi \sim 1$ and one expects a high coherence and a phase close to 180° over all wave numbers if the chemistry is fast enough. In fact a sudden drop in coherence is found at a wave number close to $k_C \equiv k_R^{3/2}\varepsilon^{-1/2}$ which corresponds to the *Corrsin* [3.94, 95] wave number at which reactive loss balances spectral transfer. In the above $k_R \equiv k_1 + k_3\{[\overline{O_3}] + [\overline{NO}]\}$ where k_1 is the photolytic rate constant for NO_2, k_3 is the bimolecular rate constant for the forward reaction in (3.98), and $[\overline{O_3}]$ and $[\overline{NO}]$ are the mean concentrations of these reactants; and ε is the dissipation rate of velocity fluctuation energy. The critical Corrsin wave number is found by setting the Onsager time scale

$$\begin{aligned} \tau_T &= [k^3 \Phi_{11}(k)]^{-1/2} \\ &\approx \varepsilon^{-1/3} k^{-2/3} \end{aligned} \tag{3.99}$$

equal to the chemical time scale k_R^{-1}. In the above $\Phi_{11}(k)$ is the velocity spectrum over wave number k and the latter result is for the inertial subrange. These results suggest thst the nonequilibrium effects manifest themselves at high wave numbers resulting in a lack of coherence there. This is of course if k_R^{-1} lies within the range of time scales corresponding to the large eddies $\tau_E \equiv (U'k_E)^{-1}$ and the smallest eddies $\tau_K \equiv (\nu/\varepsilon)^{1/2}$, the latter being the Kolmogorov time scale. It should be noted that these results and conclusions can only be considered preliminary and need further investigation and confirmation.

3.5.2 Nondimensional Numbers

The above results suggest that an appropriate classification of nonpremixed reactant flows will depend on the Damköhler numbers

$$D_E \equiv k_R \tau_E = k_R/(k_E U') \approx k_R l_0/U' \tag{3.100}$$

and

$$D_K = k_R \tau_K = k_R \nu^{1/2}/\varepsilon^{1/2}, \tag{3.101}$$

where k_R is the equivalent linear reaction rate defined in the manner discussed above and by *Bilger* [3.58]. The last form for D_E is that given in (1.71). Since $\tau_E > \tau_K$, $D_E > D_K$. We may identify the following regimes

$D_K \gg 1$	fast chemistry, equilibrium flow,	(3.102a)
$D_K < 1 < D_E$	moderately fast chemistry, spectral effects,	(3.102b)
$D_E \ll 1$	slow chemistry, no effects of turbulence on reaction rate.	(3.102c)

These classifications appear physically plausible, but at this stage must be regarded as preliminary.

Donaldson and *Hilst* [3.80] suggested that the regimes are determined by a parameter N

$$N \equiv \frac{2D}{\lambda^2 k_\alpha \bar{Y}_\alpha} = \frac{0.1 q^{1/2}}{\Lambda k_\alpha \bar{Y}_\alpha} (S_c)^{-1} \tag{3.103}$$

with $\Lambda \approx l_0 \approx k_E^{-1}$ and $q = \overline{u_i^2}$, $q^{1/2} \sim 1.6U'$. Here λ is a Taylor microscale formed from the covariance between the reactants and their joint dissipation and k_α is a bimolecular reaction rate constant. It can be seen that for $S_c \sim 1$, $N \sim (6D_E)^{-1}$. $N \ll 1$ corresponds to $D_E \gg 1$ which is ascribed [3.80] to fast chemistry mixing control conditions. This is in qualitative agreement with the criteria given above.

Seinfeld [3.98] quoting *Lamb* [3.99] derived the criterion

$$k\langle c \rangle_{\max} \tau_e \ll 1$$

for the mixing effects to be unimportant. Here k is the effective bimolecular rate constant, $\langle c \rangle_{\max}$ the maximum concentration of any reactant in the flow at the point, and τ_e is the eddy lifetime. The latter can be estimated as $\tilde{q}/\varepsilon$ so that this criterion is essentially the same as (3.102c) above.

The perturbation analysis of [3.76] leads to the definition of a nondimensional number

$$N_{\mathrm{p}} \equiv \frac{\xi' \xi_{\mathrm{s}} k (Y_{\mathrm{s}}^{\mathrm{f}})^{n-1}}{\bar{\chi} P(\xi_{\mathrm{s}}/\xi')},$$

where the k and n are those of (3.76) and $Y_{\mathrm{s}}^{\mathrm{f}}$ is the mass fraction of the appropriate reactant that would obtain in a stoichiometric mixture if chemical reaction is frozen. The perturbation of the chosen reactant from equilibrium is given by $(4N_{\mathrm{p}})^{-1/n}$. This nondimensional number has a much better theoretical basis than those quoted above. For a given chemical system and flow geometry N_{p} will be related to D_{E} but it gives a more precise accounting for the effects of the chemistry and flow conditions.

3.6 Turbulence Structure and Modeling

3.6.1 Use of Favre Averaging

The considerable simplification that results from the use of Favre averaging in the conservation equations is emphasized in Chap. 1 and by implication throughout most of this chapter. Many of the equations that appear here could equally well have appeared in conventional average form but equations such as (3.6, 7, 50–53, 66) would be complicated considerably by the presence of correlations involving the density fluctuation. Many authors neglect these correlations, not so much because they *are* negligible, but because they do not fully understand the methodology involved with Favre averaging. Often the dropping of these terms means that the equations written are in fact for Favre averages although they are written as conventional time averages. Erroneous results are then obtained since the mean density is incorrectly computed, as discussed in Sect. 3.1.8.

A common reservation is that Favre averages are of little use for comparison with experimental data. The type of average obtained by many experimental techniques is open to question. It is argued in [3.100] that measurements such as composition from sampling probes[4] and velocity from pitot probes may be closer to Favre averages than conventional averages. Even laser-Doppler velocimeter results can be biased due to the effects of fluctuations in the seeding density [3.18, 100]. More importantly, if the difference between the conventional and Favre average is significant, then the density fluctuation correlations are significant $(\tilde{Q}-\bar{Q}=\overline{\varrho' Q'}/\bar{\varrho})$ and should be included in the conservation equations. Conservation equations for $\overline{\varrho' u'}$ and $\overline{\varrho' v'}$ can be written

4 The results of *Kennedy* and *Kent* [3.55], however, indicate that sample probe measurements are close to the conventional average.

and modeled [3.39, 79]. These can be used in the solution of the conventionally averaged equations and also to obtain Favre averages from the resulting conventionally averaged velocities. Alternatively they can be used to obtain conventional averages from the Favre-averaged results of the Favre-averaged equations. For scalar quantities both Favre and conventional averages can be obtained through the use of (3.14–21). In some closure methods where the pdf of the conserved scalar is not used a model equation for the covariance with the density fluctuation would need to be used.

Another reservation commonly held is that the mathematical models of turbulence developed for conventional averages, such as the gradient transport modeling of turbulent fluxes, cannot be simply carried over into the Favre averaging methodology. The response to this is that Favre averaging is relevant only in variable density flows, and that in such flows the common models are suspect. The foundations for techniques such as the eddy viscosity modeling of the shear stresses lie in empiricism and dimensional reasoning and not in the mechanistic physical models often used to support them. In the presence of large density gradients and fluctuations new dimensionless quantities such as density ratios appear and new empiricism may be needed. Modeling of the Favre-averaged first moments and turbulence kinetic energy and dissipation has been used successfully in a range of variable-density flows [3.25, 31, 102] with more rigor than has conventionally averaged modeling.

Yet another cause of reservation relating to Favre averaging is the disappearance of the gravitational term in (1.52) for the Favre-averaged turbulent kinetic energy. With $l=3$, $g_3=-\varrho g$ and $\overline{u''_3 g_l}=-\overline{g\varrho u'_3}/\bar{\varrho}=0$. Thus buoyancy effects on the turbulent kinetic energy, which are real, must be buried in other terms. In fact, they are buried in the velocity pressure gradient term which now has a component involving the mean pressure gradient [3.23]

$$-\overline{u''_l \frac{\partial p}{\partial x_l}} = -\overline{u''} \frac{\partial \bar{p}}{\partial x_l} - \frac{\partial}{\partial x_l}(\overline{p' u''_k})$$
$$+\overline{p' \varrho Dv/Dt}\,, \qquad (3.104)$$

where $v \equiv 1/\varrho$ is the specific volume and D/Dt the total (Stokes) derivative. The mean pressure gradient will have a component which arises from the gravitational body force

$$\frac{\partial \bar{p}}{\partial x_3} = \left(\frac{\partial \bar{p}}{\partial x_3}\right)_{\text{hydrodynamic}} - \bar{\varrho} g\,. \qquad (3.105)$$

The hydrodynamic component is that balancing the inertial and viscous terms in the momentum equation and that often set to zero in fire plume problems. The split indicated by (3.105) results in the conventional generation term $\overline{\varrho' u'}\, g$ in the turbulence kinetic energy equation. A careful treatment of mean pressure gradients is thus needed in buoyant flows and in particular it may be prudent to include the radial or cross-stream momentum equation.

We conclude from these considerations that the objections to Favre averaging are not substantiated and the advantages of such averaging make it attractive for predictive methods involving variable densities.

3.6.2 Turbulence Models

The modeling, in uniform density flows, of the turbulent fluxes and other terms such as the velocity and scalar dissipation and the pressure velocity correlations is a rapidly developing subject. No attempt is made to review these developments here. Recent reviews by *Reynolds* and *Cebeci* [3.103] and *Launder* [3.104] provide excellent entries into this field. Most of this work is concerned with the modeling of turbulence in thin shear layers. There has been unfortunately little work on turbulence modeling in flows with fully three-dimensional mean flows, with recirculation, or with time dependence; these flow geometries are of relevance to the flows in many combustion systems.

It is a primary aim of turbulence modeling research to develop models in as general a form as possible. This search for generality extends into modeling in variable density and chemically reacting turbulent flows. Thus models which are used in variable density flows should reduce to the models used in uniform density flows, without alteration of modeling coefficients, when the density is made uniform. Recently *Jones* [3.105] has given an excellent review of the bases for such modeling in variable density flows. *Edelman* and *Harsha* [3.106] gave a good outline of recent problems of interest and applications of modeling techniques.

There are several different phenomena involved in variable density flows. Turbulence fluxes and turbulence kinetic energy production in simple thin shear layers are often [3.25, 31] modeled without any special allowance made for density variation other than that implied in formulating the models in terms of Favre-averaged quantities. *Libby* and *Bray* [3.107], on the other hand, found it necessary to include a density ratio effect on eddy diffusivity and length scale ratio in order to match experimental data on premixed turbulent flame spread behind baffles in ducts. Whether there is a direct effect of the density variation on the turbulence properties, one which alters the rate at which turbulence kinetic energy is generated by the shear stress and the relationship between the turbulence shear stress and the mean gradient, is a question of some interest. This effect should not be confused with those effects involving different phenomena such as those introduced in discussing the pressure-gradient velocity-fluctuation terms (3.104) involved in the turbulent kinetic energy balance. These phenomena are perhaps best discussed in the context of addressing the age-old question about the nature of flame-generated turbulence.

3.6.3 Flame-Generated Turbulence

If one has a jet of hydrogen exhausting into a stream of air one can measure a certain level of turbulence in the mixing field. On igniting the jet the turbulence fluctuations decrease near the jet exit but downstream increase considerably, representing a generation of turbulence by the flame. However, the mean velocities have also increased and have produced increased shear rates. Measurements, such as those of *Glass* and *Bilger* [3.101], indicate that parameters such as $(\overline{u'^2})^{1/2}/U_0$ and $\overline{u'v'}/(\overline{u'^2})^{1/2}(\overline{v'^2})^{1/2}$ (where U_0 is the mean velocity increment on the centerline) are almost identical to those in a nonburning, uniform-density jet. The experimentally measured flame field may be predicted [3.25] to good accuracy using the same turbulence models that are used for predicting the nonburning flow. Thus in one sense the turbulence is flame generated and there is a turbulence/chemistry interaction – the sense that turbulence levels have increased. However it is apparent that the turbulence is ordinary shear-generated turbulence. Of greater fundamental interest would be turbulence which is not shear generated and which can be associated with more direct turbulence/chemistry interaction.

Another source of turbulence generation or annihilation is that associated with the effects of gravitational or other body forces, rotation, or mean pressure gradients. These all appear in the Favre-averaged turbulence kinetic energy equation as the term involving the velocity fluctuation and mean pressure gradient (3.104, 105). To date we have little experience in understanding and evaluating the importance of this term. The effects are subtle and complex and yet are important in many designs of practical combustion systems. Increasing the swirl in a swirl burner [3.108] can dramatically shorten the flame length, while in fire whirls [3.109] turbulence mixing is suppressed by the rotation and the flame length greatly increased. It is evident that the sign of the flux $\overline{\varrho'v'}$ can be crucial. Even in nonrotating flows it is possible that the radial pressure gradient due to the variation of $\overline{\varrho v'^2}$ across the flow can give rise to turbulence suppression on the outer edges of the flame [3.23]. Moreover, in most practical combustion systems strong axial and transverse pressure gradients are present due to the geometry and/or the heat release and can interact with density inhomogeneities to cause important changes in the mean fluxes. In free-burning fires buoyancy plays a very large part in determining the turbulent entrainment rate and hence the rate of combustion; *Tamanini* [3.28, 63] has made a start on the modeling of these effects. Much work remains to be done on the modeling of all of these effects. It is probable that modeling of the balance equations for $\overline{\varrho'u'}$ and $\overline{\varrho'v'}$ is needed to obtain accurate estimates of these for use in the Favre-averaged kinetic energy equation. It may not be sufficient to view these gravitational and mean pressure gradient terms just as sources or sinks in the turbulent kinetic energy equation. The terms appear directly in the equation for the turbulent stress tensor (1.51) and thus distort the normal anisotropy found for shear-generated turbulence. A method of handling these effects has yet to be devised.

Other sources of flame-generated turbulence may arise from the second and third terms on the right-hand side of (3.104). By its nature the pressure velocity correlation term $\partial(\overline{p' u_l''})/\partial x_l$ can only convect turbulence from one part of the flow to another and as such is unlikely to be a source of flame-generated turbulence. The remaining term $\overline{p' \varrho Dv/Dt}$ is a more likely source of velocity fluctuations. It expresses the coupling of the heat release with the pressure fluctuations. However there appears to be no mechanism which would give directionality to this term and hence vorticity to the velocity fluctuations. The fluctuations produced may thus be entirely acoustic in nature and could lead to noise such as screech but not directly to flame-generated turbulence.

It appears then that apart from mean shear induced by the flame the major source of flame-generated turbulence may be that associated with mean pressure gradients. Research is currently under way to better elucidate these effects.

3.7 Summary

The last decade has seen a rapid development of the conserved scalar/fast chemistry analysis first proposed by *Hawthorne* et al. [3.2] for turbulent nonpremixed combustion. This analysis has been extended to cover systems with reversible and complex multistep chemical mechanisms. Non-Gaussian pdfs of the conserved scalar are also incorporated. Doubts about this approach, which have arisen due to the need to predict this pdf given only its first two moments, appear not to be an important barrier; the mean density, composition, and temperature are insensitive to the form of the pdf and semi-empirical models of the pdf are being used successfully.

A very significant recent development is the extension of the conserved scalar methodology to incorporate the effects of finite rate chemistry on the main reaction. Problems are posed by the closure for the chemical production term of the second variable. These appear to be considerably ameliorated by the use of a perturbation approach particularly for flows in which the chemistry is reasonably fast. An improved insight has been obtained into the relative roles of mixing and chemical kinetics.

The formulation of appropriate nondimensional numbers which can be used to discriminate between regimes of slow and fast chemistry is still not entirely satisfactory. The spectral effects of the kinetic rate are as yet not clear. Some attempt has been made here to clarify these questions even if the answers are not available.

The use of Favre-averaged variables has clarified the nature of the problems involved in mathematically modeling turbulence in variable density flows. A considerable further effort is required to satisfactorily treat these problems and to extend currently available methods for thin shear layers to the problems involving fully three-dimensional flow fields and time dependence which are of great interest in combustion.

References

3.1 F.A.Williams: *Combustion Theory* (Addison-Wesley, Reading, Mass. 1965)
3.2 W.R.Hawthorne, D.S.Weddell, H.C.Hottel: In *Third Symposium on Combustion, Flame, and Explosion Phenomena* (Williams and Wilkins, Baltimore 1949) p. 267
3.3 K.Wohl, C.Gazley, N.Kapp: In Ref. 3.2, p. 288
3.4 M.W.Thring, M.P.Newby: In *Fourth Symposium (International) on Combustion* (Williams and Wilkins, Baltimore 1953) p. 789
3.5 H.L.Toor: AIChE J. **8**, 70 (1962)
3.6 P.A.Libby: ARS J. **32**, 388 (1962)
3.7 H.L.Toor: Ind. Eng. Chem. Fundam. **8**, 655 (1969)
3.8 D.B.Spalding: VDI Ber. **146**, 25 (1970)
3.9 D.B.Spalding: In *Thirteenth Symposium (International) on Combustion* (The Combustion Institute, Pittsburgh 1971) p. 649
3.10 J.H.Kent: Ph. D. Thesis, The University of Sydney, Sydney (1972)
3.11 J.H.Kent, R.W.Bilger: In *Fourteenth Symposium (International) on Combustion* (The Combustion Institute, Pittsburgh 1973) p. 615
3.12 J.H.Kent, R.W.Bilger: TN F-47, Mech. Engng. Dept., Sydney Univ. (1972)
3.13 R.P.Rhodes, P.T.Harsha, C.E.Peters: Acta Astron. **1**, 443 (1974)
3.14 C.H.Lin, E.E.O'Brien: J. Fluid Mech. **64**, 195 (1974)
3.15 W.B.Bush, F.E.Fendell: Acta Astron. **1**, 645 (1974)
3.16 P.A.Libby: In *Turbulent Mixing in Non-reactive and Reactive Flows*, ed. by S.N.B.Murthy (Plenum, New York, London 1974) p. 333
3.17 I.E.Alber: Ref. 3.16, p. 371
3.18 F.C.Lockwood, A.S.Naguib: Combust. Flame **24**, 109 (1975)
3.19 H.A.Becker: In *Fifteenth Symposium (International) on Combustion* (The Combustion Institute, Pittsburgh 1975) p. 601
3.20 P.A.Libby: Combust. Sci. Technol. **13**, 79 (1976)
3.21 D.B.Spalding: Combust. Sci. Technol. **13**, 3 (1976)
3.22 R.W.Bilger: Combust. Sci. Technol. **13**, 155 (1976)
3.23 R.W.Bilger: Prog. Energy Combust. Sci. **1**, 87 (1976)
3.24 D.T.Pratt: Prog. Energy Combust. Sci. **1**, 73 (1976)
3.25 J.H.Kent, R.W.Bilger: In *Sixteenth Symposium (International) on Combustion* (The Combustion Institute, Pittsburgh 1977) p. 1643
3.26 P.A.Libby: In *Studies in Convection*, Vol. 2, ed. by B.E.Launder (Academic Press, London, New York, San Francisco 1977) p. 1
3.27 S.Elgobashi: Ref. 3.26, p. 141
3.28 F.Tamanini: Combust. Flame **30**, 85 (1977)
3.29 Y.Onuma, M.Ogasawara: Combust. Flame **30**, 163 (1977)
3.30 M.Glass, J.H.Kent: In *Second Australasian Conference on Heat and Mass Transfer* (The University of Sydney, Sydney 1977) p. 445
3.31 J.Janicka, W.Kollman: In *Seventeenth Symposium (International) on Combustion* (The Combustion Institute Pittsburg 1979) p. 421
3.32 A.A.Townsend: *The Structure of Turbulent Shear Flow* 2nd ed. (Cambridge University Press, Cambridge 1976) p. 53
3.33 W.B.Bush, F.E.Fendell: Combust. Sci. Technol. **11**, 35 (1975)
3.34 H.A.Becker: In Ref. 3.26, p.45
3.35 D.B.Spalding: Chem. Eng. Sci. **26**, 95 (1971)
3.36 S.Gordon, B.J.McBride: NASA SP-273 (1971)
3.37 R.W.Bilger: Combust. Flame **30**, 277 (1977)
3.38 R.W.Bilger: Combust. Sci. Technol. **11**, 215 (1975)
3.39 W.Kolbe, W.Kollman: Acta Astron. **7**, 91 (1980)
3.40 R.G.Batt: J. Fluid Mech. **82**, 53 (1977)
3.41 A.Roshko: AIAA J. **14**, 1349 (1976)

3.42 K.R.Sreenivasan, R.A.Antonia, S.E.Stephenson: AIAA J. **16**, 869 (1978)
3.43 K.S.Venkataramani, N.K.Tutu, R.Chevray: Phys. Fluids **18**, 1413 (1975)
3.44 P.Anderson, J.C.Larue, P.A.Libby: Technical Report UCSD-9-PU Project Squid Headquarters, Purdue University (1977)
3.45 K.R.Sreenivasan, R.A.Antonia: AIAA J. **16**, 861 (1978)
3.46 A.D.Birch, D.R.Brown, M.G.Dodson, J.R.Thomas: In Ref. 3.31, p. 307
3.47 R.K.Gould, R.D.Thorpe, P.J.Howard: In Ref. 3.31, p. 299
3.48 J.C.Larue, P.A.Libby: Phys. Fluids **17**, 1956 (1974)
3.49 H.Q.Danh, R.A.Antonia: Phys. Fluids **20**, 1050 (1977)
3.50 R.A.Stanford, P.A.Libby: Phys. Fluids **17**, 1353 (1974)
3.51 R.A.Antonia: In Ref. 3.30, p. 13
3.52 R.W.Bilger, R.A.Antonia, K.R.Sreenivasan: Phys. Fluids **19**, 1471 (1976)
3.53 S.Corrsin, A.L.Kistler: NACA Report 1244 (1975)
3.54 I.Ebrahimi, R.Kleine: In Ref. 3.25, p. 1711
3.55 I.M.Kennedy, J.H.Kent: In Ref. 3.31, p. 279
3.56 R.W.Bilger: AIAA J. **15**, 1056 (1977)
3.57 P.A.Libby: Phys. Fluids **19**, 494 (1976)
3.58 R.W.Bilger: Combust. Sci. Technol. **19**, 89 (1979)
3.59 J.C.Hill: Annu. Rev. Fluid Mech. **8**, 135 (1976)
3.60 R.P.Rhodes, P.T.Harsha: AIAA Paper No. 72–68 (1972)
3.61 S.H.Stårner: Private communication (1978)
3.62 J.H.Kent, R.W.Bilger: TN F-41, Mech. Engng. Dept., University of Sydney (1972)
3.63 F.Tamanini: In Ref. 3.31, p. 1075
3.64 C.H.Gibson, P.A.Libby: Combust. Sci. Technol. **6**, 29 (1972)
3.65 R.W.Bilger: Combust. Flame **26**, 115 (1976)
3.66 F.A.Williams: In Ref. 3.16, p. 189
3.67 J.W.Moseley: B.E.Thesis, Department of Mech. Engng., The University of Sydney (1978)
3.68 C.T.Bowman: Prog. Energy Combust. Sci. **1**, 33 (1975)
3.69 R.W.Bilger, R.E.Beck: In Ref. 3.19, p. 541
3.70 K.N.C.Bray, J.B.Moss: Acta Astron. **4**, 291 (1977)
3.71 W.E.Kaskan, G.L.Schott: Combust. Flame **6**, 73 (1962)
3.72 G.Dixon-Lewis, F.A.Goldsworthy, J.B.Greenberg: Proc. R. Soc. A**346**, 261 (1975)
3.73 P.Hutchinson, E.E.Khalil, J.H.Whitelaw: In *Turbulent Combustion*, ed. by L.A.Kennedy, Vol. 58 of Progress in Astronautics and Aeronautics (American Institute of Aeronautics and Astronautics, New York 1978) p. 211
3.74 R.Borghi: Adv. Geophys. **18**B, 349 (1974)
3.75 F.C.Lockwood: Combust. Flame **29**, 111 (1977)
3.76 R.W.Bilger: Combust. Sci. Technol. **22**, 251 (1980)
3.77 C.duP.Donaldson: Proceedings Turbulent Shear Flow AGARD Meeting, London (AGARDOGRAPH No. 93) (1971) pp. B.1–B.24
3.78 C.-H.Lin, E.E.O'Brien: Astron. Acta **17**, 771 (1972)
3.79 A.K.Varma, P.J.Mansfield, G.Sandri: *Second Order Closure Modeling of Variable Density Turbulent Flows* A.R.A.P. Report No. 388, Aeronautical Research Associates of Princeton, Inc, Princeton N. J. (March, 1979)
3.80 C.duP.Donaldson, G.R.Hilst: Environ. Sci. Technol. **6**, 812 (1972)
3.81 C.Bonniot, R.Borghi, P.Magre: In Ref. 3.73, p. 331
3.82 C.duP.Donaldson: In Ref. 3.16, p. 131
3.83 C.duP.Donaldson, A.K.Varma: Combust. Sci. Technol. **13**, 55 (1976)
3.84 A.K.Varma, E.S.Fishburne, C.duP.Donaldson: In Ref. 3.73, p. 117
3.85 D.J.Kewley: In *Proc. Intern. Symposium on Gasdynamic Lasers* (DFVLR Press, Goettingen 1976) p. 212
3.86 D.J.Kewley: In *Sixth Australian Conference on Hydraulics and Fluid Mechanics* (Institution of Engineers, Australia, Canberra 1977) p. 49
3.87 R.Curl: AIChE J. **9**, 175 (1963)
3.88 R.C.Flagan, J.P.Appleton: Combust. Flame **23**, 249 (1974)

3.89 K.W.Mao, H.L.Toor: AIChE J. **16**, 49 (1970)
3.90 G.Vassilatos, H.L.Toor: AIChE J. **11**, 666 (1965)
3.91 D.B.Spalding: In Ref. 3.31, p. 431
3.92 I.M.Kennedy: Ph. D. Thesis, The University of Sydney (1979)
3.93 J.S.Bendat, A.G.Piersol: *Random Data: Analysis and Measurement Procedures* (Wiley-Interscience, New York 1971) p. 43
3.94 S.Corrsin: J. Fluid Mech. **11**, 407 (1961)
3.95 S.Corrsin: Phys. Fluids **7**, 1156 (1964)
3.96 D.J.Kewley, R.W.Bilger: TN ER-30, Mech. Engng. Dept., University of Sydney (1978)
3.97 R.W.Bilger: Atmos. Environ. **12**, 1109 (1978)
3.98 J.Seinfeld: *Air Pollution* (McGraw-Hill, New York 1975) p. 271
3.99 R.G.Lamb: Atmos. Environ. **7**, 235 (1973)
3.100 R.W.Bilger: In *Experimental Diagnostics in Gas Phase Combustion Systems*, ed. by B.T.Zinn, C.T.Bowman, D.L.Hartley, E.W.Price, J.G.Skifstad, Vol. 53 of Progress in Astronautics and Aeronautics (American Institute of Aeronautics and Astronautics, New York 1977) p. 49
3.101 M.Glass, R.W.Bilger: Combust. Sci. Technol. **18**, 165 (1978)
3.102 P.A.Libby: "Free Turbulent Shear Flows", NASA SP-321 (1973) p. 421
3.103 W.C.Reynolds, T.Cebeci: In *Turbulence*, Topics in Applied Physics, Vol. 12, ed. by P.Bradshow (Springer, Berlin, Heidelberg, New York 1976) p. 193
3.104 B.E.Launder: In Ref. 3.103, p. 232
3.105 W.P.Jones: *Prediction Methods for Turbulent Flows*, Lecture Series at the von Karman Institute for Fluid Dynamics, 1979 (to be published)
3.106 R.B.Edelman, P.T.Harsha: Prog. Energy Combust. Sci. **4**, 1 (1978)
3.107 P.A.Libby, K.N.C.Bray: AIAA J. **15**, 1186 (1977)
3.108 N.A.Chigier: Prog. Energy Combust. Sci. **2**, 97 (1976)
3.109 H.W.Emmons, S.J.Ying: In *Eleventh Symposium (International) on Combustion* (The Combustion Institute, Pittsburgh 1967) p. 475

4. Turbulent Flows with Premixed Reactants

K. N. C. Bray

With 16 Figures

This chapter considers turbulent flows in which the reactants have been fully mixed prior to reaction. Application to combustion is emphasized both because many practical combustion systems require the fuel and oxidizer to be premixed, and also because the premixed flame provides a convenient test for contemporary ideas about turbulent reacting flows. Only gaseous species are considered.

The rates of the chemical kinetic processes leading to combustion are strongly dependent on temperature. Consequently, the propagation of a premixed laminar flame requires thermal conduction and diffusion from the hot products to preheat the reactive mixture. In a turbulent flame, these molecular processes are augmented both indirectly, by distortion of flame surfaces, leading to an increase in their area, and directly, by turbulent mixing. The result is that the mean rate of heat release is generally more strongly influenced by the turbulence than by chemical kinetic factors so that premixed turbulent combustion is primarily a complex fluid mechanical problem. However, ignition and flame quenching provide examples of situations in which both chemical kinetics and fluid mechanics are likely to be important.

Premixing leads to a significant simplification in the analysis as the composition of the flow is essentially uniform, in terms of the elements involved, i.e., the Z_i's of (1.19) are constants. It also causes complications; the scalar thermodynamic variables of temperature, density, and composition often fluctuate strongly within this type of flame, between values characteristic of the unburned and fully burned mixtures. These intense scalar fluctuations pose theoretical and experimental difficulties, some of which have not yet been solved.

Following a review of knowledge concerning premixed turbulent combustion, this chapter concentrates on theoretical models which are based on prior specification of a probability density function for the fluctuating thermodynamic state of the mixture. It is argued that such models provide the best available compromise between complexity and generality of application.

4.1 Introductary Remarks

If reactants are to be premixed, a characteristic mixing time must be small in comparison with a time typifying the rate of chemical reaction at the condition under which mixing occurs, so that mixing can be accomplished before significant reaction begins. Assuming that the relevant diffusion coefficients are all approximately equal, the subsequent reaction will then occur in a mixture of almost constant and uniform elemental composition.

For the present study, premixed combustion provides the most important example. The chemical reaction time under ambient conditions can often be extremely large so that premixed reagents may be stored indefinitely. However, this time falls rapidly with increasing temperature and, at temperatures typical of combustion, it can become small in comparison with the characteristic times of other contributing processes. In practice, premixed combustion is almost always turbulent. Well-known examples include combustion in spark-ignition engines, in jet engines with reheat or with premixing and prevaporization to reduce pollutant emissions, and in tunnel burners in industry. Another important example is provided by vapor cloud explosions resulting from the leakage of fuel into the atmosphere where it mixes with air and can subsequently ignite. In all of these cases, the velocity field is closely coupled to the composition and temperature fields, so that turbulent motion and combustion interact strongly with each other.

Isothermal premixed turbulent flows involving homogeneous chemical reactions in liquids, for example, are generally less complex; see Sect. 1.2. If mixing is essentially complete before the commencement of the reaction, then the subsequent chemistry takes place in a uniform isothermal mixture, and the reaction is independent of the turbulent flow. There are exceptions, particularly if reactions are allowed to occur at surfaces. Examples may be found in electrochemistry involving premixed liquid reagents flowing over electrodes; situations can arise where reaction intermediates, which are produced in reactions at the electrodes, must diffuse into the bulk liquid before the reaction can reach completion (see, for example, [4.1]). The turbulent velocity field then strongly influences the progress of the chemical reaction. It is, however, unlikely that the reaction will have any significant effect on the turbulence.

Scientific investigation of turbulent premixed flames is generally recognized to have begun in 1940 with *Damköhler*'s classical theoretical and experimental study [4.2]. Reviews of knowledge concerning turbulent flames, emphasizing premixed reactants, were published in 1961 by *Lewis* and *von Elbe* [4.3] and in 1965 by *Williams* [4.4]. More recent reviews of the subject may be found in the work of *Andrews* et al. [4.5], *Abdel-Gayed* and *Bradley* [4.6], and *Libby* and *Williams* [4.7], while turbulent combustion in spark-ignition engines has been reviewed by *Tabaczynski* [4.8].

Damköhler [4.2] first put forward the idea that sufficiently large-scale turbulence merely wrinkles a premixed laminar flame without significantly modifying its internal structure, while sufficiently small-scale turbulence primarily affects the transport processes inside the flame. Limiting regimes are thus identified in which the microscale of the turbulence is either very large or very small in comparison with the thickness of a laminar flame, and these concepts have influenced almost all later work on the subject.

There is much indirect experimental evidence to support the existence of a wrinkled laminar flame regime in premixed turbulent combustion. Within this regime, thin laminar flames separate zones of unburned and fully burned mixture. Passage of such zones past a fixed point in the flow field causes large fluctuations to occur in the temperature, density, and composition of the gas and, for the reasons explained in Sect. 1.14, the time-averaged reaction rates are strongly affected. However, sufficiently intense turbulence tears the wrinkled flame sheet into pieces and can also quench the combustion in some of the resulting flame fragments. The laminar flame is known to be unstable to a broad spectrum of disturbances. It is, therefore, possible that in the presence of strong turbulence, instantaneous burning zones bear little or no resemblance to laminar flames. For example, *Chomiak* [4.9] suggested that these zones are associated with a fine vortex structure whose transverse dimension is comparable to the Kolmogorov microscale of the turbulence. Conditions under which such complications may arise are not yet firmly established.

In *Damköhler*'s other limiting regime, where the turbulence scale is smaller than the laminar flame thickness, the turbulent fluctuations in temperature, etc., are expected to be relatively weak.

A related problem concerns the influence of heat release and combustion-generated fluctuations in density on the structure and properties of the turbulent velocity field. The existence of so-called flame-generated turbulence has long been a topic of controversy and it is not yet clear whether conventional empirical descriptions of turbulence production and turbulent transport in cold nonreacting flows can be applied without modification in flames. The combustion-turbulence interaction is discussed in a recent review by the present author [4.10].

We shall look first at experimental information about the structure and propagation of premixed turbulent flames. Although, these flames have been studied in the laboratory for many years, so that a vast body of experimental data on them exists, there are still serious gaps and inconsistencies in the description which emerges. Section 4.3 discusses the premixed laminar flame which is relevant, both as a simpler flow resembling the turbulent flame and, more importantly, because it appears in wrinkled laminar flame models of turbulent flames. Since the laminar flame is subject to instabilities which influence its behavior in a turbulent flow, a rigorous and general treatment of this subject is difficult. Accordingly, the instability mechanisms are briefly reviewed in Sect. 4.3. Theoretical studies of premixed turbulent flames are reviewed in Sect. 4.4, starting with wrinkled laminar flame theories and continuing with combustion controlled by turbulent transport. Section 4.5 then presents a unified analysis of the subject, based on the work of *Bray* and *Moss* [4.11, 12], which is capable of describing both wrinkled laminar flames and regimes of combustion dominated by turbulent transport. The chapter ends with a discussion of outstanding problems and possible approaches to them.

4.2 Review of Experiment

Premixed turbulent flames are studied in the laboratory in a number of alternative configurations (see, for example, [4.4]). Open unconfined flames may be stabilized on Bunsen-type burners with approach flow turbulence generated by the pipe flow within the burner tube, or alternatively, by flow through a screen or perforated plate. An oblique flame is generally produced but some experiments involve nominally flat turbulent flames, stabilized on open burners with the time-averaged flame perpendicular to the flow. Confined flames may be stabilized on rods, vee gutters, and pilot flames, generally in ducts of constant cross-sectional area. Unsteady spherical flames are produced by a spark or other ignition source in a combustible mixture, either in a flow, or a closed reaction vessel, or in a simulation of an unconfined explosion.

We shall first consider the structure of the turbulent flame and then turn to its speed of propagation.

4.2.1 Flame Structure

Experiments which investigate the structure of turbulent flames are of interest, both because they test the accuracy of theoretical predictions and also for the insight which they can provide into the fundamental processes involved. Crucial questions, which must eventually be resolved experimentally, include the range of validity of wrinkled laminar flame descriptions, the existence and extent of flame-generated turbulence, and the applicability of conventional empirical descriptions of turbulent transport. Measurements of mean velocity, composition, and temperature, and the fluctuation intensities of these quantities, are all required, together with their covariances, length scales, probability density functions, etc. Unfortunately, comprehensive and accurate measurements are scarce and clear answers to the above questions are not yet available, although it may be expected that this situation will improve with the development of nonintrusive optical measuring techniques of high spatial and temporal resolution.

We consider open flames first. Figure 4.1 shows short duration photographs which illustrate the instantaneous structure of such flames. Figure 4.1a in particular includes characteristic cusp-shaped regions, whose shape is very similar to that predicted for the unstable laminar flame (see Sect. 4.3). Instantaneous flame shapes, deduced from such pictures [4.13], are reproduced in Fig. 4.2, and show a continuous wrinkled flame surface. The depth of the wrinkles grows with height above the burner and leads to thickening of the time-average flame structure. Studies such as this leave little doubt that, under these conditions, when the turbulence intensity is low, the instantaneous flame is laminar. However, *Fox* and *Weinberg*'s measurements show that, near the tips of the wrinkles, and as a consequence of the instability of the laminar flame,

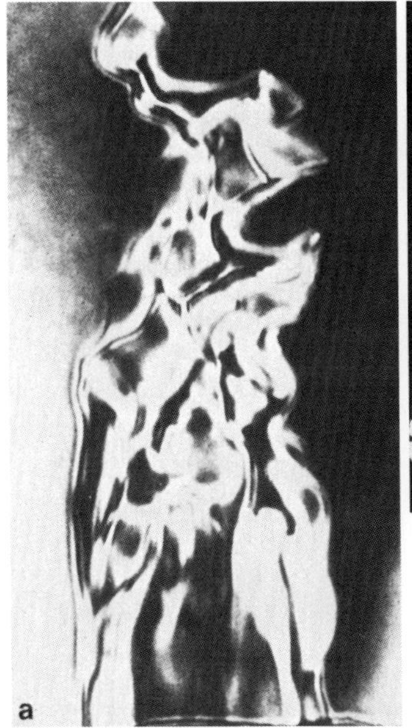

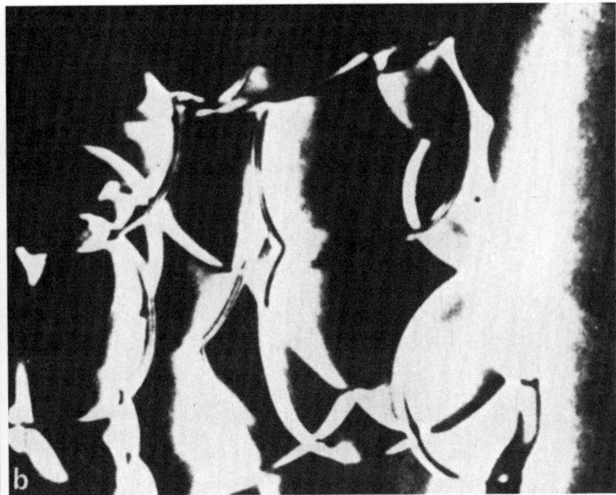

Fig. 4.1a and b. Short duration schlieren photographs of open turbulent flames [4.13]

the instantaneous flame speed is greater than that of an undisturbed laminar flame.

By seeding a combustible mixture with small inert particles, *Grover* et al. [4.14] obtained instantaneous photographs in which the boundary between the burned and unburned regions is clearly visible. They also find this boundary to be a continuous wrinkled surface. The same technique gives information about the flow velocity. In the unburned gas, the mean stream lines curve outwards towards the flame, while this outward flow is more pronounced in the combustion products which then mix with ambient air. The thickness of the turbulent flame zone appears to be related to the scale of the turbulence. *Vinckier* and *van Tiggelen* [4.15], using an opposed jet burner, found that the mean turbulent flame thickness, determined photographically, is comparable to the integral scale of the cold flow turbulence at the location of the flame, or two or three times larger than the Taylor microscale (see Sect. 1.16 for definitions of these scales).

As the wrinkled laminar flame moves in the turbulent field, the thermodynamic state of the gas at a fixed point in the flow fluctuates between values corresponding to unburned and fully burned mixture. This is illustrated in the ionization probe measurements of *Suzuki* et al. [4.16]. Measurements by *Yoshida* and *Tsuji* [4.17] and *Yoshida* and *Günther* [4.18] using fine wire thermocouples, whose frequency response has been corrected electronically, are also consistent with a wrinkled laminar flame description. The probability density function of temperature is found to be bimodal with well-defined peaks at the temperatures of unburned and fully burned mixture, and its shape is not

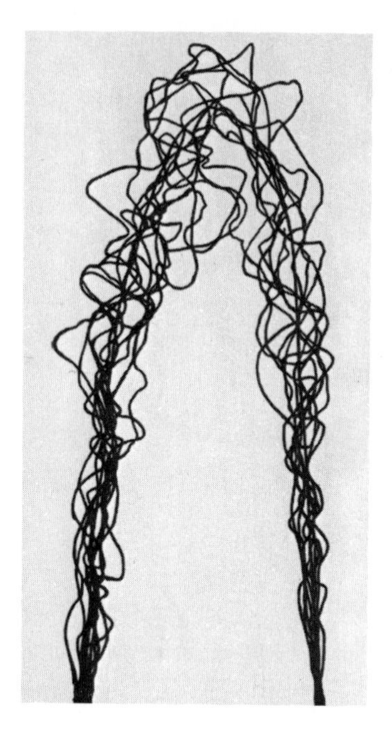

Fig. 4.2. Superimposed contours of instantaneous flame boundaries [4.13]

influenced by changes in the turbulence level. Maximum rms temperature fluctuations in excess of 600 K are observed. This is close to the maximum of 740 K which would occur if the flame were made up only of packets of unburned and fully burned mixture. The microlength scale of the temperature fluctuations is also reported [4.17].

Fluctuating velocity components are of particular interest for modeling studies. Microphone probes allow fluctuating dynamic pressure to be measured [4.19], but cannot distinguish between fluctuations in velocity and density. A sharp increase is observed in the intensity of dynamic pressure fluctuations in the flame zone. The velocity field of a turbulent Bunsen flame has also been mapped by laser anemometry [4.20, 21]. The mean flow stream lines confirm the trends determined earlier by other methods. The axial component of mean velocity along the center line remains almost constant, with height above the burner but, away from the center line, the axial mean velocity increases with height. The radial outflow component increases with distance from the center line, reaching a peak outside the flame. Both the axial and radial components of turbulent velocity fluctuation show a complex variation with position including peaks and troughs in the flame zone. There are thus indications of both generation and removal of turbulence within the flame. The Reynolds shear stress decays with increasing height above the burner, from that corresponding to an initial pipe flow profile.

We now turn to ducted flames. Figure 4.3 shows time exposure and instantaneous photographs of a flame stabilized on a small circular cylinder in

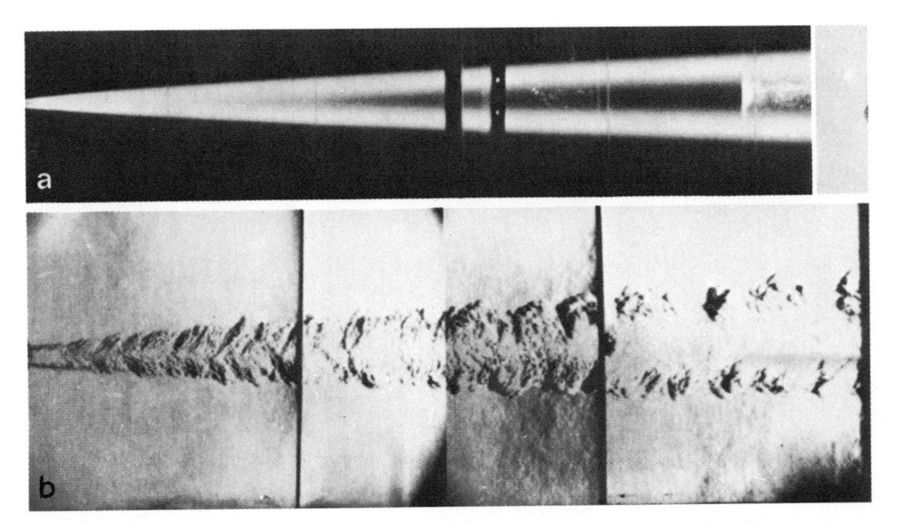

Fig. 4.3a and b. Time exposure (**a**) and instantaneous schlieren (**b**) photographs of flame stabilized in a duct [4.22]

a high-speed ducted flow [4.22]. Under these conditions the time-average flame is highly oblique and almost planar. Composition measurements on similar flames by *Howe* et al. [4.23] show that time-average concentrations of hydrocarbon fuel, carbon monoxide, hydrogen, and nitrogen are all uniquely correlated with the oxygen concentrations, independently of flow velocity or position in the flame. The same data allow profiles of time-averaged reaction rate to be calculated. It is found [4.23] that the maximum value of this rate is only 5 to 10% of the overall rate in a liminar flame of the same composition. Root mean square temperature fluctuations reach a peak value of about 600 °C within the flame [4.24] but fall to a low level on the center line. The probability density function of temperature is bimodal, with peaks at temperatures close to those of unburned mixture and combustion product. These various observations suggest a flow which consists predominantly of packets of unburned and fully burned gas. The mean flame angle is almost independent of approach flow velocity, temperature, and composition; thus the character of the flame is strongly influenced by the nature of the flow but relatively insensitive to chemical kinetics.

The mean axial velocity field of ducted flames [4.23] involves considerable acceleration resulting from gas expansion due to heat release. Typically [4.23] the axial velocity of the unburned gas doubles before it is all entrained into the flame, and the velocity at the center line doubles again. Large mean velocity gradients are, therefore, produced. The streamlines in the unburnt gas are deflected away from the flame [4.25].

Similar trends were found by *Moreau* and *Boutier* [4.26, 27] using laser anemometry in a ducted flow which is ignited and stabilized by a parallel flow of hot gas at one edge of the combustion chamber. An almost planar oblique

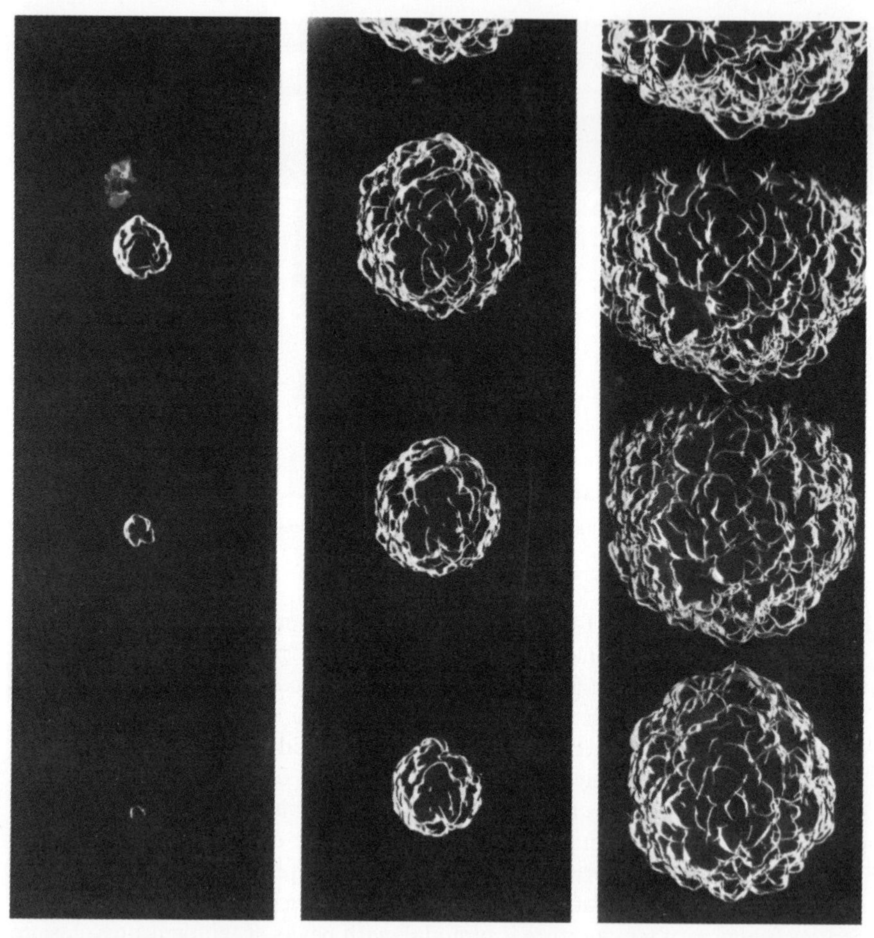

Fig. 4.4. Growth of a spherical flame in an isotropic turbulent stream [4.39]. Frame width 5 in.; time between exposures 1/300 s

turbulent flame is obtained. The rms of the axial component of fluctuating velocity rises in the flame, reaching a value about three times that measured in the cold flow, but the relative turbulence intensity decays downstream of the flame. The transverse component of the fluctuating velocity appears to be less strongly influenced by the presence of the flame. These findings are consistent with results of earlier studies by *Westenberg* and *Rice* [4.28] using a helium tracer technique, which indicate an increase of no more than about 50% in the transverse component of the fluctuating velocity within the flame zone, followed by a decay in the burned gas. In agreement with earlier workers (e.g., [4.25]), *Moreau* and *Boutier* [4.27] attributed the growth of axial turbulence in the flame zone mainly to the mean velocity gradient resulting from combustion. The production of turbulence energy by shear depends [see (1.52)] on the product of the mean velocity gradient and the Reynolds stress. Unfortunately, *Moreau* and *Boutier* [4.27] found the Reynolds stress in their flame too small to measure. Nevertheless, such stresses provide the most plausible mechanism for the modest growth in turbulence which they observe.

Another important category of laboratory flame is the unsteady spherical flame propagating from a central ignition source, which is usually a spark. Figure 4.4 illustrates this type of flame. Turbulence can be generated ahead of the flame by flowing the combustible mixture through a suitable grid or perforated plate, in which case the flame kernel is convected downstream in the turbulent flow. Alternatively, the experiment may be conducted in a closed pressure vessel, with turbulence produced either from fans or by means of a grid which is moved through the gas just before ignition. In these spherical flames, there can be no generation of turbulence by the mean velocity gradient mechanism mentioned above because, by symmetry, the shear components of Reynolds stress are zero. Measurements of the turbulence level within the flame would, therefore, be of particular interest. To date only the propagation speed has been measured.

Spark-ignited, unsteady turbulent flames are also used [4.29, 30] to investigate the effects of turbulence on the minimum ignition energy and quenching distance for premixed reactants. Of particular relevance in the present context is the observation [4.29] that an increase in turbulence intensity leads to an increase in the minimum energy required in order to obtain ignition. The effect can be quite large, as minimum ignition energies have been measured [4.29] which are up to about six times that required in the same mixture at a low level of turbulence. Such measurements suggest that in sufficiently intense turbulence a packet of burning gas, torn away from a propagating flame by the turbulent motion, may be quenched before it is completely burned.

4.2.2 Flame Speed

Many ambiguities and uncertainties arise in the experimental determination of the turbulent flame speed S_T. It is convenient to define S_T as the mean mass flux per unit area, in a coordinate system fixed to the time-average motion of the flame, divided by the unburned gas density ϱ_0. The area chosen is the smoothed out surface of the time-averaged flame zone. However, because the time-averaged flame is generally relatively thick, and also curved, the choice of a flame surface close to the inner or outer boundary of the burning zone can have a significant effect on the predicted value of S_T. Some workers use the above definition to deduce a mean value of S_T for the whole flame, while others report local values, which often vary by a factor of two or three from one point in the flame to another (see, for example, [4.14]). For a mean S_T, only the mass flow rate of fuel-air mixture and the area of a suitably defined time-average flame surface are required. In order to find a local value of S_T the magnitude and direction of the local time-averaged velocity and the local density must be determined at the time-averaged flame surface. However, heat release leads to expansion of the burned gas, causing curvature of the mean stream-lines, acceleration, and a fall in density. In most of the experiments reported in the literature, the information required for an accurate determination of local turbulent flame speeds is simply not available.

In the case of the unconfined spherical flame, the observed motion of the flame surface is the sum of effects due to expansion of the burned gas, causing an outward flow ahead of the flame, and turbulent flame propagation through this radially expanding medium. The flame speed S_T may be found by subtracting the outward flow velocity just ahead of an appropriate flame surface where the velocity is determined by calculation or by direct measurement [4.31]. In either case, an additional uncertainty is introduced. Alternatively, the flow velocity may be eliminated through the use of two flames propagating towards each other [4.31]. At the instant when the flames meet, the flow velocity ahead of each is zero.

In view of the many cited difficulties, experimental measurements of S_T have a large factor of uncertainty, which we can typify by the discrepancy of a factor of two or three between mean and local flame speed values from the same experiment, in the work of *Grover* et al. [4.14]. However, the following trends emerge:

I) S_T is almost always greater than the laminar flame speed S_L.

II) It increases with increasing intensity of turbulence ahead of the flame. Many, though not all, experimenters find the relationship to be approximately linear.

III) Some experiments (e.g., [4.32]) show S_T to be insensitive to the scale of the approach flow turbulence, although *Smith* and *Gouldin* [4.33] observed that S_T increases with turbulence macroscale in open flames, while *Ballal* and *Lefebvre* [4.34] found a complex variation of S_T with scale in ducted flames.

IV) In open flames, the variation of S_T with composition is generally much the same as for S_L, with a well-defined maximum close to stoichiometric. This has led many experimenters to present turbulent flame speed data as the ratio S_T/S_L.

V) Very large values of S_T may be observed in ducted burners at high approach flow velocities. In these circumstances [4.22, 25] S_T increases in proportion to the approach flow velocity, but is insensitive to approach flow turbulence and composition. It is believed that these effects result from the dominant influence of turbulence generated within the flame by the large velocity gradients. However, it is indicative of the state of the subject that the measurements of *Lefebvre* and *Reid* [4.35] did not conform to this description, showing a strong dependence of S_T on composition, similar to that for a laminar flame. It is possible that the pilot flame stabilization in their experiment produced smaller velocity gradients than were present in *Wright* and *Zukoski*'s rod-stabilized flames [4.22].

Ambitious correlations of flame speed data have been attempted by *Andrews* et al. [4.5, 31] and more recently by *Abdel-Gayed* and *Bradley* [4.6]. In *Andrews* et al. S_T/S_L is correlated with a turbulence Reynolds number characteristic of the unburned gas,

$$R_\lambda = \frac{u'_0 \lambda_0}{\nu_0},$$

where u'_0 is the root mean square turbulence velocity, λ_0 is the Taylor microscale of the turbulence, and ν_0 is the kinematic viscosity. Plotted in this way [4.31], data from a large number of experiments reported in the literature show a variation of a factor of four or more in values of S_T/S_L at fixed values of R_λ. Data from any one experiment generally show a much smaller scatter. Within this scatter, S_T/S_L is found to vary roughly in proportion to R_λ. However, this variation appears to be mainly due to the u'_0 factor in R_λ. As noted above, the evidence concerning variation of S_T/S_L with turbulence length scale is contradictory.

Abdel-Gayed and *Bradley* [4.6] constructed a more complex correlation, in which S_T/S_L is regarded as a function of two parameters, a turbulence Reynolds number

$$R_l = \frac{u'_0 l_0}{\nu_0},$$

where l_0 is the integral scale of the turbulence in the unburned gas, and the velocity ratio S_L/u'_0. They correlate a large amount of experimental data from all available sources to obtain a family of curves. At fixed R_l, S_T/S_L decays approximately hyperbolically with increasing S_L/u'_0, but S_T/S_L increases with R_l, if S_L/u'_0 is held constant. The correlation covers Reynolds numbers up to 3000 and values of S_T/S_L up to about 25. Although the scatter of individual data points about the mean curves is generally smaller than in the earlier correlation of *Andrews* et al. [4.31], it can still involve a factor of two or more in S_T/S_L at fixed S_L/u'_0 and R_l.

Part of this scatter can no doubt be attributed to uncertainty in determination of the turbulence parameters appearing in the correlation, many of which had to be estimated by *Abdel-Gayed* and *Bradley*. However, as pointed out by *Libby* et al. [4.36], there appear also to be serious disagreements either between different sets of experimental data or in their interpretation by *Abdel-Gayed* and *Bradley*. *Libby* et al. picked two extensive sets of experimental data, which approximately define two extremes in the correlation, and replot them as S_T/S_L vs u'_0/S_L. Each data set yields an approximate straight line, with quite small scatter, but the magnitudes of S_T/S_L at given u'_0/S_L differ by a factor of up to about three. Each set contains data covering a wide range of Reynolds numbers R_l, and these ranges overlap in the two sets of experiments. No consistent Reynolds number trend is apparent in either set of experimental results considered separately. However, when the two apparently inconsistent groups of data are taken together, and given equal weight, the Reynolds number dependence reported by *Abdel-Gayed* and *Bradley* [4.6] is obtained. If the arguments of *Libby* et al. [4.36] are accepted, experimental data on turbulent flame speeds is as well correlated by the simple formula

$$\frac{S_T}{S_L} = A + B\frac{u'_0}{S_L} \tag{4.1}$$

and the coefficients A and B are not strongly dependent on the turbulence Reynolds number.

It must be concluded that no unique relationship has yet been found between S_T and quantities such as u'_0, S_L, and turbulence Reynolds numbers. This is partly because of experimental uncertainties and errors and, possibly, also because S_T is dependent upon the hydrodynamics of the flow field in which the flame propagates. This flow field varies from one type of experiment to another. The uncertainties and conflicting trends, which are evident in the experimental literature, pose serious problems for the theoretician, who must validate his predictions on the basis of comparison with experiment.

4.3 The Premixed Laminar Flame

Premixed gaseous reactants can be caused to burn in stationary approximately planar laminar flames stabilized on a variety of burners. In the absence of significant unsteadiness, heat loss, or flame front curvature, the laminar flame speed S_L is a physicochemical and chemical kinetic property of the combustible mixture, which is independent of the method of stabilization. An essential part of the flame mechanism is the upstream laminar diffusion into the cold unburned gas mixture of heat and reactive species generated in the main reaction zone. If mean properties are assumed within the flame, and the rate of energy release due to combustion is equated to the rate at which heat is added to the incoming gas by conduction, an approximate expression can readily be found [4.4] for a characteristic flame thickness, namely

$$\delta_L \approx (k/c_p w)^{1/2}, \tag{4.2}$$

where k is the thermal conductivity, c_p the specific heat at constant pressure, and w the mass of reactant converted to product per unit volume per second. More detailed analysis shows that δ_L is the characteristic scale of the thermal preheat region of an unstrained laminar flame, while intense chemical heat release is confined to a narrower zone at the hot side of the flame. Similarly the undisturbed laminar flame speed is [4.4]

$$S_L \approx k/c_p \varrho_0 \delta_L, \tag{4.3}$$

where ϱ_0 is the density of the unburned gas. Typical values at atmospheric pressure yield flame thicknesses of a fraction of a millimeter and speeds of ten to a few hundred centimeters per second. The low velocities imply that laminar flames are almost isobaric. For a stationary flame aligned normal to the flow, continuity gives

$$m=\varrho_0 S_\mathrm{L}=\varrho_\infty u_\infty , \tag{4.4}$$

where m is the mass flux through the flame, per unit area, and ϱ_∞ and u_∞ are the density and flow velocity, respectively, downstream of the flame. The density ratio ϱ_0/ϱ_∞ is typically in the range from five to ten.

A definition of the flame speed as the mass flux through the flame, per unit area of flame, divided by the unburned gas density ϱ_0, is useful for non-stationary or oblique flames.

If we consider a stationary plane normal flame and impose a constant velocity in a direction parallel to the flame surface, a plane oblique flame is obtained. Because of the increase in velocity demanded by continuity, (4.4), a streamline through such an oblique flame is deflected towards the direction of the normal to the flame surface. A consequence of the imposed tangential velocity is that points in the oblique flame surface move along this surface. If the flame surface is also curved, adjacent points travelling along it may either move further apart (which is known as flame stretch) or they may come closer together (flame compression).

The simplest curved configuration is the spherical flame. If the direction of propagation is outwards, and the radius of the flame is not large in comparison with its thickness, the rate of increase of flame area, or flame stretch, becomes significant. A unit area of spherical flame surface must then preheat a larger area of cold gas than in the case of the planar flame in order to sustain its propagation. The speed and structure of the flame are both influenced by this process [4.3, 37, 38].

Similarly, an oblique flame is curved if the velocity U of the approach flow varies in a direction y perpendicular to the direction of the approach flow. It may be shown [4.39] that the quantity

$$K_1=\frac{\delta_\mathrm{L}}{U}\frac{\partial U}{\partial y} \tag{4.5}$$

which is known as the Karlovitz flame stretch factor, is approximately equal to the ratio of the flame thickness δ_L to its radius of curvature. *Karlovitz* et al. [4.40] and others (see, for example, [4.3]) argued that excessive stretching, as indicated by the factor K_1, can lead to local quenching of the reaction. These ideas lead [4.3] to proposed criteria for conditions under which flames will blow off from stabilizing devices. *Klimov* [4.41] and more recently *Williams*

[4.42] analyzed the propagation of a laminar flame in a shear flow, with velocity gradient $\partial U/\partial y$, in terms of a more general stretch factor

$$K_2 = \frac{\delta_L}{S_L} \frac{1}{A} \frac{dA}{dt}, \tag{4.6}$$

where A is the area of an element of flame surface, dA/dt is its rate of increase, and δ_L/S_L is a measure of the transit time of the gas passing through the flame. Stretch ($K_2 > 0$) is found to reduce both the flame thickness and reactant consumption per unit area of flame, while large stretch ($K_2 \gg 0$) may lead to extinction. On the other hand, compression ($K_2 < 0$) increases flame thickness and reactant consumption per unit area. These findings are relevant to laminar flames in turbulent flow; see Sect. 4.4.

The spontaneous instability of a laminar flame is due to two separate destabilizing effects [4.43]. The Lewis number, $L = \varrho D c_p/k$, of the reactant which limits the reaction (that is, the reactant whose disappearance finally brings the reaction to an end) is found to be an important parameter. The first destabilizing effect, which is a hydrodynamic mechanism due to expansion of the gas passing through the flame front, was described in the early studies of *Landau* [4.44] and *Markstein* [4.45] and occurs for all values of L. It leads to stationary, irregular wrinkles on the flame front but the flame continues to propagate in a laminar regime. Figure 4.5 shows a proposed [4.46] shape of the stationary wrinkled flame produced in this way, together with streamlines. According to *Sivashinsky* [4.46] the flame contour has sharp points or cusps at the hot gas side, and is curved on the cold gas side. Streamline divergence indicates that the unburned gas slows down ahead of the curved flame segments, allowing them to propagate forward.

The second destabilizing effect results from an interaction between the diffusion and heat conduction processes within the flame front structure and occurs only if $L < 1$. It leads to the formation of a regular cellular structure in the surface of the flame (see, for example, photographs by *Markstein* [4.48] and by *Palm-Leis* and *Strehlow* [4.37]). The flame shape closely resembles that illustrated in Fig. 4.5, the main difference being that the cell size is now determined by the flame rather than by the wavelength of the initially imposed disturbance [4.47]. According to *Sivashinsky* [4.43, 49] the diffusion-thermal effect is unlike the *Landau* destabilizing effect due to expansion in that it does not lead to a stationary flame surface. He predicts an essentially nonstationary structure in which the flame surface breaks up continuously into new cells in a seemingly chaotic manner which he describes as the self-turbulization of the laminar flame.

Another result of stability analysis, relevant to turbulent combustion, is the demonstration (see, for example, [4.50]) that some types of perturbations to the planar laminar flame influence its speed much more strongly than its structure. The perturbations are represented by terms in the describing equations that are

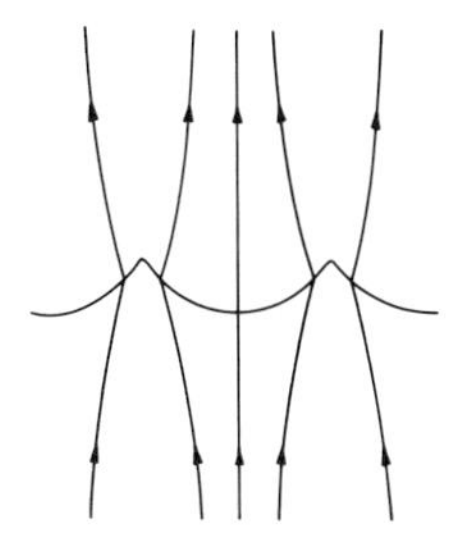

Fig. 4.5. Corrugated laminar flame shape and streamlines (after [4.46])

of order $1/N$ relative to the leading terms, where $N = T_a/T_\infty = E/R^0 T_\infty \gg 1$ is the nondimensional activation energy. Such terms yield only small perturbations to the structure of the flame, in an appropriate coordinate system, but can lead to changes of order unity in the flame speed. We shall make use of this observation in Sect. 4.5.6.

We conclude from this brief review that the laminar flame is subject to various forms of instability within a wide range of wave numbers and that the type of instability is influenced by the composition of the mixture through the Lewis number L of the reactant which limits the reaction. In general, therefore, a laminar flame propagating in a turbulent combustible mixture will be distorted both by turbulent convection and also as a result of its own instability. The curved flame imposes velocity changes which influence the turbulent velocity field. The velocity field in turn shears and stretches the flame, modifying its thickness and propagation speed and perhaps even extinguishing it locally. The interaction between these various processes complicates the theories of turbulent flames, to which we now turn our attention.

4.4 Review of Theory

In this section, we first consider the various regimes of premixed turbulent combustion and the conditions under which a laminar flame can propagate in a turbulent fluid. Following a brief discussion of the existence of a turbulent flame speed, we proceed, in Sect. 4.4.3, to discuss wrinkled laminar flame theories. In these theories, molecular transport provides the only significant transport mechanism within the narrow combustion zone or zones, although turbulence plays an important part by extending such zones. We shall then turn in Sect. 4.4.4 to theories in which combustion is considered to be controlled more directly by turbulent transport, while molecular processes are confined to their traditional role through the smallest scales of turbulence. Once the laminar flame is abandoned as the only mode of combustion we shall meet the problems of accommodating the effects of fluctuating thermodynamic state on the mean rate of reaction.

4.4.1 Regimes of Turbulent Flame Propagation

As explained in Sect. 1.16, a description of the structure of a turbulent velocity field may be presented in terms of two quantities, which are measures of the intensity and scale of the turbulence. A root mean square velocity fluctuation u' is chosen for the former. The turbulence length scale may be the integral scale l, which characterizes the large eddies; the Taylor microscale λ, from the mean rate of strain; or the Kolmogorov microscale l_k, which typifies the smallest dissipative eddies. The intensity and length scales can be combined in turbulence Reynolds numbers: $R_l = u'l/v$, $R_\lambda = u'\lambda/v$, or $R_k = u'l_k/v$, where v is the kinematic viscosity and $R_k^4 \approx R_\lambda^2 \approx R_l$ [see (1.69, 70)]. It is assumed that typical values can be chosen to evaluate these and other parameters, although we must not forget that v can increase by at least an order of magnitude, as a result of combustion.

Characteristic length and time scales of the combustion reaction may be found from the speed and thickness of the undisturbed laminar flame, although the discussion in Sect. 4.3 about the effects of stretch and shear on the laminar flame must be kept in mind. For the present purpose, we take the laminar flame speed to be $S_L \approx (vw/\varrho)^{1/2}$ and the laminar flame thickness to be $\delta_L \approx (v\varrho/w)^{1/2}$ where w is a measure of the reaction rate within the laminar flame. A ratio of characteristic flow and chemical lengths or time scales is sometimes termed a

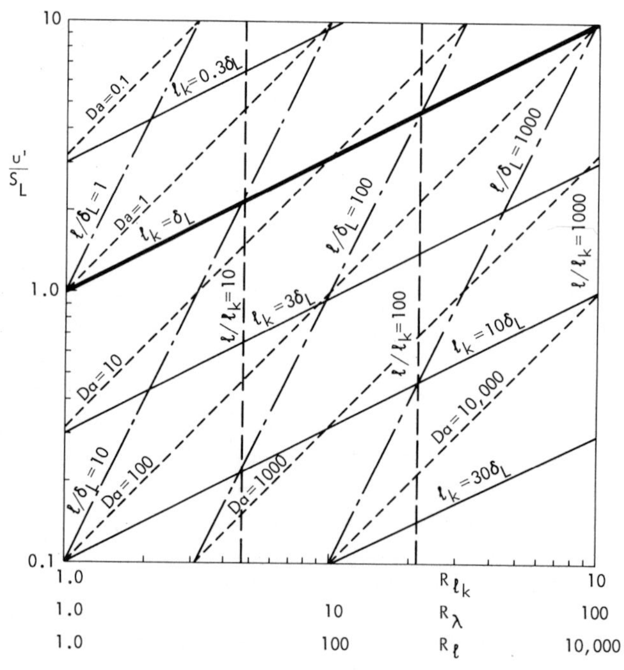

Fig. 4.6. Characteristic parameters of premixed turbulent combustion. The Klimov/Williams criterion is satisfied above the line $l_k = \delta_L$

Damköhler number; see Sect. 1.16. If the characteristic chemical time in (1.71) is evaluated from the laminar flame as δ_L/S_L, then a Damköhler number may be defined as $D_a = lS_L/\delta_L u'$. When D_a is much greater than unity, a fast chemistry regime is entered in which S_L/S is much smaller than l/u'. We shall see later, in Sect. 4.5.3, that the formulation of the problem is then greatly simplified.

These definitions and equations are sufficient to relate the Damköhler number D_a and the length ratios l/δ_L, l_k/δ_L and l/l_k to u'/S_L and the three turbulence Reynolds numbers. This is done in Fig. 4.6. In the right-hand half of this figure $R_\lambda > 100$, ensuring the separation of length scales which is a characteristic of high Reynolds number behavior. The largest Damköhler numbers are found in the bottom right-hand corner of the figure.

We now turn to the important question of whether or not a laminar flame can exist in a turbulent flow. We have seen earlier that if allowance is made for flame front curvature effects, a laminar flame is stable to disturbances of sufficiently short wavelength but that intense shear can lead to extinction. *Klimov* [4.41] and *Williams* [4.42, 51] showed, from solutions of the laminar flame equations in an imposed shear flow, that a conventional propagating laminar flame may exist only if the stretch factor K_2 defined in (4.6) is less than a critical value of the order of unity. Modeling the area change term in (4.6) as

$$\frac{1}{A}\frac{dA}{dt} \approx \frac{u'}{\lambda}$$

and using relationships quoted above, we can deduce [4.42] that $K_2 \approx (\delta_L/l_k)^2$. Thus the criterion to be satisfied if a laminar flame is to exist in a turbulent flow is that the laminar flame thickness δ_L must be less than the Kolmogorov microscale l_k of the turbulence.

A heavy line in Fig. 4.6 indicates the condition $\delta_L = l_k$. Below and to the right of this line the Klimov-Williams criterion is satisfied and wrinkled laminar flames may occur. The figure shows that this region includes both large and small values of turbulence Reynolds numbers and velocity ratio $u'/S_L \gtrless 1$, but predominantly large D_a.

Above and to the left of the heavy line in Fig. 4.6 is the region in which $l_k < \delta_L$. According to the *Klimov-Williams* criterion, the turbulent velocity gradients in this region are sufficiently intense to destroy a laminar flame. The figure shows that $u' \geqq S_L$ in this region, which predominantly includes relatively low Damköhler numbers. At the highest Reynolds numbers the region is entered only for very intense turbulence $u' \gg S_L$.

It is interesting to note that the correlation by *Abdel-Gayed* and *Bradley* [4.6] of a large number of experiments includes data on both sides of the line $l_k = \delta_L$.

The question arises, therefore, of the time-dependent structure of the turbulent heat release zone under conditions where $l_k < \delta_L$, so that laminar flames are not expected to occur. Also it must be pointed out that even when

$l_k > \delta_L$ the *Klimov-Williams* criterion states only that a laminar flame can exist. It does not prove that combustion must occur in this manner. An early proposal, due to *Summerfield* et al. [4.52], is that the turbulence convects the reaction regions, producing a distributed reaction zone and only small fluctuations in temperature and composition. This description is likely to be most nearly appropriate when the laminar flame is thick in comparison with the integral scale of turbulence, the conditions prevailing in the top left-hand corner of Fig. 4.6.

Although the thin vortex tube description of combustion at high Reynolds numbers ($R_\lambda > 100$) due to *Chomiak* [4.9] is in some ways attractive, his experimental evidence [4.53] in support of it must be regarded as inconclusive. Unfortunately, the Kolmogorov microscale and the laminar flame thickness are both small and difficult to resolve in a turbulent flame. Spark schlieren photographs are inconclusive because they involve spatial integration along the optical path, and thus cannot distinguish the image of a burning vortex tube from that of a curved laminar flamelet. The real nature of the fine structure of the flame in intense turbulence is still unknown.

On the other hand, many experiments confirm the existence of a wrinkled laminar flame regime at least for turbulence of low intensity and large scale, i.e., corresponding to the bottom right-hand corner of Fig. 4.6. The photographs of Fig. 4.1 provide an example in which the characteristic shape of the wrinkled flame is apparent.

4.4.2 Existence of the Turbulent Flame Speed

Many investigations of premixed turbulent combustion are concerned primarily with the turbulent flame speed S_T. Unlike the laminar flame, for which the speed S_L of a stationary planar wave is a physicochemical and chemical kinetic property of the gas mixture alone, S_T must be presumed to depend also on many properties of the turbulent flow field, including the method of flame stabilization. The existence of a properly defined turbulent flame speed has been questioned many times. Under restrictive assumptions, including vanishingly small heat release, *Williams* [4.54] has shown that S_T can be determined as an eigenvalue in a manner analogous to the laminar flame speed problem. More recently, subject to certain modeling assumptions, *Bray* and *Libby* [4.55] have calculated S_T from the solution of an eigenvalue problem including effects of heat release and density variation. In some circumstances, their results (reviewed in Sect. 4.5.5) show a sensitivity of S_T to initial conditions, which is interpreted as an influence of the method of flame stabilization. However, although the concept of a uniquely defined turbulent flame speed may be suspect, it remains widely used in many practical applications. As we have seen in Sect. 4.2, a large amount of experimental data is expressed in this form and theory must seek to explain such data.

4.4.3 Wrinkled Laminar Flame Theories

The basic concept, common to all wrinkled laminar flame descriptions, is that combustion occurs only in narrow zones whose instantaneous structure is that of a laminar flame. However, this concept can be applied in many different ways, some of which are illustrated below, starting with the simplest and most empirical. A wrinkled laminar flame assumption can also be incorporated into the unified model of premixed turbulent combustion which will be described in detail in Sect. 4.5.

The work of *Damköhler* [4.2] considered large-scale turbulence as simply distorting the laminar flame without affecting its structure or propagation speed. With this assumption, each unit area of laminar flame continues to consume a mass $\varrho_0 S_L$ of mixture per second, but the total flame area is greatly increased as a result of wrinkling. The turbulent flame speed S_T is then given by

$$\frac{S_T}{S_L}=\frac{A_L}{A_T}, \tag{4.7}$$

where A_L is the total area of laminar flame surface contained within an area of turbulent flame whose time average is A_T. The problem then is to determine A_L/A_T, which depends on the shape taken up by the laminar flame in response to the turbulence. This is, in fact, a very difficult problem since the flame shape can be strongly influenced by the various laminar flame instabilities discussed earlier. *Damköhler* took $A_L/A_T=1+u'/S_L$, giving [see (4.1)]

$$S_T=S_L+u', \tag{4.8}$$

where u' is the rms turbulence velocity of the unburned gas ahead of the flame. Later authors assumed various empirical flame shapes and obtained a variety of different expressions for S_T.

This simple description of the wrinkled flame was modified by *Karlovitz* et al. [4.56] and *Scurlock* and *Grover* [4.57] through the inclusion of so-called flame-generated turbulence. They postulated that the turbulence in the unburned gas ahead of the flame is augmented by additional turbulence generated within the flame; the origin of this turbulence is identified as the fluctuating orientation of the velocity of the burned gas just behind the wrinkled flame. For unconfined flames, *Scurlock* and *Grover* predicted the flame-generated turbulence to be comparable in magnitude to the approach flow turbulence. On the other hand, they proposed that the flame-generated turbulence in confined flames far outweights that in the approach flow. However, these predictions are not supported by experiment.

A different category of wrinkled laminar flame model assumes that the flame surface has been torn apart by intense turbulence. Burning then takes place in many separate eddies or parcels but, in accordance with the wrinkled flame concept, the individual flames are laminar. *Shipman* et al. [4.58, 59] provided such an example. In this model, the burning zone is characterized as parcels of unburned gas dispersed through fully burned gas, and parcels of burned gas in unburned. The parcels are assumed to be spherical and the combustion at their periphery takes place in undisturbed laminar flames. An essential feature is the splitting of these parcels into smaller but still spherical parcels to represent the tearing action of intense strain, represented in the model by the shearing effect of the mean transverse velocity gradient. The size distribution of the parcels is governed by burning and splitting where the latter is related to the gradient of the mean velocity. Comparison with experimental data from high-speed ducted flames shows [4.59] that the fluid mechanical process of parcel splitting controls the rate of heat release in regions of intense strain.

The work of *Shipman* et al. is important in the context of later developments because of its clear recognition of the influence of strain on heat release. Also it exploits the thermochemical simplicity of a system consisting predominantly of fully burned and unburned regions separated by flames of negligible volume. We shall meet these ideas again in descriptions of more recent theories.

There are fundamental difficulties associated with each of the wrinkled laminar flame theories described here. All of these difficulties can be traced back to the fact that instabilities of the laminar flame are ignored. As a result there are three shortcomings. The first is that the laminar flame speed is not a constant, as assumed, but a variable, which is strongly dependent on factors such as flame front curvature and acceleration. The second factor is that the flame shape cannot be specified arbitrarily; it results from a complex interaction between the instability of the flame and its distortion by the turbulence. Thirdly, there is the development of the turbulent velocity field within the time-average reaction zone. The theories described here cannot tell when the turbulence intensity of the unburned gas is frozen, as assumed, in effect, by *Damköhler* [4.2] or strongly augmented by the flame-generated turbulence of *Karlovitz* et al. [4.56] or when mean strain effects predominate [4.58].

More recently, important developments towards a rigorous and general statistical treatment of the wrinkled laminar flame in a turbulent approach flow were reported by *Williams* [4.42, 54] and by *Clavin* and *Williams* [4.60]. The problem considered is that of a turbulent flame which on average is perpendicular to a mean flow containing grid-generated turbulence. The most recent work [4.60] avoids the large temperature fluctuations resulting from laminar flame movement by using a coordinate system which moves in the mean flow direction with the fluctuating velocity of the laminar flame. The analysis involves a regular perturbation for large values of the ratio of turbulence scale to laminar flame thickness, together with a singular perturbation for large $N = T_a/T_\infty$ where T_a is the activation temperature. The analysis requires no empirical closure assumptions. Two key simplifications are, however, made in order to avoid effects arising from instabilities of the laminar flame. The first is an assumption that the Lewis number is unity thus eliminating cellular instability. Secondly, the density change associated with combustion is neglected; this avoids the hydrodynamic, thermal expansion type of instability. With constant density combustion, heat release has no effect on the flow, and it is possible to consider the velocity field as a given quantity, specified in advance. There is, therefore, no flame-generated turbulence considered. The authors pointed out that methods are available which permit each of these restrictions to be relaxed.

The analysis is too long to be reproduced here but some of the important results will be quoted. A key quantity is the longitudinal (x-wise) displacement α of the flame in an Eulerian reference frame, by turbulent velocity fluctuations. Thus the equation of the reactive surface in the flame is $x=\alpha(y,z,t)$, where x is in the direction of the mean flow. To first order in the ratio of length scales, $\alpha=\alpha_0$ is equal to the longitudinal displacement of fluid elements in a Eulerian frame by turbulent fluctuations. That is, $\partial\alpha_0/\partial t=u'$, where $u'(x, y, z, t)$ is the fluctuating component of the specified turbulent velocity field. *Clavin* and *Williams* showed that the turbulent flame speed S_T is related to the mean square of the transverse gradient of α_0. In terms of u', they find that

$$S_T=S_L\left[1+1/2\overline{\left(\frac{\partial}{\partial y}\int_0^t u'dt\right)^2}+1/2\overline{\left(\frac{\partial}{\partial z}\int_0^t u'dt\right)^2}\right]. \tag{4.9}$$

The term in square brackets is shown to be the average ratio of the wrinkled flame area to the cross-sectional area of the approach flow. The authors infer that, to the order to which their analysis has been carried, effects of neither strain rate nor flame curvature on the laminar flame speed influence S_T. Presumably this is a consequence of their choice of conditions which eliminate laminar flame instabilities. Equation (4.9) differs significantly from the empirical result (4.1). However, assuming isotropic turbulence and making a Taylor hypothesis (see Chap. 1), (4.9) reduces to

$$S_T=S_L\left(1+\frac{\overline{u'^2}}{S_L^2}\right), \tag{4.10}$$

which differs from a formula due to *Shelkin* [4.61] (see [4.4]) only in the absence of a factor of 2 multiplying $\overline{u'^2}/S_L^2$. From (4.10) S_T/S_L is seen to be independent of the scale of turbulence in this limit of large-scale turbulence.

The flame structure is also calculated. If Θ is a dimensionless temperature $(T_\infty-T)/(T_\infty-T_0)$, then its time average $\bar{\Theta}$ is related to x and to the probability density function for α_0. Taking this to be Gaussian, and writing $\delta=(\overline{\alpha^2})^{1/2}$, reduces this relation in the limit of large δ to the simple form

$$\bar{\Theta}=1/2\,\mathrm{erf}(x/2^{1/2}\delta). \tag{4.11}$$

A method of calculating the mean square temperature fluctuation is also provided.

Another prediction of the analysis of *Clavin* and *Williams* [4.60] has an intriguing consequence. It is not clear from turbulence theory whether or not the Eulerian mean square displacement $\overline{\alpha^2}$ will grow with time (that is, whether or not α is stationary) for flows of interest. If α is nonstationary, the result of the analysis could be consistent with a conventional gradient approximation to turbulent transport

$$\overline{u'T'} = -\nu_T d\bar{T}/dx ,$$

where ν_T is an empirical transport coefficient (see Sect. 1.13); but the turbulent flame then thickens as time increases. On the other hand, if α is assumed to be statistically stationary, the flame thickness remains constant but a negative turbulent transport coefficient ν_T is required in part of the flame. Since negative transport coefficients are physically unacceptable, it is concluded that the gradient transport closure is inapplicable in the limit considered.

The work of *Sivashinsky* [4.43, 46, 49] already referred to in Sect. 4.3 is in a sense complementary to that of *Clavin* and *Williams*. *Sivashinsky* treated the spontaneous instability and transition to turbulence (self-turbulization [4.49]) of initially laminar flames in the absence of turbulence in the approach stream. Any turbulence which occurs must, therefore, be generated within the flow field of the flame. Because spontaneous instabilities are investigated, he considers both the Lewis number L of the limiting reactant, i.e., of the reactant whose consumption brings the reaction to an end, and density variations due to heat release, which influence the diffusion-thermal (cellular) and hydrodynamic (Landau) modes of instability, respectively. The phenomena which he predicted are described [4.49] as occurring gradually; for example, well-developed turbulence is found when the radius of a spherical flame has grown to fifty times a critical radius, at which the cellular structure first begins to form. We may, therefore, expect that self-turbulization will become dominant only when the turbulence in the approach stream is of low intensity.

We must conclude from this review that the rigorous treatment of a wrinkled laminar flame in a turbulent flow is not yet developed in a form suitable for routine application, for example, in the combustion chamber of an engine. Such problems must be tackled much more empirically and, therefore, with a greater margin of uncertainty. However, the wrinkled laminar flame concept can still be helpful, as we shall see in Sect. 4.5.6.

4.4.4 Combustion Controlled by Turbulent Mixing

Turbulent mixing is important in all turbulent flames. However, we may distinguish two sets of conditions in which laminar flames will not occur and turbulent transport is clearly the most important controlling process. The first is the classical *Damköhler* [4.2] limit in which all turbulence length scales are small in comparison with the laminar flame thickness; see the top left-hand corner of Fig. 4.6. The second case is combustion with intense turbulence and moderate *Damköhler* numbers, corresponding to the top right-hand corner of Fig. 4.6. The laminar flame thickness is then smaller than the integral scale of the turbulence, so the *Damköhler* limit of small-scale turbulence is not achieved but, above the line corresponding to the *Klimov-Williams* criterion, laminar flames are not expected to occur.

The first case, in which all the scales of the turbulence are smaller than the laminar flame thickness, is difficult to achieve in practice. It is nevertheless informative to study the mechanisms involved. *Damköhler* suggested that in this case the principal effect of the turbulence is to enhance the laminar transport processes within the flame. An analogy can then be drawn with the mechanism controlling laminar flame propagation. For the latter (4.2) and (4.3) give

$$S_{\mathrm{L}}=\frac{1}{\varrho_0}\left(\frac{kw}{C_{\mathrm{p}}}\right)_{\mathrm{L}}^{1/2}, \tag{4.12}$$

where the subscript denotes mean values within the laminar flame. We, therefore, postulate

$$S_{\mathrm{T}}=\frac{1}{\varrho_0}\left(\frac{kw}{C_{\mathrm{p}}}\right)_{\mathrm{T}}^{1/2}, \tag{4.13}$$

in which the subscript refers to a turbulent flame. If effects of turbulent fluctuations in temperature, density, and composition could be neglected (a condition which, in practice, cannot be achieved), the reaction rate w would be unaffected by turbulence, i.e., $w_{\mathrm{T}}=w_{\mathrm{L}}$. In addition, if it is assumed that the Prandtl number $\mu C_{\mathrm{p}}/k$ is unity for both laminar and turbulent transport processes, (4.12) and (4.13) then give

$$S_{\mathrm{T}}/S_{\mathrm{L}}=(\nu_{\mathrm{T}}/\nu)^{1/2}, \tag{4.14}$$

where ν is the molecular kinematic viscosity, and ν_{T} is a turbulent exchange coefficient or eddy kinematic viscosity; see Sect. 1.13. If $\nu_{\mathrm{T}}\sim u'l$, then $S_{\mathrm{T}}/S_{\mathrm{L}}\sim R_l^{1/2}$.

The second, and much more practical situation, is that of intense turbulence with relatively small length scales. This can be achieved most easily in confined flames where the turbulence is augmented due to shear. The controlling processes were clearly identified by *Spalding* [4.62, 63]. A simple model [4.62] incorporating a conventional description of turbulent transport, together with an assumption that the gas mixture burns instantaneously as soon as it is entrained into the mixing layer, successfully predicts several features of such flows. Turbulent mixing is, therefore, the controlling process.

The second work [4.63] is of greater interest to us. In this *Spalding* compared two different physical models of a confined, premixed turbulent flame, using the same eddy viscosity formulation and the same finite difference numerical method for both models. The two models differ only in their treatment of chemical reaction rate term in the governing equations.

In the first model, this term, describing the time-average volumetric rate of consumption of oxidizer, is written [cf. (1.5) and (1.7)]

$$\bar{w}_{\mathrm{ox}} = K\bar{p}^2\,\bar{Y}_{\mathrm{fu}}\bar{Y}_{\mathrm{ox}}\exp(-E/R^0\bar{T}), \tag{4.15}$$

where subscripts fu and ox denote fuel and oxidizer, respectively. Equation (4.15) implies that the consumption of oxidizer is controlled by chemical kinetics. Also, because it is expressed only in terms of the mean quantities $\bar{p}$, $\bar{Y}_{\mathrm{fu}}$, $\bar{Y}_{\mathrm{ox}}$, and $\bar{T}$, it assumes that the effects of turbulent fluctuations in the thermodynamic state variables are negligible. All the additional fluctuation terms in, for example, (1.62) and (1.65) are set to zero, corresponding to the "well-mixed" limit of Sect. 1.14. This model is unsuccessful as it is unable to match even qualitative trends of experimental data for any values of the empirical constants K and E in (4.15).

The second model achieves a much more satisfactory agreement with experiment by replacing (4.15) with a so-called eddy breakup model for $\bar{w}_{\mathrm{ox}}$. This model envisages the burning zone as being made up mainly of alternating fragments of unburned gas and almost fully burned gas. In order to estimate the rate at which lumps of unburned gas are broken down into smaller ones with enough interfaces with hot gas to allow reaction, *Spalding* used an analogy with the decay of turbulence energy. This decay comes about by the breaking of large lumps of fluid into successively smaller ones by turbulent strain until viscous forces become significant. The decay rate of turbulence kinetic energy is produced by the viscous dissipation term

$$\bar{\Phi} = \overline{\tau_{lk}\frac{\partial u_l''}{\partial x_k}} \tag{4.16}$$

in (1.52) [see also (1.53)]. *Spalding* assumed that $\bar{\Phi}$ is equal to the term in (1.52) representing production of turbulence kinetic energy, namely

$$-\overline{\varrho u_k'' u_l''}\,\partial\tilde{u}_l/\partial x_k\,,$$

and postulates a similar expression for reaction rate $\bar{w}_{\mathrm{ox}}$, which is, therefore, proportional to the velocity gradient (cf. [4.58, 59]). In fact he used a conventional time average rather than the Favre average shown above.

The eddy breakup rate description is expressed in a more convenient and universal form by *Mason* and *Spalding* [4.64], who write

$$\bar{w}_{\mathrm{ox}} = C_{\mathrm{EBU}}(\overline{Y'^2})^{1/2}\,\bar{\Phi}/\bar{q}\,, \tag{4.17}$$

where $(\overline{y'^2})^{1/2}$ is the root mean square fluctuation in mass fraction, $\bar{\Phi}$ is the viscous dissipation, and $\bar{q}$ is the turbulence kinetic energy, all conventionally averaged. The eddy breakup constant C_{EBU} is regarded as a universal constant; the value $C_{EBU}=0.53$ is quoted. Although (4.17) is obtained intuitively, it has been widely and often successfully used. Since $\bar{\Phi} \sim \bar{\varrho}\bar{q}^{3/2}/l$, (4.17) describes a reaction rate which is proportional to the turbulence velocity $\bar{q}^{1/2}$ and inversely proportional to the length scale l. It would be zero in the absence of composition fluctuations. It contains no information about the chemical kinetics and relates to combustion which is entirely mixing limited.

A rigorous derivation of an equation similar to (4.17) will be presented in Sect. 4.5.3.

4.5 A Unified pdf Model

Practical combustion calculations require a unified theoretical framework which will be applicable, as far as possible, in all the regimes illustrated in Fig. 4.6. This theoretical framework should not be based upon a particular physical picture of the flame structure, such as the wrinkled laminar flame, which cannot apply universally. It must be capable of accommodating the effects of large fluctuations in thermodynamic state since both experiment and theory indicate that such fluctuations will occur in some circumstances, including most reacting flows of practical interest.

These fluctuations in the scalar variables, which specify the instantaneous thermodynamic state of the mixture, produce many additional terms in the equations for time-averaged rates of reaction. As explained in Sect. 1.14, the extra terms, which take the form of covariances of the scalar variables, can have a large effect on the behavior of the flow. In particular, the mean reaction rates cannot be regarded as functions of the time mean state variables alone. The most convenient description of the fluctuating thermodynamic state is through a probability density function (pdf). In the most general case, a joint pdf for all the state variables is required. The time-average reaction rate may be expressed in terms of this joint pdf [see (1.66)].

There are two alternative approaches to the pdf. The first is to calculate the pdf from its own balance equation as described in Chap. 5. This is an attractive possibility which provides a rational route to a detailed description of the effects of scalar fluctuations. The difficulties involved in this approach are indicated in Chap. 5. In addition, it leads to large computing requirements if practical combustion problems are to be tackled. Empirical assumptions are needed in order to model terms in the pdf balance equation and their form is not yet universally accepted, particularly for the term representing effects of molecular diffusion [4.65, 66]. The alternative approach is to specify the pdf empirically. We shall show in Sect. 4.5.3 that, in many circumstances, the time mean reaction rate is relatively insensitive to the shape of the function chosen for this purpose. Thus, an analysis which is sufficiently simple to be incorporated into practical engineering calculations can nevertheless include the important consequences of the fluctuating thermodynamic state. This is the approach which we shall describe. Note that specification of scalar pdf's does not overcome the closure problem involving turbulent transport processes.

The statistics of the fluctuating thermodynamic state are represented by a pdf which in general takes the form $P[\varrho, T, Y_i(i=1, 2, \ldots, N); \boldsymbol{x}]$ and is, therefore, a function of $N+2$ variables. The pdf P is to be determined empirically, and its form specified a priori. In appropriate circumstances, it may be derived from a physical model of the flame structure, such as the wrinkled laminar flame model, or from experimental data. *Pope* [4.67, 68] described a rational method for determining the statistically most likely pdf in terms of its moments. Alternatively, subject to appropriate physical and chemical constraints, it may be chosen for convenience: an array of Dirac delta functions, or regions of constant probability, etc.

However, the number of variables can become prohibitively large. In whatever way it is specified, the model pdf must be chosen so that its moments satisfy appropriate balance equations. With Favre averaging we have, for example,

$$\tilde{Y}_j(\boldsymbol{x}) = \frac{1}{\bar{\varrho}(\boldsymbol{x})} \int\int \ldots \int \varrho Y_j P[\varrho, T, Y_i(i=1, 2, \ldots, N); \boldsymbol{x}] \, d\varrho \, dT \cdot dY_i(i=1, 2, \ldots, N) \tag{4.18}$$

and a similar expression for $\overline{\varrho Y_j'' Y_k''}$. Partial differential equations may be written for these various moments which may be regarded as parameters of the empirically specified pdf. The larger the number of moments that are taken into account, the more accurately may we expect to determine P.

Let us look at the number of variables involved. If we choose as parameters of the empirical pdf all the first- and second-order Favre moments, i.e., $\bar{\varrho}$, $\tilde{T}$, $\tilde{Y}_i(i=1, 2, \ldots, N)$, $\overline{\varrho T'}$, $\overline{\varrho T'' Y_i''}$ $(i=1, 2, \ldots, N)$, and $\overline{\varrho Y_i'' Y_j''}$ $(i, j=1, 2, \ldots, N)$ then, since only $n-1$ of the Y_i's are independent, there is a total of $N+2$ parameters. The same number of differential equations is required in order to determine these parameters. For example, with only three species, $N=3$, the number of parameters is eleven.

To reduce this number as far as possible, use may be made of element conservation equations and equations of state. For premixed reactants, in circumstances where all species can be assumed to have the same coefficient of molecular diffusion, the element mass fractions Z_j $(j=1, 2, \ldots, M)$ can be regarded as known constants, related to the species mass fractions by (1.18). The number of independent mass fractions is then reduced to $N-M$. Two further reductions in number can be made from additional assumptions. If pressure fluctuations are neglected, then the instantaneous pressure can be replaced by its time-average value. The equation of state (1.20) then relates the instantaneous density to temperature and composition. Further, if the flow can be assumed to be adiabatic, the Mach number low, and the Prandtl and Schmidt numbers unity, then from (1.15) the specific enthalpy will be constant everywhere. Using (1.13) and (1.20) to eliminate ϱ and T, the joint pdf P is a function of only $N-M$ time-dependent composition variables. If all first and second moments of these variables are used as parameters in an empirical model of this pdf, we now find a maximum of $N-M+(N-M)^2$ such parameters.

In the simplest possible case, with three species (fuel, oxidizer, and product) and two elements, we have $N=3$ and $M=2$. Then the pdf P for the thermodynamic state is a function of only one variable. The simple version of this pdf has two parameters consisting of the first and second moments. The theory in this simple form will be developed in the following sections. It should be noted that *Pope* [4.69] found his statistically most likely pdf's to require three rather than two moments as parameters in order to match experimental data. The prediction of three moments is, however, beyond present modeling capabilities.

Analysis of chemically complex systems will be discussed in Sects. 4.5.6 and 4.5.7.

4.5.1 Simple Model: The Progress Variable c

Premixed turbulent combustion has been analyzed in terms of a pdf for a single thermochemical variable by *Bray* and *Moss* [4.11, 12], *Bray* and *Libby* [4.55], *Borghi* et al. [4.70, 71], *Lockwood* [4.72], and others.

The main assumptions of the following analysis [4.12, 55] are as follows:

I) Combustion is controlled by a single-step, irreversible chemical reaction whose rate is specified by a global reaction rate expression.

II) The specific heat at constant pressure and the molecular weight of the reacting gas mixture are both constant and independent of the progress of the reaction.

III) The Prandtl and Schmidt numbers of the mixture are unity.

IV) The flow is adiabatic and occurs at a Mach number much less than unity.

V) Pressure fluctuations are neglected.

We consider a mixture of four species: fuel ($i=1$), oxidizer ($i=2$), product ($i=3$), and inert ($i=4$) and specify that the initial mixture contains no product. The element mass fractions Z_j are related to the time-varying species mass fractions Y_i through three equations of the form (1.18). The equivalence ratio of the mixture is given by

$$\phi=\frac{Y_{1,0}/Y_{2,0}}{(Y_{1,0}/Y_{2,0})_{\text{stoic}}}=\frac{\mu_{23}Z_1}{\mu_{13}Z_2}=\frac{\mu_{11}\mu_{23}Y_{1,0}}{\mu_{22}\mu_{13}Y_{2,0}},$$

where μ_{ij} is the mass of element i in unit mass of species j, and subscripts 0 and "stoic" represent initial conditions and a stoichiometric mixture, respectively.

The caloric equation of state (1.13) may be written

$$h_0 = c_p T - Y_3 \Delta, \tag{4.19}$$

where h_0 is the enthalpy of the unburned mixture, Δ is the heat of reaction, and c_p is assumed constant. Finally, the thermal equation of state is

$$\bar{p} = \varrho R^0 T/\bar{W}, \tag{4.20}$$

where the mean molecular weight $\bar{W}$ is assumed constant.

A progress variable c for the global combustion reaction is defined from

$$Y_3 = c Y_{3,\infty}, \tag{4.21}$$

where $Y_{3,\infty}$ is the mass fraction of product in the fully burned gas whose temperature is the adiabatic flame temperature

$$T_\infty = \frac{h_0}{c_p} + \frac{\Delta}{c_p} Y_{3,\infty}$$

and whose density is ϱ_∞. Thus, $c=1$ when combustion is complete while $c=0$ in the unburned mixture which contains no product. Equations (1.18, 4.19, 20) enable all the instantaneous state variables to be uniquely related to c. The equation for c is [see (1.17)]

$$\frac{\partial}{\partial t}(\varrho c) + \frac{\partial}{\partial x_k}(\varrho u_k c) = \frac{\partial}{\partial x_k}\left(\varrho \mathscr{D} \frac{\partial c}{\partial x_k}\right) + w, \tag{4.22}$$

where w is related to the mass rate of production of product species per unit volume and time by

$$w_3 = Y_{3,\infty} w.$$

From (1.18)

$$Y_1 = \frac{1}{\mu_{11}}(Z_1 - \mu_{13} Y_{3,\infty} c), \tag{4.23}$$

$$Y_2 = \frac{1}{\mu_{22}}(Z_2 - \mu_{13} Y_{3,\infty} c), \tag{4.24}$$

and Y_4 is independent of c. Equation (4.19) enables the temperature T to be obtained from

$$T = T_0(1 + \tau c), \tag{4.25}$$

where τ is a heat release parameter defined as

$$\tau = \frac{\Delta Y_{3,\infty}}{h_0} = \frac{T_\infty}{T_0} - 1\,. \tag{4.26}$$

Values of τ of practical interest range from 4 to 9. Finally, from (4.20) the density is

$$\varrho = \varrho_0 T_0/T = \varrho_0/(1+\tau c)\,. \tag{4.27}$$

With the fluctuating thermochemical state related to c alone, Favre-averaged state variables may be expressed in terms of a pdf $P(c;\boldsymbol{x})$ through equations of the form

$$\tilde{g}(\boldsymbol{x}) = \frac{1}{\bar{\varrho}(\boldsymbol{x})} \int_0^1 \varrho(c)\, g(c)\, P(c;\boldsymbol{x})\, dc\,, \tag{4.28}$$

where $g(c)$ represents any thermodynamic state variable, and $\int_0^1 P(c;\boldsymbol{x}) = 1$. Thus, from (4.27)

$$\bar{\varrho}(\boldsymbol{x})\, \tilde{c}(\boldsymbol{x}) = \varrho_0 \int_0^1 \frac{cP(c;\boldsymbol{x})}{1+\tau c}\, dc \tag{4.29}$$

while

$$\bar{\varrho}(\boldsymbol{x}) = \varrho_0 \int_0^1 \frac{P(c;\boldsymbol{x})}{1+\tau c}\, dc\,. \tag{4.30}$$

Combining (4.29) and (4.30), we find that

$$\begin{aligned} \bar{\varrho}(\boldsymbol{x})[1+\tau\tilde{c}(\boldsymbol{x})] &= \varrho_0 \left[\int_0^1 \frac{P(c;\boldsymbol{x})}{1+\tau c}\, dc + \int_0^1 \frac{\tau c P(c;\boldsymbol{x})}{1+\tau c}\, dc \right] \\ &= \varrho_0 \int_0^1 P(c;\boldsymbol{x})\, dc = \varrho_0\,, \end{aligned}$$

giving

$$\frac{\bar{\varrho}(\boldsymbol{x})}{\varrho_0} = \frac{1}{1+\tau\tilde{c}(\boldsymbol{x})}\,. \tag{4.31}$$

The chemical kinetic rate term $w\,[\varrho,\, T,\, Y_i\ (i=1\ldots 4)]$ in (4.22) reduces to $w(c)$, whose time average is

$$\bar{w}(\boldsymbol{x}) = \int_0^1 w(c)\, P(c;\boldsymbol{x})\, dc\,. \tag{4.32}$$

Favre-averaged balance equations may now be written in terms of the progress variable c. We shall require the following [see (1.42, 46, 52)]:

Continuity

$$\frac{\partial}{\partial x_k}(\bar{\varrho}\tilde{u}_k)=0. \tag{4.33}$$

Species Conservation

$$\frac{\partial}{\partial x_k}(\bar{\varrho}\tilde{u}_k\tilde{c})=\frac{\partial}{\partial x_k}\left[\overline{\varrho\mathscr{D}\frac{\partial c}{\partial x_k}-\varrho u_k''c''}\right]+\bar{w}. \tag{4.34}$$

Species Fluctuation

$$\begin{aligned}\frac{\partial}{\partial x_k}(\overline{\varrho c''^2}\tilde{u}_k)=2&\left[\overline{c''\frac{\partial}{\partial x_k}\left(\varrho\mathscr{D}\frac{\partial c}{\partial x_k}\right)}\right]\\ &-2\overline{\varrho u_k''c''}\frac{\partial\tilde{c}}{\partial x_k}-\frac{\partial}{\partial x_k}(\overline{\varrho u_k''c''^2})+2\overline{c''w}.\end{aligned} \tag{4.35}$$

Turbulence Kinetic Energy

$$\begin{aligned}\frac{\partial}{\partial x_k}(\bar{\varrho}\tilde{u}_k\tilde{q})=-&\overline{\varrho u_k''u_l''}\frac{\partial\tilde{u}_l}{\partial x_k}-\tfrac{1}{2}\frac{\partial}{\partial x_k}(\overline{\varrho u_k''u_l''^2})\\ &-\overline{u_k''\frac{\partial p}{\partial x_k}}+\overline{u_k''\frac{\partial\tau_{kl}}{\partial x_l}},\end{aligned} \tag{4.36}$$

where, $\tilde{u}_k$ is the Favre-averaged mean velocity in direction x_k and

$$\tilde{q}=\tfrac{1}{2}\overline{\varrho u_k''u_k''}/\bar{\varrho}$$

is the Favre-averaged turbulence kinetic energy. Two distinct closure models are required in order to convert (4.33–36) into a closed set of equations. The first is a thermochemical closure which enables us to express the chemical source terms in (4.34, 35) as functions of $\tilde{c}$ and $\overline{\varrho c''^2}$. The second is a turbulent transport model; we shall provisionally use a conventional eddy viscosity formulation for this purpose (see Sect. 4.5.4), while recognizing that this may prove not to be appropriate.

4.5.2 Thermochemical Closure for $P(c;x)$

In general, the pdf $P(c;x)$ must represent gas which has not yet started to burn and gas which has burned completely as well as gas mixture which is in the

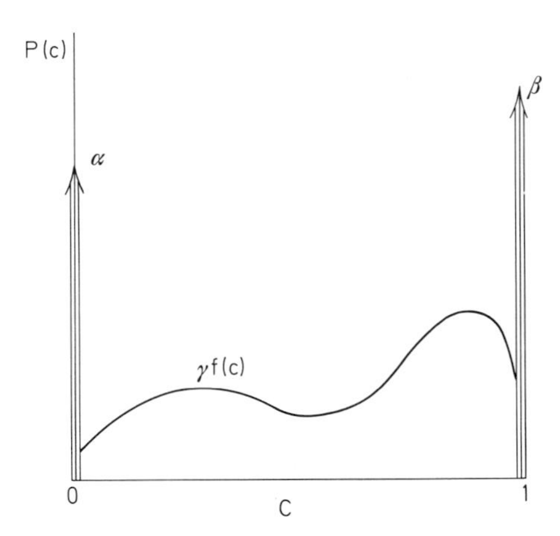

Fig. 4.7. The pdf $P(c)$ including Dirac delta functions at $c=0, 1$

process of burning. It is convenient to separate these three contributions. We therefore write

$$P(c;\boldsymbol{x})=\alpha(\boldsymbol{x})\,\delta(c)+\beta(\boldsymbol{x})\,\delta(1-c) + [H(c)-H(c-1)]\,\gamma(\boldsymbol{x})f(c;\boldsymbol{x}), \tag{4.37}$$

where $\delta(c)$ and $H(c)$ are the Dirac delta and Heaviside functions, respectively; the coefficients $\alpha(\boldsymbol{x})$, $\beta(\boldsymbol{x})$, and $\gamma(\boldsymbol{x})$ are nonnegative functions of position and $f(c;\boldsymbol{x})$ satisfies the normalization condition

$$\int_0^1 f(c;\boldsymbol{x})dc=1.$$

Thus

$$\alpha(\boldsymbol{x})+\beta(\boldsymbol{x})+\gamma(\boldsymbol{x})=1.$$

The delta functions at $c=0$ and $c=1$ may be identified (see Fig. 4.7) with packets of unburned and fully burned mixture, respectively, and $f(c;\boldsymbol{x})$ with product distributions associated with burning. The coefficients $\alpha(\boldsymbol{x})$, $\beta(\boldsymbol{x})$, and $\gamma(\boldsymbol{x})$ describe the partitioning among these three possible modes.

In line with our earlier discussion about the a priori pdf method, $P(c;\boldsymbol{x})$, and hence the burning mode pdf, $f(c;\boldsymbol{x})$, is to be represented empirically. As discussed at the beginning of this section, the functional form selected for $f(c;\boldsymbol{x})$ will be assumed to have at most two parameters, which are chosen as the Favre average $\tilde{c}$ and the Favre mean square fluctuation $\overline{\varrho c''^2}$. Thus the reacting mode pdf becomes $f(c;\tilde{c},\overline{\varrho c''^2})$.

Our next task is to express $\tilde{c}$ and $\overline{\varrho c''^2}$ together with the chemical source terms $\bar{w}$ and $\overline{c''w}$ in (4.34, 35), respectively, in terms of this new pdf. For brevity the dependence of variables on the position $\boldsymbol{x}$ will no longer be indicated. From (4.29) and (4.37)

$$\tilde{c}=\frac{1}{\bar{\varrho}}\left[\beta\varrho_\infty+\gamma\int_0^1 c\varrho(c)f(c;\tilde{c},\overline{\varrho c''^2})dc\right], \tag{4.38}$$

while the mean square fluctuation is

$$\begin{aligned}\overline{\varrho c''^2}&=\int_0^1(c-\tilde{c})^2\varrho(c)\,P(c)dc\\&=\bar{\varrho}\tilde{c}^2+\beta\varrho_\infty(1-2\tilde{c})+\gamma\int_0^1 c^2\varrho(c)f(c;\tilde{c},\overline{\varrho c''^2})\,dc\\&\quad-2\tilde{c}\gamma\int_0^1 c\varrho(c)f(c;\tilde{c},\overline{\varrho c''^2})dc.\end{aligned} \tag{4.39}$$

If β is eliminated between (4.38) and (4.39), it is found that

$$\overline{\varrho c''^2}/\bar{\varrho}=\tilde{c}(1-\tilde{c})-\gamma(1+\tau\tilde{c})\,M(\tilde{c},\overline{\varrho c''^2}), \tag{4.40}$$

where

$$M(\tilde{c},\overline{\varrho c''^2})=\int_0^1\frac{c(1-c)f(c;\tilde{c},\overline{\varrho c''^2})}{1+\tau c}dc.$$

We see that the three variables $\tilde{c}$, $\overline{\varrho c''^2}$, and γ are linked through (4.40). If the reacting mode pdf $f(c;\tilde{c},\overline{\varrho c''^2})$ can be approximated as $f(c)$, then M will be a constant and (4.40) takes a particularly simple form.

Consider next the time-average rate expression, (4.32). Since $w(c)=0$ when $c=0$ and when $c=1$, the delta functions in (3.49) make no contribution to $\bar{w}$. Accordingly we write

$$\bar{w}(\tilde{c},\overline{\varrho c''^2})=w_{\max}\gamma I_3(\tilde{c},\overline{\varrho c''^2}), \tag{4.41}$$

where $w_{\max}$ is the maximum value of $w(c)$ and

$$I_3(\tilde{c},\overline{\varrho c''^2})=\int_0^1\frac{w(c)}{w_{\max}}f(c;\tilde{c},\overline{\varrho c''^2})\,dc. \tag{4.42}$$

Similarly,

$$\overline{c''w}=w_{\max}\gamma[I_4(\tilde{c},\overline{\varrho c''^2})-\tilde{c}I_3(\tilde{c},\overline{\varrho c''^2})], \tag{4.43}$$

where

$$I_4(\tilde{c},\overline{\varrho c''^2})=\int_0^1\frac{w(c)}{w_{\max}}cf(c;\tilde{c},\overline{\varrho c''^2})\,dc. \tag{4.44}$$

Equation (4.40) enables us to eliminate γ from (4.41, 43). We then finally obtain [4.10]

$$\frac{\bar{w}}{w_{\max}}=K_1(\tilde{c},\overline{\varrho c''^2})\frac{\tilde{c}(1-\tilde{c})}{1+\tau\tilde{c}}\left[1-\frac{\overline{\varrho c''^2}}{\bar{\varrho}\tilde{c}(1-\tilde{c})}\right], \tag{4.45}$$

$$\overline{c''w}=[c_m(\tilde{c},\overline{\varrho c''^2})-\tilde{c}]\bar{w}. \tag{4.46}$$

The two new quantities in these equations are

$$K_1(\tilde{c},\overline{\varrho c''^2})=I_3(\tilde{c},\overline{\varrho c''^2})/M(\tilde{c},\overline{\varrho c''^2}) \tag{4.47}$$

and

$$c_m(\tilde{c},\overline{\varrho c''^2})=I_4(\tilde{c},\overline{\varrho c''^2})/I_3(\tilde{c},\overline{\varrho c''^2}). \tag{4.48}$$

We observe from (4.46) that c_m is the value of $\tilde{c}$ at which $\overline{c''w}$ changes sign. Note that the burning mode pdf $f(c;\tilde{c},\overline{\varrho c''^2})$ influences $\bar{w}$ and $\overline{c''w}$ only through K_1 and c_m, both of which are ratios of moments. This results in an insensitivity of the source terms to details of the specification of the pdf.

If $f(c;\tilde{c},\overline{\varrho c''^2})$ is specified, (4.45, 46) relate the source terms $\bar{w}$ and $\overline{c''w}$ to $\tilde{c}$ and $\overline{\varrho c''^2}$, thus completing the thermochemical closure of the model.

4.5.3 Effects of Turbulence on Reaction Rate

We are now in a position to investigate the effects of turbulent fluctuations on the mean reaction rate $\bar{w}$ within the accuracy of the above model. The following analysis comes from *Bray* [4.10].

It is convenient to introduce a normalized mean square fluctuation intensity g defined as

$$g=\frac{\overline{\varrho c''^2}}{\bar{\varrho}\tilde{c}(1-\tilde{c})}. \tag{4.49}$$

Equation (4.40) shows that $g<1$ but approaches unity if the burning mode probability $\gamma\to 0$. We shall determine $\bar{w}$ for different values of g.

I) *Small fluctuations*, $g\ll 1$

In this limit, the instantaneous reaction rate may be expanded in a Taylor series about $\tilde{c}$. Rather than expand $w(c)$, it is convenient to treat $\bar{\varrho}w(c)/\varrho(c)$. Thus

$$\bar{\varrho}w(c)/\varrho(c)=w(\tilde{c})+\left[\frac{d}{dc}\left(\frac{\bar{\varrho}w(c)}{\varrho(c)}\right)\right]_{\tilde{c}}(c-\tilde{c})$$
$$+\frac{1}{2}\left[\frac{d^2}{dc^2}\left(\frac{\bar{\varrho}w(c)}{\varrho(c)}\right)\right]_{\tilde{c}}(c-\tilde{c})^2+\dots.$$

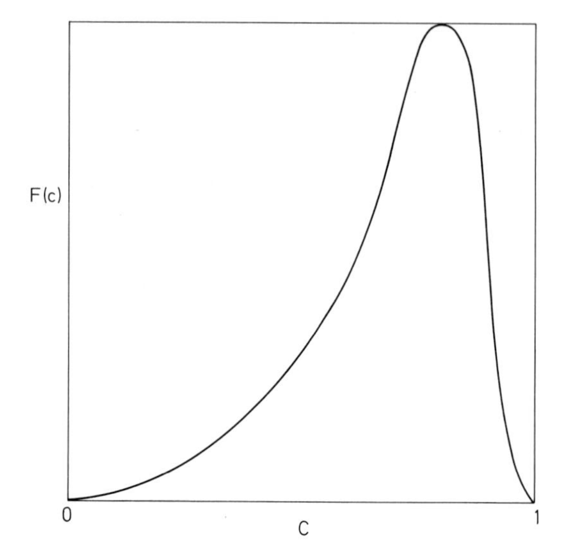

Fig. 4.8. The function $F(c)$

With Favre averaging we obtain the lowest order contributions to the source terms in the form

$$\bar{w}=w(\tilde{c})[1+gG(\tilde{c})+\ldots] \tag{4.50}$$

$$\overline{c''w}=gK(\tilde{c})+\ldots, \tag{4.51}$$

where

$$G(\tilde{c})=\frac{\tilde{c}(1-\tilde{c})}{2w(\tilde{c})}\left[\frac{d^2}{dc^2}\left(\frac{\bar{\varrho}w(c)}{\varrho(c)}\right)\right]_{\tilde{c}}$$

and

$$K(\tilde{c})=\tilde{c}(1-\tilde{c})\left[\frac{d}{dc}\left(\frac{\bar{\varrho}w(c)}{\varrho(c)}\right)\right]_{\tilde{c}}.$$

The function

$$F(c)=\frac{\bar{\varrho}w(c)}{\varrho(c)w_{\max}}$$

is sketched in Fig. 4.8. It will be seen that d^2F/dc^2 is positive, close to $c=0$ and $c=1$, so $G(\tilde{c})$ is positive, and $\bar{w}>w(\tilde{c})$ in these regions, that is, turbulent fluctuations increase the mean rate of reaction. On the other hand, near the peak in $F(c)$, $d^2F/dc^2<0$, so $\bar{w}<w(\tilde{c})$ and the fluctuations reduce the mean rate. Similarly, dF/dc and $K(\tilde{c})$ both change sign at the peak; $\overline{c''w}>0$ to the left of the peak and $\overline{c''w}<0$ to the right [cf. (4.46)].

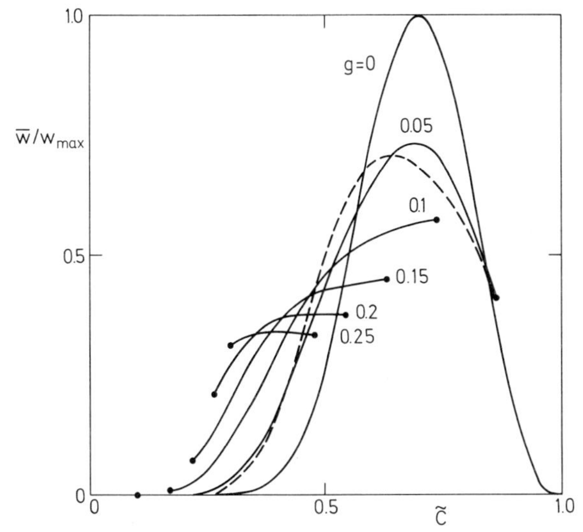

Fig. 4.9. *Full curves:* the chemical source term, w/w_{max}, evaluated from (4.45, 49) assuming a rectangular pdf without intermittency. The curves terminate at points where intermittency occurs. *Dashed curve:* linear approximation from (4.50) with $g=0.05$ [4.10]

The magnitude of $G(\tilde{c})$ is generally large in comparison with unity, particularly near $\tilde{c}=0$, 1, so the range of validity of the above linear approximation is severely limited although adequate for our purpose.

II) *Finite g*

We first consider the case where $\gamma=1$. Packets of unburned or fully burned mixture are then assumed not to occur so the composition field is without intermittency. Although this is not the normal situation in practice, it is a convenient starting point for the analysis. Cases where $\gamma<1$ are analyzed later. A rectangular pdf is assumed for simplicity, and $\bar{w}$ is calculated from (4.45). The instantaneous reaction rate is calculated from the following equation [4.12]:

$$w(c)=Bw_{max}(1-c)^3\exp[-E^*(1-c)/c], \tag{4.52}$$

where B is a normalizing constant. Numerical values assumed are $E^*=5$, $\tau=5$. Figure 4.9 shows the results, for specified values of g. The curves terminate at values of $\tilde{c}$ at which the rectangular pdf reaches one or other of the bounds $c=0, 1$; to extend the curves outside the range illustrated therefore requires the introduction of intermittency. It may be seen from Fig. 4.9 that as g increases the $\tilde{c}$ range which can be covered without intermittency becomes smaller. At $g=0.34$, this range reduces to a single value, $\tilde{c}=0.36$, at which $\bar{w}/w_{max}=0.29$ while, for $g>0.34$, no solution is possible without intermittency. Experiment [4.23] indicates peak values of $\bar{w}/w_{max}$ which are much smaller than this, so we may conclude that extensive intermittency is a normal feature of premixed turbulent combustion.

Table 4.1. Reaction rate parameters K_1 and c_m [see (4.45), 46)] for several simple reacting mode pdfs $f(c;\boldsymbol{x})$

Reacting mode pdf: $f(c;\boldsymbol{x})$	y	K_1	c_m
Uniform	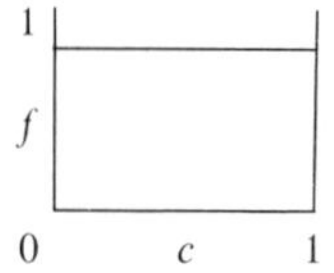	5.41	0.692
Battlement	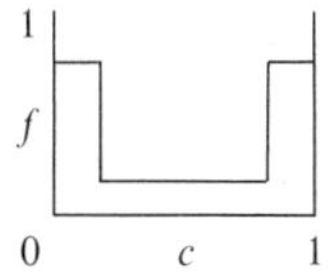	3.89	0.716
Triangular	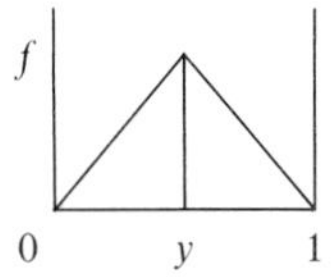 0	2.91	0.667
	0.25	3.54	0.651
	0.5	5.60	0.654
	0.75	8.57	0.689
	1.00	8.93	0.707
Delta function	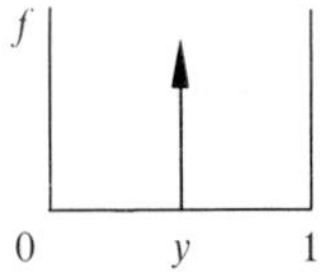 0	0	0
	0.25	0.0005	0.25
	0.5	3.72	0.5
	0.75	23.60	0.75
	1.00	0	1.0

Figure 4.9 (dashed curve) shows the linear, small perturbation approximation to $\bar{w}/w_{max}$ from (4.50), evaluated at a small fluctuation level, $g=0.05$. It will be seen that in this case the approximation compares quite well with the more accurate results.

Equations (4.45) and (4.46) can also be used to determine $\bar{w}$ and $\langle c''w\rangle$ in cases where intermittency occurs, if the burning mode pdf $f(c;\tilde{c},\overline{\varrho c''^2})$ is specified. An important question is the sensitivity of the source terms $\bar{w}$ and $\overline{c''w}$ to the specification of $f(c;\tilde{c},\overline{\varrho c''^2})$. The moment ratios K_1 and c_m which contain the sole effects of $f(c;\tilde{c},\overline{\varrho c''^2})$ are shown in Table 4.1 for several simple burning mode pdfs. As $w(c)=0$, both for unburned mixture ($c=0$) and for fully burned product, K_1 and c_m are independent of the intermittency. Apart from the Dirac delta function pdf we see that c_m for all the other pdf's lies in the narrow range $0.651\leqq c_m\leqq 0.716$. The values of K_1 involve a considerably greater spread: $2.91\leqq K_1\leqq 8.93$. However, this range is still relatively small in comparison with the uncertainty which is typical of chemical reaction rates. We conclude that a reasonable a priori specification of the pdf is likely to cause errors no greater than others inherent in the modeling of reacting turbulent flows. Clipped Gaussian and battlement pdf models have been used to estimate c_m and K_1 [4.12].

On the other hand, the table shows that, if the reacting mode pdf is represented by a single delta function, i.e., if $P(c;\mathbf{x})$ is approximated by three delta functions, at $c=0$, y, and 1, respectively, then the relevant moments are excessively sensitive to the location y.

III) *Large fluctuation,* $g \approx 1$

With g close to unity, the fluctuation level approaches the maximum possible value. Equation (4.40) may be written

$$1-g=\gamma M \frac{1+\tau\tilde{c}}{\tilde{c}(1-\tilde{c})}$$

so, as g approaches unity, γ must become small. In this case the mixture is made up of packets of burned and unburned gas, separated by narrow reaction zones. We make no assumption concerning the structure of these narrow burning zones; they may be wrinkled laminar flames or the narrow dissipative vortex filaments of *Chomiak* [4.9], or any other structures which turn out to be appropriate on the basis of experiment or further analysis. The limit $\gamma \ll 1$ leads to drastic simplifications in the model, enabling us to dispense with the composition fluctuation balance (4.35) and to express $\bar{w}$ directly in terms of $\tilde{c}$. Before discussing this, we must identify the regime of validity of the approximation $\gamma \ll 1$.

For this purpose the balance equations (4.34, 35) are written in normalized form by introducing the following nondimensional variables:

$$X_i = x_i/l_0\,;\; U_i = u_i/\bar{q}_0^{1/2}$$

$$\Omega = \varrho/\varrho_0\,;\; \mathcal{M} = \mu/\mu_0\,,$$

where l_0, $\bar{q}_0$, ϱ_0, and μ_0 are the integral scale of turbulence, turbulence kinetic energy, density, and viscosity of the unburned gas. The symbol $\bar{q}_0$ is used rather than $\tilde{q}_0$ because the Favre average reduces to the conventional one in the unburned gas where $\varrho \equiv \varrho_0$. These variables are appropriate, since the turbulent flame thickness and normal propagation speed are known from experiment to be of order l_0 and $\bar{q}_0^{1/2}$, respectively. Equations (4.34) and (4.35) may then be written

$$\frac{\partial}{\partial X_k}(\bar{\Omega}\tilde{U}_k\tilde{c}) = \frac{1}{R_l}\frac{\partial}{\partial X_k}\overline{\left(\mathcal{M}\frac{\partial c}{\partial X_k}\right)} - \frac{\partial}{\partial X_k}(\overline{\Omega U_k'' c''}) + D_{\mathrm{a}}\frac{\bar{w}}{w_{\max}} \tag{4.53}$$

and

$$\frac{\partial}{\partial X_k}(\overline{\Omega c''^2}\tilde{U}_k) = \frac{2}{R_l}\overline{\left[c''\frac{\partial}{\partial X_k}\left(\mathcal{M}\frac{\partial c}{\partial X_k}\right)\right]} - 2\overline{\Omega U_k'' c''}\frac{\partial\tilde{c}}{\partial X_k}$$
$$-\frac{\partial}{\partial X_k}(\overline{\Omega U_k'' c''^2}) + 2D_{\mathrm{a}}\frac{\overline{c''w}}{w_{\max}}. \tag{4.54}$$

Two parameters in these equations are a turbulence Reynolds number

$$R_l = \varrho_0 \bar{q}_0^{1/2} l_0 / \mu_0$$

and a Damköhler number

$$D_\mathrm{a} = w_\mathrm{max} l_0 / \varrho_0 \bar{q}_0^{1/2} .$$

It is noted, in passing, that these are similar to two of the parameters appearing in Fig. 4.6.

Using (4.40, 41, 43) we may write (4.53, 54) as

$$\frac{\partial}{\partial X_k}(\bar{\Omega}\tilde{U}_k\tilde{c}) = \frac{1}{R_l}\frac{\partial}{\partial X_k}\overline{\left(\mathcal{M}\frac{\partial c}{\partial X_k}\right)} - \frac{\partial}{\partial X_k}(\overline{\Omega U_k'' c''}) + \gamma D_\mathrm{a} I_3 \tag{4.55}$$

and

$$\frac{\partial}{\partial X_k}[\bar{\Omega}\tilde{U}_k\tilde{c}(1-\tilde{c}) - \gamma\bar{\Omega}\tilde{U}_k M(1+\tau\tilde{c})]$$

$$= \frac{2}{R_l}\left[\overline{c''\frac{\partial}{\partial X_k}\left(\mathcal{M}\frac{\partial c}{\partial X_k}\right)}\right]$$

$$-\overline{2\Omega U_k'' c''}\frac{\partial\tilde{c}}{\partial X_k} - \frac{\partial}{\partial X_k}(\overline{\Omega U_k'' c''^2}) + 2\gamma D_\mathrm{a} I_3(c_\mathrm{m} - \tilde{c}) . \tag{4.56}$$

From the first of these, (4.55), we deduce that γD_a is of order unity. Thus, the limit $\gamma \ll 1$ accompanies a large Damköhler number $D_\mathrm{a} \gg 1$. Therefore, the chemical time ϱ_0/w_max is much smaller than the turbulence time $l_0/\bar{q}_0^{1/2}$, and the rate of combustion is controlled mainly by turbulent mixing rather than by chemical kinetics. The large fluctuation analysis, as such, sets no constraint on the turbulence Reynolds number R_l. However, existing empirical models of turbulent transport generally assume that it is large. The regime of validity of the model is then $D_\mathrm{a} \gg 1$, $R_l \gg 1$; see Fig. 4.6.

In a formal expansion in powers of $1/D_\mathrm{a}$ [4.36, 55] the lowest order equations contain γ only in combination with D_a. Thus, in this approximation, γ is retained in the chemical source terms on the right-hand sides of (4.55, 56), but the term containing γ is omitted from the left-hand side of (3.68). In terms of the dimensional variables (4.56) becomes

$$(1-2\tilde{c})\frac{\partial}{\partial x_k}(\bar{\varrho}\tilde{u}_k\tilde{c}) = 2\left[\overline{c''\frac{\partial}{\partial x_k}\left(\varrho\mathcal{D}\frac{\partial c}{\partial x_k}\right)}\right]$$

$$-\overline{2\varrho u_k'' c''}\frac{\partial\tilde{c}}{\partial x_k} - \frac{\partial}{\partial x_k}(\overline{\varrho u_k'' c''^2}) + 2\bar{w}(c_\mathrm{m} - \tilde{c}) + \mathrm{O}(\gamma) .$$

Combining this with the species balance, (4.55), gives an equation for $\bar{w}$

$$(1-2c_{\mathrm{m}})\bar{w}=2\left[\overline{c''\frac{\partial}{\partial x_k}\left(\varrho\mathscr{D}\frac{\partial c}{\partial x_k}\right)}\right]$$
$$-(1-2\tilde{c})\frac{\partial}{\partial x_k}\left(\overline{\varrho\mathscr{D}\frac{\partial c}{\partial x_k}}\right)-\overline{2\varrho u_k''c''}\frac{\partial\tilde{c}}{\partial x_k}$$
$$-\frac{\partial}{\partial x_k}(\overline{\varrho u_k''c''^2})+(1-2\tilde{c})\frac{\partial}{\partial x_k}(\overline{\varrho u_k''c''})+\mathrm{O}(\gamma)\,. \tag{4.57}$$

It will be shown later that the last three terms in (4.57) sum to zero. However, further analysis is required first.

If we return to the pdf $P(c)$ given by (4.37), the condition $\gamma \ll 1$ leads to a pdf consisting mainly of the two delta functions with $\alpha+\beta\simeq 1+\mathrm{O}(\gamma)$; (4.38) then allows us to express α and β in terms of $\tilde{c}$, giving

$$\alpha=\frac{1-\tilde{c}}{1+\tau\tilde{c}}+\mathrm{O}(\gamma)\,, \tag{4.58}$$

$$\beta=\frac{(1+\tau)\tilde{c}}{1+\tau\tilde{c}}+\mathrm{O}(\gamma)\,. \tag{4.59}$$

To this accuracy the burning mode pdf $f(c)$ no longer enters these equations.

Similar arguments apply to the joint pdf for velocity and concentration [4.73]. For simplicity, we consider only a single velocity component denoted by u and express the pdf $P(u, c; \boldsymbol{x})$ in a form analogous to (4.37), namely

$$P(u,c;\boldsymbol{x})=\alpha(\boldsymbol{x})P_{\mathrm{r}}(u,0;\boldsymbol{x})\,\delta(c)+\beta(\boldsymbol{x})P_{\mathrm{p}}(u,1;\boldsymbol{x})\,\delta(1-c)$$
$$+\gamma(\boldsymbol{x})[H(c)-H(c-1)]\,P_{\mathrm{uc}}(u,c;\boldsymbol{x})\,, \tag{4.60}$$

where P_{r}, P_{p}, and P_{uc} are joint pdfs for reactant, product, and burning modes, respectively, and, as before,

$$\alpha(\boldsymbol{x})+\beta(\boldsymbol{x})+\gamma(\boldsymbol{x})=1\,.$$

The following normalization conditions apply:

$$\int_{-\infty}^{\infty} du\int_0^1 dcP(u,c;\boldsymbol{x})=1$$

$$\int_{-\infty}^{\infty} duP_{\mathrm{r}}(u,0;\boldsymbol{x})=1$$

$$\int_{-\infty}^{\infty} duP_{\mathrm{p}}(u,1;\boldsymbol{x})=1$$

$$\int_{\infty}^{\infty} du\int_0^1 dcP_{\mathrm{uc}}(u,c;\boldsymbol{x})=\int_0^1 dcf(c;\boldsymbol{x})=1\,,$$

ensuring that the integral of (4.60) over all velocities u recovers the one-dimensional pdf of (4.37).

In the limit $\gamma \ll 1$, the joint pdf for the burning mode, P_{uc}, makes only a small contribution to the moments. Thus, for example

$$\bar{\varrho}(\boldsymbol{x})\tilde{u}(\boldsymbol{x}) = \int_{-\infty}^{\infty} du\, u \int_{0}^{1} dc\, \varrho(c) P(u, c; \boldsymbol{x})$$

$$= \alpha(\boldsymbol{x})\varrho_0 \int_{-\infty}^{\infty} du\, u P_r(u, 0; \boldsymbol{x})$$

$$+ \beta(\boldsymbol{x})\varrho_\infty \int_{-\infty}^{\infty} du\, u P_p(u, 1; \boldsymbol{x}) + \mathrm{O}(\gamma). \tag{4.61}$$

Let

$$\bar{u}_r(\boldsymbol{x}) = \int_{-\infty}^{\infty} du\, u P_r(u, 0; \boldsymbol{x}), \tag{4.62}$$

$$\bar{u}_p(\boldsymbol{x}) = \int_{-\infty}^{\infty} du\, u P_p(u, 1; \boldsymbol{x}), \tag{4.63}$$

which may be identified as the *conditioned* mean velocities of the reactant and product packets, respectively, at position $\boldsymbol{x}$ in the flame. Thus, for example, $\bar{u}_r(\boldsymbol{x})$ is a mean velocity, calculated from the fluctuating velocity at $\boldsymbol{x}$, but averaged only over those periods of time for which $c=0$; see Sect. 1.15. If these expressions are substituted into (4.61) and use is made of (4.58, 59), it is readily found that

$$\tilde{u} = (1-\tilde{c})\bar{u}_r + \tilde{c}\bar{u}_p + \mathrm{O}(\gamma), \tag{4.64}$$

where the dependence on position $\boldsymbol{x}$ is no longer shown explicitly.

Similarly, the mean flux of product can be expressed in terms of the conditioned velocities as

$$\overline{\varrho u'' c''} = \bar{\varrho}\tilde{c}(1-\tilde{c})(\bar{u}_p - \bar{u}_r) + \mathrm{O}(\gamma). \tag{4.65}$$

These and similar expressions provide a valuable insight into the processes occurring in the large fluctuation limit. They also suggest a description of turbulent transport [4.74] which avoids the conventional eddy viscosity assumption. We shall return to this formulation in Sect. 4.5.5.

Of particular relevance to the present analysis is the result

$$\overline{\varrho u'' c''^2} = \bar{\varrho}\tilde{c}(1-\tilde{c})(\bar{u}_p - \bar{u}_r)(1-2\tilde{c}) + \mathrm{O}(\gamma)$$

which is obtained in the same manner as (4.61). Combined with (4.65) it gives

$$\overline{\varrho u'' c''^2} = \overline{\varrho u'' c''}(1-2\tilde{c}) + \mathrm{O}(\gamma). \tag{4.66}$$

It is important to note that (4.66) is a direct consequence of the assumption $\gamma \ll 1$. No model of turbulent transport is assumed.

Substitution of (4.66) into the rate expression (4.57), which is derived on the assumption of $\gamma \ll 1$, shows that the sum of the last three terms on the right-hand side of (4.57) is zero. Thus,

$$\bar{w} = \frac{1}{2c_{\mathrm{m}} - 1} \left\{ (1 - 2\tilde{c}) \frac{\partial}{\partial x_k} \overline{\left[\varrho \mathscr{D} \frac{\partial c}{\partial x_k} \right]} - \overline{2c'' \frac{\partial}{\partial x_k} \left(\varrho \mathscr{D} \frac{\partial c}{\partial x_k} \right)} \right\} \tag{4.67}$$

which relates $\bar{w}$ to the averages of molecular diffusion terms alone. The first term inside the brace on the right of (4.67) is $(1-2\tilde{c})$ times the molecular diffusion term in (4.34). As shown in (4.55) this term becomes small a high turbulence Reynolds number R_l. The other term is minus the molecular diffusion contribution to the species fluctuation balance (4.35); it is usual to split this term into two parts,

$$\overline{2c'' \frac{\partial}{\partial x_k} \left[\varrho \mathscr{D} \frac{\partial c}{\partial x_k} \right]} = 2 \frac{\partial}{\partial x_k} \overline{\left[c'' \varrho \mathscr{D} \frac{\partial c}{\partial x_k} \right]} - \overline{2 \varrho \mathscr{D} \frac{\partial c}{\partial x_k} \frac{\partial c''}{\partial x_k}} .$$

The second contribution, which is usually the larger, is known as the scalar dissipation, and is denoted by $\bar{\chi}$. Thus,

$$\bar{\chi} \equiv \overline{2 \varrho \mathscr{D} \frac{\partial c}{\partial x_k} \frac{\partial c''}{\partial x_k}} . \tag{4.68}$$

It represents the rate at which concentration fluctuations, in this case, fluctuations in c, are reduced by molecular diffusion. Equation (4.67) may now be written

$$\bar{w} = \frac{1}{2c_{\mathrm{m}} - 1} \left[\bar{\chi} + (1 - 2\tilde{c}) \frac{\partial}{\partial x_k} \overline{\left(\varrho \mathscr{D} \frac{\partial c}{\partial x_k} \right)} - 2 \frac{\partial}{\partial x_k} \overline{\left(c'' \varrho \mathscr{D} \frac{\partial c}{\partial x_k} \right)} \right] . \tag{4.69}$$

The chemical kinetic rate does not appear in this expression, which corresponds, therefore, to the mixing limited reaction rate. The only assumption made in obtaining (4.69) is that the burning mode probability γ is small. It is thus a rigorous consequence of the initial postulates of the present analysis in the limit $\gamma \ll 1$. At high turbulence Reynolds numbers $R_l \gg 1$ the first term on the right-hand side is the only important one and (4.69) takes the very simple form

$$\bar{w} = K_2 \bar{\chi} , \tag{4.70}$$

where $K_2 = 1/(2c_{\mathrm{m}} - 1)$. The coefficient K_2 is listed in Table 4.2 for several simple assumed forms for the burning mode pdf, and is seen to be generally quite insensitive to the shape chosen, as required in the a priori pdf approach.

Table 4.2. Reaction rate parameters K_2 and c_m [see (4.70)] for several simple reacting mode pdfs $f(c; \boldsymbol{x})$

Reacting mode pdf: $f(c; \boldsymbol{x})$	y	K_2	c_m
Uniform		2.61	0.692
Battlement		2.32	0.716
Triangular	0	2.99	0.667
	0.25	3.31	0.651
	0.5	3.25	0.654
	0.75	2.64	0.689
	1.00	2.41	0.707
Delta function	0	-1	0
	0.25	-2	0.25
	0.5	$\pm\infty$	0.5
	0.75	2	0.75
	1.00	1	1.0

The delta function is again an exception which tests excessively the concept of insensitivity.

Thus, the problem of modeling the mean reaction rate $\bar{w}$ in mixing limited combustion with $D_a \gg 1$ and $R_l \gg 1$ is exchanged for the problem of describing the scalar dissipation function $\bar{\chi}$. It may be anticipated [4.74] that a proper representation of $\bar{\chi}$ will include effects of the reaction and heat release which occur in turbulent flames. However, such a representation, fully validated, is not yet available. Meanwhile, we tentatively adopt a model currently used in cold, nonreacting, constant density flows; the scalar dissipation may then be represented empirically [4.75] by

$$\bar{\chi} = C\tilde{q}^{1/2}\overline{\varrho c''^2}/l_c, \tag{4.71}$$

where C is a modeling constant and l_c is a length scale of the composition field. Substitution of (4.71) into (4.70) yields a reaction rate expression similar to the eddy breakup model of *Mason* and *Spalding* [4.64]; see (4.17). Finally, since $\gamma \ll 1$, (4.40) simplifies to

$$\overline{\varrho c''^2} = \bar{\varrho}\tilde{c}(1-\tilde{c}) + O(\gamma).$$

That is, the Favre-averaged scalar fluctuations have their maximum possible intensity for a given value of $\tilde{c}$. This gives

$$\bar{w}=K_2 C\bar{\varrho}\tilde{q}^{1/2}\tilde{c}(1-\tilde{c})/l_c \tag{4.72}$$

for the mixing limited reaction rate. To this approximation the species fluctuation balance (4.35) is no longer required. Reaction proceeds most rapidly in regions where the turbulence is of high intensity and the length scale of the composition field is small.

Predictions employing (4.72) are discussed in Sect. 4.5.5. First it is necessary to discuss the second area requiring modeling, namely that dealing with turbulent transport.

4.5.4 Turbulent Transport Model

There are many fundamental ways in which combustion can affect the physics and, hence, the modeling of turbulent transport, for example, through production of density variations, buoyancy effects, dilation due to heat release; and through its influence on molecular transport and on instability effects, etc. These influences are not well understood; by and large, modellers have followed *Spalding*'s "Art of Partial Modelling" [4.76] in assuming that their empirical closures and model equations can be carried over unchanged from constant density unreacting flow problems to turbulent combustion. This assumption is hard to justify theoretically when it is recalled that many closures originated in simple dimensional analysis, while combustion introduces additional dimensionless groups such as Damköhler numbers, density ratio, etc. Also the exact, unclosed equations [4.77, 78] contain many extra terms which are not presented in the constant density case (see Chap. 1).

Extreme situations can be found where a large effect on turbulent transport is readily observed experimentally. For example, baffle-stabilized, ducted, premixed flames can be caused to "buzz", with the result that the flame spreading rate is increased. Thus, a combustion instability leads to a sharp increase in turbulent transport with, in this case, pressure fluctuations playing a significant part in the interaction.

For the present purposes, we shall follow normal practice and assume a conventional eddy viscosity gradient transport model of turbulent transport (see Sect. 1.13) in which the turbulent flux of a scalar quantity g is written

$$\overline{\varrho u''_m g''}=-\bar{\varrho}\nu_T\partial\tilde{g}/\partial x_m\,, \tag{4.73}$$

where the eddy kinematic viscosity ν_T is approximated by

$$\nu_T=a\tilde{q}^{1/2}l\,; \tag{4.74}$$

a is an empirical constant and l is a length scale associated with the large eddies.

The turbulent kinetic energy $\tilde{q}$ and the scale l of the turbulence must, of course, be calculated. A popular approach [4.79] is to calculate $\tilde{q}$ and l [or alternatively, the viscous dissipation function,

$$\varepsilon = \overline{(\tau_{kl} \partial u_k''/\partial x_l)}/\bar{\varrho}$$

which is related to l] from two differential equations, both of which require modeling. However, drastic and fundamentally unjustified modeling assumptions are required in order to derive a useable balance equation for either l or ε. Accordingly, in the calculations which are reviewed in the following section, the consequences of a much simpler algebraic specification of the length scale variation are explored. Three alternative empirical equations have been used

$$\begin{aligned} \frac{l}{l_0} &= 1 && \text{I} \\ \frac{l}{l_0} &= \left(\frac{\bar{q}_0}{\tilde{q}}\right)^{1/2} && \text{II} \\ \frac{l}{l_0} &= 1 + \tau\tilde{c}\,. && \text{III} \end{aligned} \tag{4.75}$$

The first assumes that the length scale remains frozen during combustion while the second holds constant the characteristic Reynolds number $\tilde{q}^{1/2} l/\nu_{\mathrm{T}}$, and the third assumes that l expands with the time-average dilatation and so keeps $\bar{\varrho} l$ constant. In some cases (see, for example, [4.55]) useful information can be obtained from the theory without any specification of a length scale.

As the scalar dissipation function $\bar{\chi}$ must be calculated from (4.71), we also require an expression for the length scale l_c of the composition fluctuations. It is assumed, again in the interest of simplicity, that l_c/l is a constant; more elaborate assumptions are not justified at present.

The turbulence kinetic energy (TKE) $\tilde{q}$ is obtained from a balance equation, (4.36). As written, this equation is essentially exact for low Mach number flows in the absence of buoyancy, as considered here, so it automatically includes the effects of any mechanism of flame-generated turbulence. Whether the net effect of burning and flow is to generate additional turbulence or to remove turbulence energy depends [4.80, 81] on the balance between positive and negative terms. However, several of these terms must be represented empirically; these terms are now identified and discussed. We have

$$\underset{\text{(I)}}{\frac{\partial}{\partial x_k}(\bar{\varrho}\tilde{u}_k\tilde{q})} = \underset{\text{(II)}}{-\overline{\varrho u_k'' u_l''}\frac{\partial \tilde{u}_l}{\partial x_k}} \underset{\text{(III)}}{- \tfrac{1}{2}\frac{\partial}{\partial x_k}(\overline{\varrho u_k'' u_l''^2})} + \underset{\text{(IV)}}{\frac{\partial}{\partial x_k}(\overline{u_l''\tau_{lk}})} \underset{\text{(V)}}{- \overline{\tau_{kl}\frac{\partial u_k''}{\partial x_l}}} \underset{\text{(VI)}}{- \overline{u_k''\frac{\partial p}{\partial x_k}}}\,. \tag{4.76}$$

I) *Convection* of TKE in the mean flow with velocity $\tilde{u}_k$.
II) When $m \neq n$, these terms are generally positive, so there is *production* of TKE due to the work done against the Reynolds stresses $\overline{\varrho u_k'' u_l''}$, which are modelled

from a gradient transport assumption. This is the main source of TKE in shear flows. The three normal components $k=l$ represent effects of *dilatation*, which is absent in constant density flows. Their effect is approximated [4.80] by

$$-\overline{\varrho u_k''^2}\frac{\partial \tilde{u}_l}{x_l} = -\varepsilon_d \bar{\varrho}\tilde{q}\frac{\partial \tilde{u}_l}{\partial x_l}, \tag{4.77}$$

where ε_d is a positive constant which is expected to be of order unity. Since expansion due to heat release makes $\partial \tilde{u}_l/\partial x_l$ positive in flames, this term results in a removal of TKE.

III) *Turbulent diffusion* of TKE, normally modelled by an eddy viscosity assumption.

IV) *Molecular diffusion* which will be negligible except at low turbulence Reynolds numbers.

V) *Viscous dissipation* represents removal of TKE due to viscous effects, which must be modelled. Conventional expressions introduce a length scale which may be calculated from (4.75).

VI) Within the assumptions stated above, this *pressure-velocity correlation* term represents the second possible source of flame-generated turbulence. Most studies suggest that it is small and we shall assume it to be zero ([4.80, 82] – but see Sect. 4.6).

A turbulent transport model based upon these assumptions is employed in Sect. 4.5.5. Evidence regarding the limitations of this transport model is reviewed in Sect. 4.6.

4.5.5 Application to Planar Turbulent Flames

The model described above is applied to an infinite planar turbulent combustion wave. Details may be found elsewhere [4.36, 55, 73]. Here we shall briefly present some results of the calculations.

Three flame configurations are considered: I) a *normal* flame, in which the time-average approach flow is perpendicular to the planar flame, so that no time-average shear occurs: II) an *unconfined oblique* flame obtained by adding a constant transverse velocity to the normal flame solution, leading to deflection of the mean flow but again no shear; and III) a *confined oblique* flame, in which mean shear occurs and leads [see (4.76)] to generation of additional turbulence. The model employed is the simple thermochemical model of Sect. 4.5.1, in terms of the progress variable c, and the mean heat release rate $\bar{w}$ is evaluated for the mixing limited case ($\gamma \ll 1$), using (4.72). Turbulent transport is modelled as described in Sect. 4.5.4.

The calculation includes predictions of the flame structure. Figure 4.10 shows the variation of $\tilde{c}$ and $Q=\tilde{q}/q_0$ through normal or unconfined oblique flames, using the three alternative length scale models of (4.75). It will be seen that the turbulence decays due to the dilatation effect. The flame thickness is of the order of the initial length scale l_0 and is influenced by the choice of the

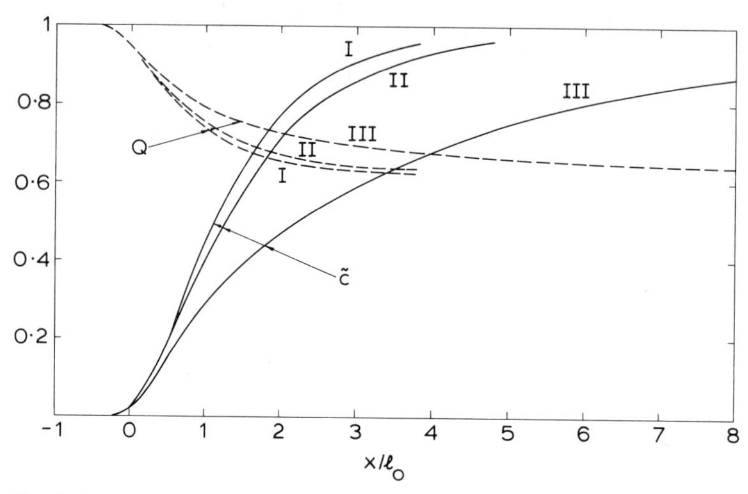

Fig. 4.10. The distributions of progress variable and turbulence kinetic energy for various models of the length scale [see (4.75)] for normal flames with $\tau=4$, $\tilde{c}_0=0.02$ [4.55]

length scale model. Model III, in which l grows by a factor $1+\tau$, gives the greatest thickness. Figure 4.11 gives similar information for confined oblique flames at various angles θ to the flow, using length scale model II. In the upstream part of the flame Q always decreases due to dilatation but it then increases again as a result of the production of turbulence energy due to shear, the predominant effect at small angles θ. The turbulence kinetic energy generated by the shear mechanism within the flame is then much greater than that in the approach stream. This suggests a theoretical limit, which is referred to as *strong interaction*, because the combustion then strongly interacts with the turbulence field.

The same model [4.55] also predicts the turbulent flame speed S_T. In the case of normal and unconfined oblique flames S_T is found from

$$\frac{S_T}{S_L}=\left[\frac{\Phi}{\beta(1+Q_\infty)}\right]^{1/2}\frac{\bar{q}_0^{1/2}}{S_L} \tag{4.78}$$

which may be compared with (4.1). Note that $\bar{q}_0^{1/2}$ is approximately proportional to the rms velocity component u'_0 appearing in (4.1). In (4.78) $\bar{q}_0$ is the TKE ahead of the flame, β and $Q_\infty=\tilde{q}_\infty/\bar{q}_0$ are eigenvalues, determined in the numerical solution, and

$$\Phi=\frac{aCl}{(2c_m-1)l_c} \tag{4.79}$$

is a modeling parameter. Figure 4.12 shows the two eigenvalues and $S_T/\bar{q}_0^{1/2}$ as functions of the heat release parameter $\tau=(T_\infty/T_0)-1$. It turns out that S_T is

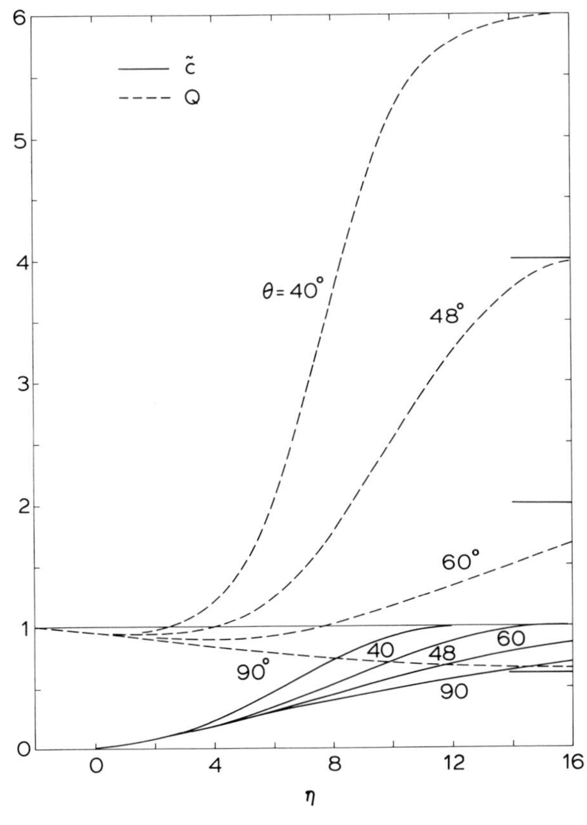

Fig. 4.11. Distributions of progress variable and turbulence kinetic energy vs $\eta = S_T x / \bar{q}_0^{1/2} a l_0$ for various values of the flame angle θ with $\tau = 4$, $\tilde{c}_0 = 0.02$ [4.55]

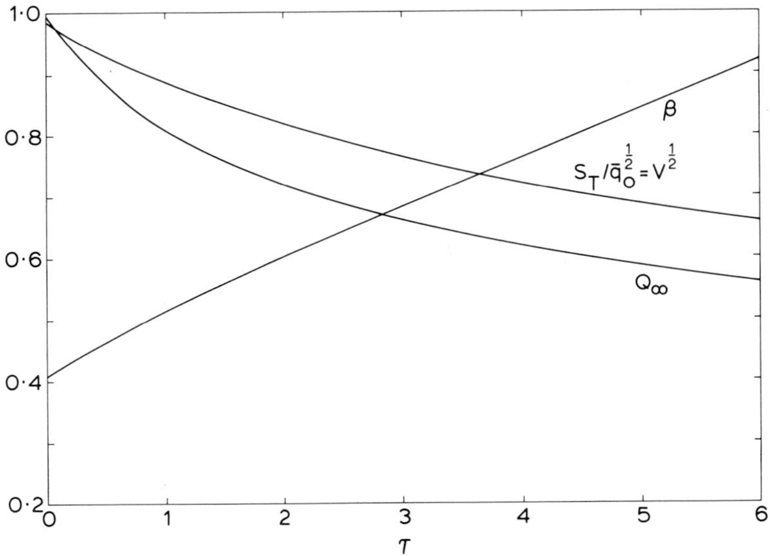

Fig. 4.12. Eigenvalues and flame speed for normal flames with $\tilde{c}_0 = 0.02$ (from [4.55])

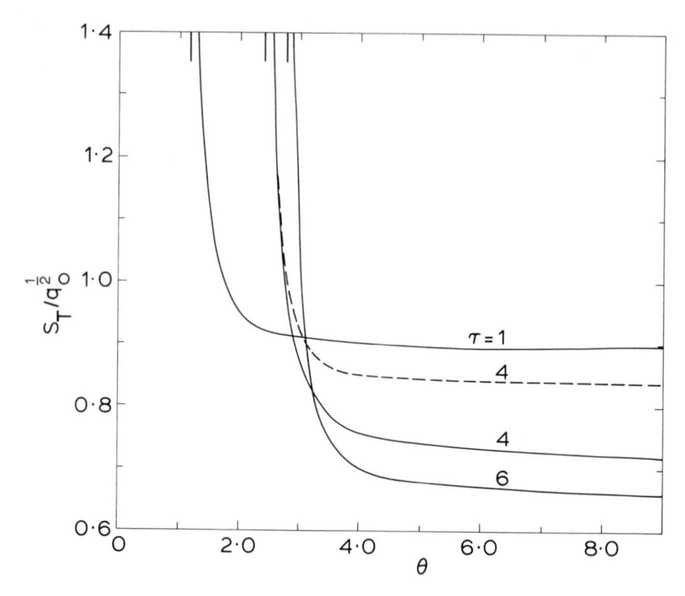

Fig. 4.13. Variation of flame speed with flame angle [from [4.55] (——) $\tilde{c}_0 = 0.02$; (– – –) $\tilde{c}_0 = 0.01$]

independent of the model employed for the length scale l. However, S_T does vary quite strongly with the initial value of the progress variable $\tilde{c}_0$ at which the reaction is allowed to commence. At $\tau = 4$, changing $\tilde{c}_0$ from 0.01 to 0.02 decreases $S_T/\bar{q}_0^{1/2}$ from 0.84 to 0.66. This "cold boundary problem" is interpreted [4.55] as an indication of dependence on the method of flame stabilization.

In the case of confined oblique flames, the additional turbulence, generated by shear within the flame, leads to an increase in S_T with decrease in flame angle θ, as shown in Fig. 4.13. As θ approaches the value corresponding to the strong interaction limit, defined above, the ratio $S_T/\bar{q}_0^{1/2}$ is predicted to tend towards infinity. At the same time the solution becomes independent of the initial condition $\tilde{c}_0$ indicating (see [4.22]) that the propagation of highly oblique confined flames is insensitive to the method of flame stabilization.

A comparison [4.55] of predictions from this model with experimental trends shows satisfactory agreement, within the inherent limitations of both the model and the experimental data. Comments made at the end of Sect. 4.2.2 about the present disarray of the experimental evidence should again be noted.

It will be recalled that the model used to obtain the above results assumes mixing limited heat release under conditions where the Damköhler number D_a and the turbulence Reynolds number R_l of Sect. 4.5.3 are both very large. Molecular transport and finite rate chemical kinetics are both neglected and consequently the laminar flame speed S_L does not appear in the model equations, although experiment shows it to be important. To overcome this discrepancy, *Libby* et al. [4.36] carried out a linear perturbation analysis involving an expansion in powers of $1/D_a$ and $1/R_l$. For normal and unconfined oblique flames they find that

$$\frac{S_T}{S_L} - 1 = 1.14 \frac{\bar{q}_0^{1/2}}{S_L} \left[1 + \left(126 - 8.33 \frac{\bar{q}_0}{S_L^2} \right) \frac{1}{R_l} + \dots \right] \tag{4.80}$$

which shows similar trends to those obtained in the empirical correlation of *Abdel-Gayed* and *Bradley* [4.6], but see the discussion of this correlation in Sect. 4.2. For confined flames in the strong interaction limit, the analysis gives the flame angle θ as

$$\theta = \theta_0 \left[1 + \frac{1}{R_l} (130 - 0.133 V_0^2 / S_L^2) + \dots \right], \tag{4.81}$$

where V_0 is the velocity of the unburned gas and θ_0 is a constant. For typical values the positive and negative terms tend to cancel, and θ departs little from θ_0. This may perhaps explain the insensitivity of the observed flame angle [4.22] to a wide variety of conditions.

We now return to results of the simpler, mixing limited model which assumes $D_a \gg 1$ and $\gamma \ll 1$. *Libby* and *Bray* showed [4.73] that if $\gamma \ll 1$ many fluctuating state variables can readily be calculated and in particular Favre and conventional Reynolds time averages can be compared. The Reynolds average of c is

$$\bar{c} = \int_0^1 c P(c) dc \,. \tag{4.82}$$

Substituting for $P(c)$ from (4.37), and taking $\gamma \ll 1$, we find

$$\bar{c} = \beta + \mathrm{O}(\gamma)$$

which becomes, when β is evaluated from (4.59)

$$\bar{c} = \left(\frac{1+\tau}{1+\tau\tilde{c}} \right) \tilde{c} + \mathrm{O}(\gamma) \,. \tag{4.83}$$

A similar calculation yields

$$\bar{T} = \frac{1 + 2\tau\tilde{c} + \tau^2\tilde{c}}{(1+\tau\tilde{c})^2} \tilde{T} + \mathrm{O}(\gamma) \,. \tag{4.84}$$

Figure 4.14 shows $\bar{c}/\tilde{c}$ and $\bar{T}/\tilde{T}$ plotted against $\tilde{c}$, and illustrates the remarkable difference between the two mean values of c, which occurs in the upstream part of the flame, where $\tilde{c}$ is small. Also $\bar{T}/\tilde{T}$ reaches a peak value in excess of two for

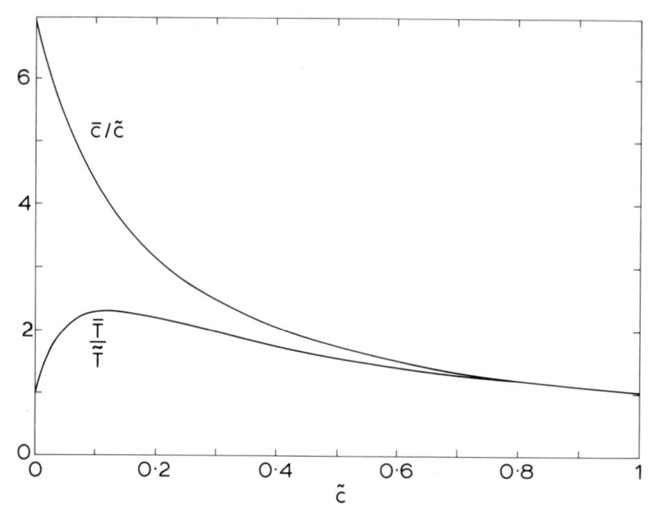

Fig. 4.14. Ratios of Reynolds to Favre means from (4.83, 84) with $\tau = 6$ [4.73]

values of τ of practical interest. Clearly, these large differences must be taken into account if a valid comparison is to be made between theory and experiment.

If Reynolds fluctuations are denoted by a single prime so that, for example, $c' = c - \bar{c}$, straightforward analysis yields the following results [4.73]:

$$\overline{c'^2} = \frac{1+\tau}{(1+\tau\tilde{c})^2} \frac{\overline{\varrho c''^2}}{\bar{\varrho}} + O(\gamma), \tag{4.85}$$

$$\overline{T'^2} = \frac{1+\tau}{(1+\tau\tilde{c})^2} \frac{\overline{\varrho T''^2}}{\bar{\varrho}} + O(\gamma), \tag{4.86}$$

$$[\overline{\varrho'^2}]^{1/2} = \tau\bar{\varrho}\left[\frac{\tilde{c}(1-\tilde{c})}{1+\tau}\right]^{1/2} + O(\gamma) \tag{4.87}$$

which are plotted in Fig. 4.15. Once again the difference between Reynolds and Favre averages is large, particularly in the upstream part of the flame. The trend of the variation of scalar fluctuation intensities is similar to that predicted by *Clavin* and *Williams* [4.60] from a quite different starting point.

In order to compare velocities, we must use results obtained from the joint pdf $P(u, c)$ of (4.60). The analysis of Sect. 4.5.3 allows us to relate the diffusion flux $\overline{\varrho u'' c''}$ to conditioned velocities $\bar{u}_r$ and $\bar{u}_p$ through (4.65). The planar flame calculations reviewed in this section assume an eddy viscosity transport model for which

$$\overline{\varrho u'' c''} - \bar{\varrho} v_T d\tilde{c}/dx\,. \tag{4.88}$$

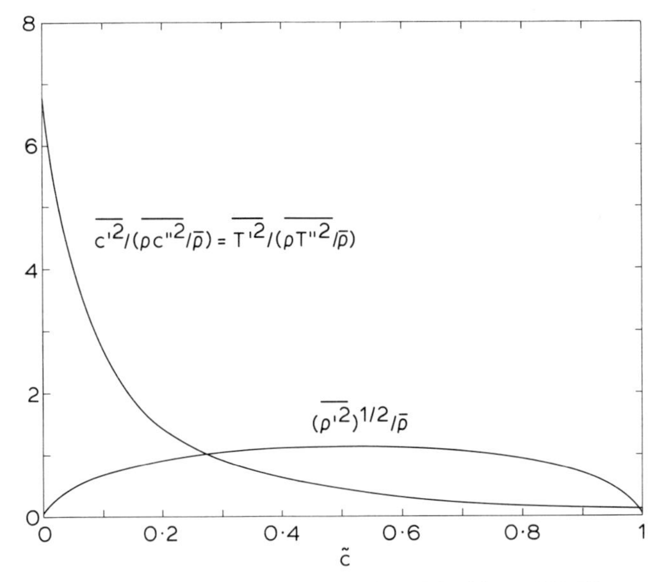

Fig. 4.15. Fluctuation intensities from (4.85–87) with $\tau = 6$ [4.73]

Equating this to (4.65) gives the following results which are consistent with the present model:

$$\bar{u}_r = \tilde{u} + \frac{v_T}{1-\tilde{c}} \frac{d\tilde{c}}{dx} + O(\gamma) \tag{4.89}$$

and

$$\bar{u}_p = \tilde{u} - \frac{v_T}{\tilde{c}} \frac{d\tilde{c}}{dx} + O(\gamma). \tag{4.90}$$

The Reynolds mean velocity is

$$\begin{aligned} \bar{u} &= \int_{-\infty}^{\infty} du\, u \int_0^1 P(u,c)\, dc \\ &= \frac{1-\tilde{c}}{1+\tau\tilde{c}} u_r + \frac{(1+\tau)\tilde{c}}{1+\tau\tilde{c}} \bar{u}_p + O(\gamma). \end{aligned}$$

Using (4.89) and (4.90) to eliminate $\bar{u}_r$ and $\bar{u}_p$, we find that

$$\bar{u} = \tilde{u} - \frac{\tau v_T}{1+\tau\tilde{c}} \frac{d\tilde{c}}{dx} + O(\gamma). \tag{4.91}$$

The ratio $\bar{u}/\tilde{u}$ may then be calculated for planar flames. It is found [4.73] that $\bar{u}/\tilde{u}$ is less than unity within the flame, the minimum value being about 0.82.

However, different results are obtained ([4.83]; see also Sect. 4.6) if the present eddy viscosity model is replaced by "second-order closure" assumptions.

The present calculations, in which the reacting mixture is made up of packets of fully burned and unburned gas, provide an upper bound to scalar fluctuation effects and show that these effects can be of crucial importance. They also emphasize that a large and significant difference can exist between the Favre and conventional Reynolds average of the same variable. Comparisons between model predictions of flame structure and experiment, therefore, involve an additional uncertainty, of up to almost an order of magnitude; this uncertainty will be resolved only when both modeller and experimenter clearly state whether Favre or conventional averaging is used. Unfortunately, in many cases, it is difficult to determine what type of average is effectively formed by some instruments in combustion experiments. There is no such excuse for the theoretician, yet the neglect of density fluctuations effectively mixes the two types of average [4.73].

4.5.6 Laminar Flamelet pdf Model

As indicated in Sect. 4.4, *Damköhler*'s [4.2] concept of the wrinkled laminar flame has influenced a majority of theoretical studies of premixed turbulent combustion. Nevertheless, two alternative approaches may be found in the literature. The first, based on a rigorous perturbation of the laminar flame solution, is at present too complex for practical application, for example, in engines. The second approach is more empirical. It uses equations for the surface area and propagation speed of a laminar flame, which is assumed to be undisturbed by the turbulent flow. This approach is fundamentally unsatisfactory, because the shape and propagation speed of a laminar flame are both sensitive to the hydrodynamic and diffusion-thermal instabilities reviewed in Sect. 4.3.

As noted earlier (Sect. 4.3), the structure of a laminar flame is much less influenced by the effects of disturbances than are the flame shape and speed. This suggests that a turbulent flame model which assumes the instantaneous flame structure to consist of packets of fully burned and unburned gas, separated by narrow reaction zones with the structure of undisturbed laminar flames, will be less sinsitive to laminar flame instability effects and extend the range of earlier empirical models based on wrinkled flames. We shall see that this assumption allows effects of multiple species and reactions of arbitrary complexity to be included in the a priori pdf analysis. It is assumed throughout the present section that these reactions reach completion within the narrow laminar reaction zones. Complete thermochemical equilibrium, therefore, exists in the fully burned gas packets. A relaxation of this equilibrium assumption is discussed at the end of Sect. 4.5.7. The present treatment may be compared with the analysis of diffusion flames with fast chemistry in Sect. 3.1.3 (see also Sect. 3.1.9).

The starting point for this study is a redefinition of the progress variable c as a normalized temperature, so that

$$c=\frac{T-T_0}{T_\infty-T_0}, \tag{4.92}$$

where T_0 and T_∞ are the temperature upstream and downstream of an undisturbed laminar flame. A calculation for such a flame then yields all species mass fractions and state variables which can be expressed numerically as functions of c, so that

$$Y_i=Y_i(c)\,i=1,2,\ldots,N, \tag{4.93}$$

$$\varrho=\varrho(c) \tag{4.94}$$

[cf. (3.11)]. The chemical kinetics and other information incorporated in this calculation can be highly complex and realistic (see, for example, [4.84]). If it is assumed for the turbulence model that $c_{pi}=c_p(i=1,2,\ldots,N)$ and that $\partial p/\partial t=0$, then the equation for c reduces to (4.22); see (1.16) and the discussion in Sect. 1.6. The laminar flame calculation, therefore, yields $w(c)$ also.

Neglecting pressure fluctuations the equation of state (1.20) gives

$$\frac{\varrho_0}{\varrho}=(1+\tau c)\frac{W_0}{W(c)}, \tag{4.95}$$

where $\tau=(T_\infty/T_0)-1$, as before, and $W(c)$ is the mean molecular weight of the mixture, whose initial value is W_0. Although there is no conceptual difficulty in incorporating the variation in $W(c)$, we shall assume here that $W_0/W(c)=1$ so that (4.27) is recovered. The equations of Sects. 4.5.1–4.5.5 may then be applied without change. We assume also that a pdf $P(c)$ exists and can be decomposed into the three modes: unburned, fully burned, and burning, according to (4.37). The burning mode pdf $f(c)$ can be calculated directly from the assumption that all burning takes place in undisturbed laminar flames. If the distribution of normalized temperature through an undisturbed laminar flame is expressed in the form

$$c=c(x/\delta_L),$$

where x is distance perpendicular to the flame, and δ_L is the flame thickness, then [4.12]

$$f(c)=\left[\frac{dc}{d(x/\delta_L)}\right]^{-1}. \tag{4.96}$$

Note that the pdf thus calculated is influenced by the value chosen for the laminar flame thickness.

Since the laminar flamelets have been assumed to be thin, we have $\gamma \ll 1$ and all the results of the large fluctuation analysis of Sect. 4.5.3 apply without modification. In particular, $\bar{w}$ is related to $\bar{\chi}$ by (4.70) and if $\bar{\chi}$ is modelled, e.g., as in (4.71), $\bar{w}$ may be obtained from (4.72). The only differential equation to be solved for a scalar variable is that [(4.34)] for the Favre mean normalized temperature $\tilde{c}$. As before, all of this is insensitive to the shape chosen for the burning mode pdf, $f(c)$, and hence to the value of δ_L in (4.96).

The individual species mass fractions may now be calculated from

$$\bar{\varrho}\tilde{Y}_i = \int_0^1 Y_i(c)\varrho(c)P(c)dc$$

or using (4.37, 58, 59, 95),

$$\tilde{Y}_i = (1-\tilde{c})Y_{i,0} + \tilde{c}Y_{i,\infty} + \frac{1}{\bar{\varrho}}\gamma\int_0^1 \varrho(c)Y_i(c)f(c)dc\,, \tag{4.97}$$

where $Y_{i,0}$ and $Y_{i,\infty}$ are the values of Y_i corresponding to T_0 and T_∞, respectively. There are two cases to consider. For species occurring in significant concentration in reactants and/or products, we can neglect the last term and write

$$\tilde{Y}_i = (1-\tilde{c})Y_{i,0} + \tilde{c}Y_{i,\infty} + \mathrm{O}(\gamma)\,. \tag{4.98}$$

On the other hand, for reaction intermediates, which the laminar flame calculation shows to be absent in both reactants and products, the last term in (4.97) is the only contribution. From (4.41, 70) we have

$$\gamma = \frac{K_2\bar{\chi}}{w_{\max}I_3}$$

leading to

$$\tilde{Y}_i = \frac{K_2\bar{\chi}}{w_{\max}I_3}(1+\tau\tilde{c})\int_0^1 \frac{Y_i(c)f(c)}{1+\tau c}dc \tag{4.99}$$

for such reaction intermediates. It is necessary to specify $f(c)$ in order to evaluate this equation. Equation (4.96) must be used for this purpose and the results will, therefore, be influenced by value chosen for the laminar flame thickness δ_L. According to (4.99), all reaction intermediates will have the same normalized mass fraction profiles, $\tilde{Y}_i/Y_{i,\max}$, through the flame.

4.5.7 A Joint pdf Model for Consecutive Reactions

In more complex systems, the probability density function appearing in (1.66) cannot be reduced to a function of a single scalar variable. Examples include cases where coupled chemical reactions involving finite reaction rates are to be described, and also where partially premixed, nonadiabatic, or high Mach number flows occur. As many practical situations involve at least one of these complicating factors, methods of analysis are needed. A joint pdf is then required.

In this section, we shall describe a simple decomposition of the joint pdf which occurs [4.85] if the chemical reactions are assumed to occur consecutively, irreversibly, and at different rates. More general approaches to the joint pdf problem are briefly reviewed in Sect. 4.6. For the present purpose the reaction scheme is assumed [4.85] to be of the form

$$X_1 \xrightarrow{k_A} X_2 \xrightarrow{k_B} X_3 . \tag{4.100}$$

It is further assumed that reaction A is much faster than reaction B, so species X_1 and X_3 do not instantaneously coexist at any point in the flow field. This restriction allows the joint pdf to be simplified. The case where reaction B is much faster than A can be treated in the same way.

In reactions (4.100), X_1, X_2, and X_3 may represent either pure species or more commonly mixtures of species in fixed proportions. A practical and simple example is provided by a two-step description of hydrocarbon oxidation: the hydrocarbon C_nH_m is first rapidly burned to form a mixture containing CO, which is then oxidized more slowly in the second reaction. However, see comments in Sect. 1.5 about the possible dangers of global reaction schemes. The fast reaction A is represented as

$$C_nH_m + rO_2 + tN_2 \xrightarrow{k_A} (n-s)CO + \frac{m}{2} H_2O + sCO_2 + tN_2 , \tag{4.101}$$

where

$$r = \frac{n+s}{2} + \frac{m}{4} .$$

The slower reaction B, CO oxidation, is also represented globally; it is written

$$2CO + O_2 \xrightarrow{k_B} 2CO_2 . \tag{4.102}$$

For combustion of a stoichiometric mixture of hydrocarbon and oxygen, with nitrogen as a dilutent, we thus have

$$\begin{aligned} X_1 &= C_nH_m + [r + \tfrac{1}{2}(n-s)]\,O_2 + tN_2 \\ X_2 &= (n-s)\,CO + \frac{m}{2} H_2O + sCO_2 \\ &\quad + \tfrac{1}{2}(n-s)\,O_2 + tN_2 \\ X_3 &= nCO_2 + \frac{m}{2} H_2O + tN_2 . \end{aligned} \tag{4.103}$$

Mean molecular weights may then be calculated for the mixtures, X_k. For X_1 for example

$$W_1 = \frac{W(C_nH_m) + [r + \frac{1}{2}(n-s)]\,W(O_2) + tW(N_2)}{1 + r + \frac{1}{2}(n-s) + t}.$$

We now consider the thermochemistry of the mixture of three species X_i whose mass fractions are Y_i $(i=1, 2, 3)$ and whose heats of formation are $\Delta_1 = 0$, $\Delta_2 = -\Delta_A$, $\Delta_3 = -\Delta_B$. The flow is assumed adiabatic with enthalpy h_0 and the specific heat c_p constant. Since X_1 and X_3 do not coexist locally, we can identify at any instant two distinct regimes, A and B.

Regime A where reaction A occurs with $Y_3 = 0$ and

$$h_0 = c_p T - \Delta_A Y_2 \tag{4.104}$$

so the temperature range extends from an initial value $T_0 = h_0/c_p$ when $Y_1 = 1$, $Y_2 = Y_3 = 0$, to an intermediate value $T_1 = (h_0 + \Delta_A)/c_p$, at which point $Y_2 = 1$ and $Y_1 = Y_3 = 0$. Thus, a progress variable for reaction A can be defined as

$$c_A = (T - T_0)/(T_1 - T_0) = (c_p T - h_0)/\Delta_A \tag{4.105}$$

in the range (0, 1), such that the three mass fractions are

$$\begin{aligned} Y_1 &= 1 - c_A \\ Y_2 &= c_A \\ Y_3 &= 0. \end{aligned} \tag{4.106}$$

Similarly, reaction B occurs in *regime B* where $Y_1 = 0$ and

$$h_0 = c_p T - \Delta_A - (\Delta_B - \Delta_A)\,Y_3. \tag{4.107}$$

The temperature range in this regime is from T_1 to the final adiabatic flame temperature, $T_\infty = (h_0 + \Delta_B)/c_p$, so the progress variable for reaction B is

$$\begin{aligned} c_B &= (T - T_1)/(T_\infty - T_1) \\ &= \frac{c_p T - h_0 - \Delta_A}{\Delta_B - \Delta_A}. \end{aligned} \tag{4.108}$$

The mass fractions are related to c_B by

$$\begin{aligned} Y_1 &= 0 \\ Y_2 &= 1 - c_B \\ Y_3 &= c_B. \end{aligned} \tag{4.109}$$

The two time-dependent progress variables c_A and c_B obey equations of exactly the same form as (4.22) with two chemical source terms, w_A and w_B, respectively, where $w_A = -w_1$ and $w_B = w_3$.

Having expressed the mass fractions in terms of the progress variables, c_A or c_B, we may put the time-dependent gas density and reaction rates in terms of the same variables. In regime A, the equation of state (1.20) gives

$$\varrho = \frac{\bar{p}}{R^0}[T_0 + c_A(T_1 - T_0)]\left(\frac{c_A}{W_2} + \frac{1-c_A}{W_1}\right) = \varrho(c_A) \tag{4.110}$$

and the reaction rate

$$w_A = w_A(\varrho, T, Y_1, Y_2) = w_A(c_A). \tag{4.111}$$

Similarly in regime B

$$\varrho = \frac{\bar{P}}{R^0}[T_1 + c_B(T_\infty - T_1)]\left(\frac{c_B}{W_3} + \frac{1-c_B}{W_2}\right) = \varrho(c_B) \tag{4.112}$$

and

$$w_B = w_B(\varrho, T, Y_2, Y_3) = w_B(c_B). \tag{4.113}$$

Thus, all possible instantaneous states of the mixture have been uniquely expressed in terms of c_A and c_B.

In view of the consecutive nature of the chemistry proposed, the joint pdf $P(c_A, c_B)$ is confined to the planes $c_B = 0$ (regime A) and $c_A = 1$ (regime B) as illustrated in Fig. 4.16. Thus,

$$\begin{aligned} P(c_A, c_B) = {} & \delta(c_B)\{\alpha\delta(c_A) + [H(c_A) - H(c_A - 1)]\gamma f(c_A)\} \\ & + \delta(1 - c_A)\{\beta\delta(c_B) + [H(c_B) - H(c_B - 1)]\psi g(c_B) + \phi\delta(1 - c_B)\} \end{aligned} \tag{4.114}$$

which may be compared with (4.37). The five coefficients represent the probabilities of the corresponding modes, as follows:

α : unburned mixture

γ : reaction A in progress

β : reaction A completed

ψ : reaction B in progress

ϕ : final all burned state.

The burning mode pdfs $f(c_A)$ and $g(c_B)$ are normalized so that

$$\int_0^1 f(c_A)dc_A = \int_0^1 g(c_B)dc_B = 1$$

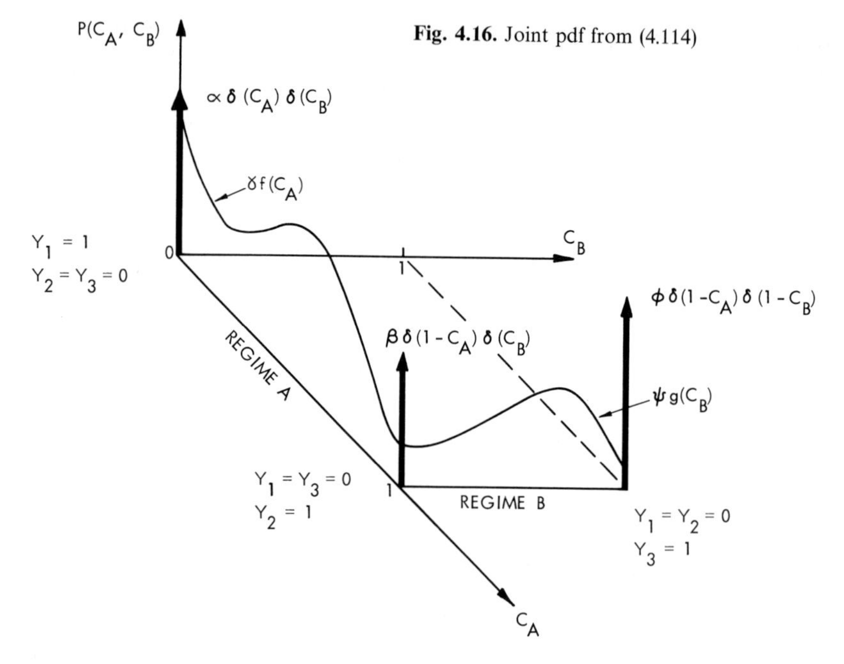

Fig. 4.16. Joint pdf from (4.114)

whence

$$\alpha+\beta+\gamma+\psi+\phi=1. \tag{4.115}$$

Four relationships are, therefore, required to determine the coefficients in the joint pdf. These are specified as follows:

$$\bar{\varrho}\tilde{c}_A=\varrho_1\beta+\varrho_\infty\phi+\gamma\int_0^1 c_A\varrho(c_A)\,f(c_A)dc_A +\psi\int_0^1 \varrho(c_B)\,g(c_B)dc_B, \tag{4.116}$$

$$\bar{\varrho}\tilde{c}_B=\varrho_\infty\phi+\psi\int_0^1 c_B\varrho(c_B)\,g(c_B)dc_B, \tag{4.117}$$

$$\overline{\varrho c_A^2}=\bar{\varrho}\tilde{c}_A^2+\overline{\varrho c_A''^2}=\varrho_1\beta+\varrho_\infty\phi +\gamma\int_0^1 c_A^2\varrho(c_A)\,f(c_A)dc_A+\psi\int_0^1 \varrho(c_B)\,g(c_B)dc_B, \tag{4.118}$$

$$\overline{\varrho c_B^2}=\varrho\tilde{c}_B^2+\overline{\varrho c_B''^2} =\varrho_\infty\phi+\psi\int_0^1 c_B^2\varrho(c_B)\,g(c_B)\,dc_B, \tag{4.119}$$

where $\varrho_1 = \varrho(c_A = 1)$, evaluated from (4.110). Equations (4.115–119) permit α, β, γ, ψ, and ϕ to be found in terms of $\tilde{c}_A$, $\tilde{c}_B$, $\overline{\varrho c_A''^2}$, and $\overline{\varrho c_B''^2}$ which must be calculated from balance equations; see (4.34, 35).

In view of the assumption that species X_1 and X_3 do not coexist, it is pertinent to look particularly at the case where reaction A is fast. If the Damköhler number for this reaction is much greater than unity then, as in Sect. 4.5.3, the reaction zones will be thin and γ will be small. With $\gamma \ll 1$, the right-hand sides of (4.116, 118) are equal, so that

$$\overline{\varrho c_A''^2} = \bar{\varrho}\tilde{c}_A(1-\tilde{c}_A) + \mathrm{O}(\gamma) \tag{4.120}$$

which is of the same form as (4.40) with $\gamma \ll 1$. Consequently, the analysis of Sect. 4.5.3 with $\gamma \ll 1$ may be applied directly. In particular from (4.70)

$$\bar{w}_A = K_2 \bar{\chi}_A, \tag{4.121}$$

where

$$\bar{\chi}_A = \overline{2\varrho\mathscr{D}\frac{\partial c_A}{\partial x_m}\frac{\partial c_A''}{\partial x_m}}$$

and $K_2 = 1/(2c_\mathrm{m} - 1)$ with in this case

$$c_\mathrm{m} = \frac{\int_0^1 \frac{c_A w_A(c_A)}{w_{A,\max}} f(c_A) dc_A}{\int_0^1 \frac{w_A(c_A)}{w_{A,\max}} f(c_A) dc_A}.$$

As before (see Tables 4.1, 4.2) c_m and K_2 are insensitive to the shape of $f(c_A)$. The model for reaction A is completed by assuming [cf. (4.71, 120)]

$$\bar{\chi}_A = C_A \tilde{q}^{1/2} \bar{\varrho}\tilde{c}_A(1-\tilde{c}_A)/l_A, \tag{4.122}$$

where C_A is a constant and l_A is a length scale of the scalar field. As suggested earlier, the validity of this equation in these circumstances is unproven; see Sect. 1.13.

Turning to reaction B, we recognize that there will, in general, be contributions to w_B from the delta function at $c_B = 0$, whose weighting is β, and from the distribution $0 < c_B \leqq 1$, weighting ψ. Thus,

$$\frac{\bar{w}_B}{w_{B,\max}} = \beta W_0 + \psi I_3, \tag{4.123}$$

where $W_0 = w_B(c_B = 0)/w_{B,\max}$ and

$$I_3 = \int_0^1 \frac{w_B(c_B)}{w_{B,\max}} g(c_B) dc_B.$$

If the weights β and ψ are expressed in terms of $\tilde{c}_A$, $\tilde{c}_B$, and $\overline{\varrho c_A''^2}$ (4.123) becomes

$$\frac{\bar{w}_B}{w_{B,\max}}=K_1\frac{\bar{\varrho}}{\varrho_1}\tilde{c}_B(1-\tilde{c}_B)\left[1-\frac{\overline{\varrho c_B''^2}}{\bar{\varrho}\tilde{c}_B(1-\tilde{c}_B)}\right]$$
$$+W_0\frac{\bar{\varrho}}{\varrho_1}\left\{(\tilde{c}_A-\tilde{c}_B)-K_3\tilde{c}_B(1-\tilde{c}_B)\left[1-\frac{\overline{\varrho c_B''^2}}{\bar{\varrho}\tilde{c}_B(1-\tilde{c}_B)}\right]\right\}, \tag{4.124}$$

where $K_1=I_3/M$ and $K_3=N/M$ are ratios of moments of the pdf $g(c_B)$ with

$$M=\int_0^1 c_B(1-c_B)\frac{\varrho(c_B)}{\varrho_1}g(c_B)dc_B$$
$$N=\int_0^1 (1-c_B)\frac{\varrho(c_B)}{\varrho_1}g(c_B)dc_B\,.$$

As before K_1 and K_3 are insensitive to the shape chosen for $g(c_B)$. If the initial reaction rate $w_B(c_B=0)$ is sufficiently small, the term on the right-hand side of (4.124) containing W_0 may be neglected. Equation (4.124) is then the same as (4.45) and Fig. 4.9 applies.

The modeling of sequential reactions A and B is now complete. *Champion* et al. [4.86] use a similar technique to model propane-air combustion, but incorporate more realistic chemical kinetics by introducing a partial equilibrium assumption.

Nitric oxide production may be treated similarly [4.87], so long as it is assumed that the nitric oxide is formed only in the products of combustion. The combustion and nitric oxide reactions may then be regarded as sequential. If c is the progress variable for the combustion reaction, and Y_{NO} is the nitric oxide mass fraction, then the joint pdf is written

$$P(c,Y_{NO})=P(Y_{NO}|c)P(c)\,,$$

where $P(Y_{NO}|c)$ is the conditional pdf for Y_{NO} given c. With nitric oxide formed only in the combustion product, the instantaneous formation rate is $w_{NO}(c=1, Y_{NO})$. Its average is simply

$$\bar{w}_{NO}=\beta\int_0^{Y_{NO,e}} w_{NO}(c=1,Y_{NO})\,P(Y_{NO}|c=1)\,dY_{NO}\,, \tag{4.125}$$

where $w_{NO}(c=1, Y_{NO})$ is the local rate of nitric oxide formation when $c=1$ (i.e., when $T=T_\infty$), β is the weight factor for fully burned mixture [see (4.37)] given approximately by

$$\beta=\frac{(1+\tau)\tilde{c}}{1+\tau\tilde{c}}+\mathrm{O}(\gamma)$$

if $\gamma \ll 1$, and $Y_{NO,e}$ is the equilibrium value of Y_{NO}. Equation (4.125) allows $\bar{w}_{NO}$ to be calculated if the conditional pdf is specified.

If $Y_{NO}/Y_{NO,e}$ is sufficiently small to allow $w_{NO}(c=1, Y_{NO})$ to be linearized then is is easy to show [4.87] that with Favre averaging,

$$\bar{w}_{NO} = w_{NO,e} \frac{(1+\tau)\tilde{c}}{1+\tau\tilde{c}} \left[1 - K\tilde{c}\left(\frac{1+\tau}{1+\tau\tilde{c}}\right) \frac{\tilde{Y}_{NO}}{\tilde{Y}_{NO,e}}\right] \tag{4.126}$$

for the kinetic scheme [4.88, 89]

$$N + NO \overset{k_1}{\rightleftarrows} 1\, N_2 + O$$

$$N + O_2 \overset{k_2}{\rightleftarrows} 2\, NO + O$$

$$N + OH \overset{k_3}{\rightleftarrows} 3\, NO + H\,,$$

where

$$w_{NO,e} = 2\varrho_\infty k_1 Y_{N,e} Y_{NO,e}$$

and

$$K = \frac{k_1 Y_{NO,e}}{k_2 Y_{O_2,e} + k_3 Y_{OH,e}}$$

are constants, subscript e denotes equilibrium, and the rate coefficients k_1, k_2, and k_3 are evaluated at temperature T_∞. The pdf $P(Y_{NO}|c=1)$ does not appear in (4.126).

We now discuss circumstances in which the consecutive reaction model of this section can be combined with the laminar flamelet approach of Sect. 4.5.6. To do so we must relax the assumption (Sect. 4.5.6) that complete thermochemical equilibrium exists in the fully burned gas packets. It is replaced by the following two assumptions:

1) all thermally significant reactions reach completion within the thin laminar flamelets;

2) a slower and essentially isothermal chemical reaction is allowed to occur only within the burned gas.

The chemical kinetic mechanism is then divided into two consecutive steps; the first complex reaction occurs within the laminar flamelet and reaches completion before the second slower reaction begins. Thus, the analysis of the present section can be applied.

This combined model is capable, for example, of accommodating a multi-reaction mechanism of hydrocarbon combustion within the laminar flamelets, followed by a conventional mechanism [4.88, 89] for nitric oxide formation in the burned gas packets.

4.6 Discussion and Concluding Remarks

In this chapter we have been largely concerned with two fundamental problems

I) turbulent reaction rates: the effects of turbulence-induced fluctuations in thermochemical state on time-average reaction rates, and
II) turbulent transport in reacting flows: applicability of empiricism developed for constant density flows and the influence of density fluctuations and flame-generated turbulence.

Experiments confirm that large scalar fluctuations do occur in premixed turbulent flames, and that mean reaction rates are affected. However, details of the instantaneous flame structure, permitting assessment of the applicability of descriptions such as *Damköhler*'s [3.2] wrinkled laminar flame or *Chomiak*'s [3.9] dissipative vortex are not available from experiment.

Fortunately, as we have seen, an analysis can be constructed which does not depend critically on this information. The a priori pdf method provides a simple description of the effects of turbulence on a single, global combustion reaction, which explains the main trends of much experimental data, including the appearance of a region of mixing limited reaction. The wrinkled laminar flame version of this analysis (Sect. 4.5.6) can incorporate chemical schemes of unlimited complexity and yet relates all species concentrations, including reaction intermediates, to the mean local temperature through the results of separate calculations for an undisturbed laminar flame [(4.97, 98)]. If post-flame chemistry is incorporated via one or more sequential reactions (Sect. 4.5.7) a complete description of the turbulent chemistry becomes possible in appropriate circumstances but predictions of this treatment await experimental test.

However, none of the methods described in Sect. 4.5 is adequate to treat more realistic situations, in which chemical kinetic rate expressions cannot be reduced to functions of a single progress variable, and where reactions do not occur consecutively. A joint pdf is thus required for the relevant concentrations and thermodynamic state variables. The a priori pdf method then involves postulating a general empirical expression for a joint pdf whose shape depends on a suitable number of parameters. The parameters are related to moments of the joint pdf representing mean concentrations, fluctuation intensities, and covariances, which are determined from balance equations similar to (4.34) and (4.35).

The joint pdf is, however, restricted to a finite domain in the phase space of the random variables [4.90, 91]. In the case of a premixed flame with a two-step reaction the domain of the two-dimensional pdf is limited [4.90, 91] by paths which represent pure mixing and pure reaction, respectively. Circumstances arise where these two paths are close to each other and the available domain is then reduced to a narrow corridor. *Champion* [4.92] exploited this fact in an analysis of turbulent combustion of premixed propane and air.

Another joint pdf method has been proposed by *Lockwood* [4.72] who treats the case of a partially premixed flow in which the mixture state is regarded as a function of two time-dependent variables: a measure of fuel-air mixing, the mixture fraction f (cf. Chap. 3), and the reactedness r, which is analogous to our progress variable c. Both f and r range from zero to unity. *Lockwood* assumed that the joint pdf $P(f,r)$ can be factored so that

$$P(f,r)=P_{\mathrm{f}}(f)\,P_{\mathrm{r}}(r)\,. \tag{4.127}$$

The time-averaged value of any fluid property $g(f,r)$ is then obtained as

$$\bar{g}=\int_0^1\int_0^1 g(f,r)\,P_{\mathrm{f}}(f)\,P_{\mathrm{r}}(r)\,df\,dr\,.$$

However, since the pdf given by (4.127) ensures that the covariance $\overline{f'r'}$ is identically zero, this approach, although algebraically convenient, cannot account properly for the fluctuation effects illustrated by (1.57) and must be regarded as unrealistic.

The "typical eddy" treatment of *Donaldson* and co-workers [4.93, 94] postulated a joint pdf consisting of Dirac delta functions at fixed locations $Y_s(s=1,2,\ldots)$ in composition space. The strengths α_s of these delta functions are parameters, variable in physical space, to be determined from appropriate moments. A simple description of turbulent chemistry results, in which a mean reaction rate is

$$\bar{w}_i=\sum_s \alpha_s w_i(Y_s)\,.$$

Coupled reactions in turbulent flow can be described in this manner [4.95] and the method can in principle be extended to kinetic schemes of arbitrary complexity. However, as discussed at the beginning of Sect. 4.5, the number of differential equations to be solved increases rapidly with number of species, and it seems unlikely that realistic descriptions of the complex chemical kinetics of combustion reactions will be modelled in this way.

Our earlier discussion of problems where the pdf is a function of a single progress variable shows that moments of delta function pdfs can be strongly dependent on the location of the single delta function (see Tables 4.1 and 4.2). A typical eddy joint pdf may contain seven delta functions [4.94], but at least three of these are in states where no reaction occurs, so typically only one delta function per composition coordinate leads to reaction, as in Tables 4.1 and 4.2. Typical eddy predictions must, therefore, be expected to be sensitive to the locations of the constituent delta functions, particularly in the practically important case where the chemistry is fast. A possible cure for this difficulty lies in the use of alternative "building blocks" in probability space, which are more widely distributed than the delta functions of the typical eddy. However, the number of parameters and balance equations are still prohibitively large, if a realistic description of, say, hydrocarbon combustion is required. See Chap. 3 for related discussion of the "typical eddy" treatment in the context of nonpremixed combustion.

It has often been pointed out that, from the point of view of molecules taking part in a chemical reaction, turbulent combustion might be more readily described as a Lagrangian process rather than a Eulerian one. Stirred reactor models [4.96, 97] identify time as the independent variable. By calculating the time histories of postulated "gas particles", under the influence of empirical mixing rate expressions, they can accommodate effects of complex chemical kinetic schemes in the presence of unmixedness and scalar fluctuations. However, to do so they must sacrifice spatial information and trade one set of problems for another set.

Several flow field models combine Lagrangian and Eulerian features with the aim of accommodating complex kinetics. One such method is *Spalding*'s "*escimo*". Formidable problems remain to be overcome before this technique can be applied to practical problems. Another method, due to *Felton* et al. [4.98], combined a finite-difference prediction of the three-dimensional flow pattern in a gas turbine combustor with a network of interconnected stirred and plug-flow reactors. In the stirred reactors the "gas particles" have an exponential distribution of residence times whereas, for the plug flow reactors, all particles entering together also leave together. Evaporation of the polydispersed liquid fuel is represented in the reactors, together with a detailed multi-reaction chemical kinetic scheme. In such models, a problem always arises in the relationship between the Lagrangian and Eulerian descriptions. Also, as pointed out elsewhere [4.10], an a priori pdf approach can often provide an equivalent description in a more straightforward manner.

We conclude that much work remains to be done before the effects of coupled chemical reactions or nonadiabatic flow in turbulent flames can be predicted with confidence. Meanwhile, the treatment of important practical problems such as hydrocarbon emissions from engines remains highly empirical.

The second problem, which has occupied a smaller part of this chapter, because it is less well understood, is the effect of combustion on turbulent transport. As discussed earlier, *Sivashinsky* [4.43, 49] studied the process of transition of a laminar flame towards turbulence, in relatively simple configurations which eliminate time-mean shear. His analysis implies that the flame does generate turbulence in these circumstances and that the mechanism depends on the Lewis number of the limiting reactant. The consequences of this generation must appear in the turbulence kinetic energy balance, (4.76), and the most likely source is the velocity/pressure gradient correlation, term (VI). No theoretical studies are known of transition in flames with mean shear.

For fully developed turbulence there appear to be no experimental indications that the Lewis number of the limiting reactant plays the same important role as in transitional flames. Experiment shows relatively modest changes in turbulence velocities in flames, including growth by factors of two or three; (4.74) implies that such growth will influence turbulent fluxes in the same ratio. There is no direct experimental support for the generation of intense turbulence as proposed, for example, by *Scurlock* and *Grover* [4.57]. Thus, in many circumstances, the growth of turbulence kinetic energy in a flame appears to be mainly due to a balance between generation due to shear and removal due to dissipation and dilatation [see the TKE balance equation (4.76)]. However, term (vi) in this equation is

$$-\overline{u_k''\frac{\partial p}{\partial x_k}} = -\overline{u_k''}\frac{\partial \bar{p}}{\partial x_k} - \overline{u_k''\frac{\partial p'}{\partial x_k}} \tag{4.128}$$

and the contribution $-\overline{u_k''}(\partial\bar{p}/\partial x_k)$, which would be zero in constant density flows, can be shown to represent a positive source of TKE in a one-dimensional turbulent flame. No experimental evidence is known from premixed flames in externally imposed pressure gradients but experiments [4.99] on turbulent diffusion flames in pressure gradients suggest that this term may significantly influence the TKE and hence the turbulent transport[1].

Situations can be found in which other pressure-related processes become important. For example, baffle-stabilized, premixed flames in ducts can exhibit a violent low-frequency instability known as "buzz", in which the flame shape appears to be distorted by large coherent structures [4.100], and the mean spreading rate is greatly increased. Pressure and velocity fluctuations are highly correlated [4.100] and the increase in turbulence may be traced to term (vi) in (4.76). The incorporation of these effects into turbulent combustion models will be difficult to achieve; see *Marble* and *Candel* [4.101] for a recent analysis, in which the turbulent flame is treated as a moving flame sheet. A full description will allow for interactions between turbulence, combustion, and duct acoustics.

Even if flows with large-scale instabilities and acoustic coupling are excluded from consideration, there is still the basic question of whether empirical equations or models developed for constant density, nonreacting flows can be applied without modification to turbulent combustion. Models are generally required for quantities such as the turbulent eddy viscosity ν_T [see (4.74)] and the scalar dissipation function $\bar{\chi}$, which plays an important part in turbulent chemistry [see (4.70)]. There is a widely held belief that the use of Favre averaging automatically takes account of all the effects of density fluctuations and chemical reactions, and, therefore, allows empiricism from constant density flows to be exploited without change. While this may indeed be true in some cases, it can be justified only by experiment, and the experimental evidence is incomplete. Meanwhile, as emphasized elsewhere in this volume, skepticism is called for regarding the application to combustion of constant density flow models, with or without Favre averaging. The following two examples illustrate the growing evidence for this skepticism.

---

1 A recent unpublished study by *Bray*, *Moss*, and *Libby* of a planar premixed flame suggests that the flame induced mean pressure gradient leads to significant generation of turbulence via the same mechanism.

In a recent theoretical study, *Borghi* and *Dutoya* [4.102] investigated the influence of premixed combustion reactions on turbulent transport. They set up a balance equation for the turbulent mass flux $\overline{\varrho u_m'' Y_i''}$, model the chemical reaction term $\overline{\varrho u_m'' w_i}$, and looked at the case of quasi-homogeneous turbulence; a gradient transport expression is recovered. From this analysis they predicted an increase in the eddy diffusion coefficient, on the fresh gas side of the flame as a direct result of the chemical reaction and a larger decrease in the eddy diffusion coefficient on the burned gas side. *Borghi* and *Dutoya* also formulated and solved a model balance equation for the scalar dissipation function $\bar{\chi}$ in a premixed turbulent flame. They presented their results in terms of a Taylor microscale for composition fluctuations and predicted that this length scale will be increased by a factor of about two as a result of combustion. These effects are absent in constant density nonreacting flows and cannot be predicted unless the constant density flow models are modified.

The second example suggests that the conventional gradient transport expression (Sect. 1.13) may not even predict the sign of the turbulent transport flux correctly in premixed flames, i.e., that countergradient diffusion may occur. Evidence of this comes from a "second-order closure" model of a one-dimensional flame [4.83] employing a balance equation for the flux $\overline{\varrho u'' c''}$, with closure empiricism derived from (4.62–66). At small values of the heat release parameter τ in (4.25), the new model predicts a flux distribution in satisfactory agreement with that obtained earlier from a gradient transport assumption. However, with larger τ and hence larger density fluctuations, significant countergradient diffusion is predicted within the flame. The effect is shown to be caused by a term $\overline{c''} d\bar{p}/dx$ in the balance equation for $\overline{\varrho u'' c''}$, which is zero in the absence of density fluctuations. This prediction is confirmed by a recent preliminary experimental study of turbulent transport in open premixed flames [4.103], which provides direct experimental evidence of countergradient diffusion.

We therefore conclude that the gradient transport assumption (Sect. 1.13), and other empiricism derived from cold constant density turbulent flows, may lead to serious errors if applied to combustion.

Premixed turbulent combustion is often regarded as a relatively simple testbed, on which new ideas about turbulent reacting flows can be assessed and validated for application to more complex combustion systems. As we have seen, however, it provides formidable challenges not only for modellers but, particularly, for those concerned with experiments. Can the contradictions, which are so apparent in the experimental literature, be resolved? Can experiments be made which will provide a genuine and searching test of the claims of the many competing models? To provide affirmative answers to these questions will call for significant effort.

References

4.1 R.E.W.Jansson, M.Fleischmann: AIChE Symposium Series: Electro-organic Synthesis Technology. AIChE National Meeting; Atlanta, Ga. (1978)

4.2 G. Damköhler: Z. Elektrochem. **46**, 601–626 (1940)
4.3 B. Lewis, G. von Elbe: *Combustion, Flames, and Explosions of Gases*, 2nd ed. (Academic, New York 1961)
4.4 F. A. Williams: *Combustion Theory* (Addison-Wesley, Reading, Mass. 1965)
4.5 G. E. Andrews, D. Bradley, S. B. Lwakabamba: Combust. Flame **24**, 285–304 (1975)
4.6 R. Abdel-Gayed, D. Bradley: In *Sixteenth Symposium (International) on Combustion* (The Combustion Institute, Pittsburgh 1977) pp. 1725–1735
4.7 P. A. Libby, F. A. Williams: Annu. Rev. Fluid Mech. **8**, 351–376 (1976)
4.8 R. J. Tabaczynski: Prog. Energy Combust. Sci. **2**, 143–165 (1976)
4.9 J. Chomiak: Combust. Flame **15**, 319–321 (1970)
4.10 K. N. C. Bray: In *Seventeenth Symposium (International) on Combustion* (The Combustion Institute, Pittsburgh 1979) pp. 223–233
4.11 K. N. C. Bray, J. B. Moss: "A Unified Statistical Model of the Premixed Turbulent Flame", University of Southampton Report AASU No. 335 (1974)
4.12 K. N. C. Bray, J. B. Moss: Acta Astron. **4**, 291–320 (1977)
4.13 M. D. Fox, F. J. Weinberg: Proc. R. Soc. London A **268**, 222–239 (1962)
4.14 J. H. Grover, E. N. Fales, A. C. Scurlock: In *Ninth Symposium (International) on Combustion* (Academic, New York 1963) pp. 21–35
4.15 J. Vinckier, A. van Tiggelen: Combust. Flame **12**, 561–568 (1968)
4.16 T. Suzuki, T. Hirano, H. Tsuji: In Ref. 4.10, pp. 289–297
4.17 A. Yoshida, H. Tsuji: In Ref. 4.10, pp. 945–956
4.18 A. Yoshida, R. Günther: "Experimental Investigation of Thermal Structure of Turbulent Premixed Flames", American Institute of Aeronautics and Astronautics, 18th Aerospace Sciences Meeting, Pasadena, Calif. (Jan. 1980)
4.19 Y. Mizutani, T. Nakayama, T. Yuminaka: Combust. Flame **25**, 5–14 (1975)
4.20 R. Kleine, R. Günther: "Flow Field and Turbulent Flame Velocity in Bunsen Flames", *Proc. Deuxième Symp. Européen sur la Combustion* (The Combustion Institute, Section Française 1975) pp. 617–622
4.21 F. Durst, R. Kleine: Gas Wärme Int. **22**, 484–492 (1973)
4.22 F. H. Wright, E. E. Zukoski: In *Eighth Symposium (International) on Combustion* (Williams and Wilkins, Baltimore 1962) pp. 933–943
4.23 N. M. Howe, Jr., C. W. Shipman, A. Vranos: In Ref. 4.14, pp. 36–47
4.24 K. J. Lewis, J. B. Moss: In Ref. 4.10, pp. 267–277
4.25 G. C. Williams, H. C. Hottel, A. C. Scurlock: In *Third Symposium on Combustion and Flame and Explosion Phenomena* (Williams and Wilkins, Baltimore 1949) pp. 21–40
4.26 P. Moreau: AIAA 15th Aerospace Sciences Mtng., Los Angeles, Calif. (1977)
4.27 P. Moreau, A. Boutier: In Ref. 4.6, pp. 1747–1756
4.28 A. A. Westenberg, J. L. Rice: Combust. Flame **3**, 459–465 (1959)
4.29 D. R. Ballal, A. M. Lefebvre: Proc. R. Soc. London A **357**, 163–181 (1977)
4.30 R. G. Abdel-Gayed, D. Bradley, M. McMahon: In Ref. 4.10, pp. 245–254
4.31 G. E. Andrews, D. Bradley, S. B. Lwakabamba: In *Fifteenth Symposium (International) on Combustion* (The Combustion Institute, Pittsburgh 1975) pp. 655–664
4.32 F. W. Bowditch: In *Fourth Symposium (International) on Combustion* (Williams and Wilkins, Baltimore 1953) pp. 674–682
4.33 K. O. Smith, F. C. Gouldin: "Experimental Investigation of Flow Turbulence Effects on Premixed Methane-Air Flames" in *Turbulent Combustion*, Progress in Astronautics and Aeronautics, Vol. 58, ed. by L. A. Kennedy (American Institute of Aeronautics and Astronautics, 1978)
4.34 D. R. Ballal, A. H. Lefebvre: Proc. R. Soc. London A **344**, 217–234 (1975)
4.35 A. H. Lefebvre, R. Reid: Combust. Flame **10**, 355–366 (1966)
4.36 P. A. Libby, K. N. C. Bray, J. B. Moss: Combust. Flame **34**, 285–301 (1979)
4.37 A. Palm-Leis, R. A. Strehlow: Combust. Flame **13**, 111–129 (1969)
4.38 G. Dixon-Lewis, I. G. Shepherd: In Ref. 4.31, pp. 1483–1491
4.39 R. A. Strehlow: *Fundamentals of Combustion* (International Textbook, Scranton, Penn. 1968)

4.40 B.Karlovitz, D.W.Denniston,Jr., D.H.Knapschaefer, F.E.Wells: In Ref. 4.32, pp. 613–620
4.41 A.M.Klimov: Zh. Prikl. Mekh. Tekh. Fiz. **3**, 49–58 (1963)
4.42 F.A.Williams: "A Review of Some Theoretical Considerations of Turbulent Flame Structure" in *Analytical and Numerical Methods for Investigation of Flow Fields with Chemical Reactions, Especially Related to Combustion* (AGARD Conf. Proc. 164, NATO, Paris 1975) pp. II 1-1 to II 1–25
4.43 G.I.Sivashinsky: Acta Astron. **4**, 1177–1206 (1977)
4.44 L.D.Landau: Acta Physicochim. URSS **19**, 77 (1944)
4.45 G.H.Markstein: J. Aeronaut. Sci. **18**, 199–209 (1951)
4.46 G.I.Sivashinsky: Acta Astron. **3**, 889–918 (1976)
4.47 T.Mitani: "Studies of Premixed Hydrogen Flames", Ph.D. Thesis, University of California, San Diego (1979)
4.48 G.H.Markstein: *Nonsteady Flame Propagation* (Pergamon, Oxford 1964)
4.49 G.I.Sivashinsky: Acta Astron. **6**, 569–592 (1979)
4.50 J.Buckmaster: Combust. Flame **28**, 225–239 (1977)
4.51 F.A.Williams: Combust. Flame **26**, 269–270 (1976)
4.52 M.Summerfield, S.H.Reiter, V.Kebely, R.W.Mascolo: Jet Propul. **25**, 377 (1955)
4.53 J.Chomiak: In Ref. 4.6, pp. 1665–1673
4.54 F.A.Williams: J. Fluid Mech. **40**, 401–421 (1970)
4.55 K.N.C.Bray, P.A.Libby: Phys. Fluids **19**, 1687–1701 (1976)
4.56 B.Karlovitz, D.W.Denniston, F.E.Wells: J. Chem. Phys. **19**, 541–547 (1951)
4.57 A.C.Scurlock, J.H.Grover: In Ref. 4.32, pp. 645–658
4.58 N.M.Howe,Jr., C.W.Shipman: In *Tenth Symposium (International) on Combustion* (The Combustion Institute, Pittsburgh 1965) pp. 1139–1149
4.59 B.S.Cushing, J.E.Faucher, S.Gandbhir, C.W.Shipman: In *Eleventh Symposium (International) on Combustion* (The Combustion Institute, Pittsburgh 1967) pp. 817–824
4.60 P.Clavin, F.A.Williams: J. Fluid Mech. **90**, 589–604 (1979)
4.61 K.I.Shelkin: Zh. Tekh. Fiz. (USSR) **13**, 520–530 (1943), [English translation, NACA Tech. Memo. No. 1110 (1947)]
4.62 D.B.Spalding: In Ref. 4.59, pp. 807–815
4.63 D.B.Spalding: In *Thirteenth Symposium (International) on Combustion* (The Combustion Institute, Pittsburgh 1971) pp. 649–657
4.64 H.B.Mason, D.B.Spalding: In *Combustion Institute European Symposium*, ed. by F.J.Weinberg (Academic, New York 1973) pp. 601–606
4.65 C.Dopazo: Combust. Flame **34**, 99–101 (1979)
4.66 S.B.Pope: Combust. Flame **34**, 103–105 (1979)
4.67 S.B.Pope: Philos. Trans. R. Soc. London A **291**, 529–568 (1979)
4.68 S.B.Pope: J. Non-Equilib. Thermo. **4**, 309–320 (1979)
4.69 S.B.Pope: "Probability Distributions of Scalars in Turbulent Shear Flow", in *Proceedings Second Symposium on Turbulent Shear Flows*, ed. by L.J.S.Bradbury et al., Imperial College, London, July, 1979
4.70 R.Borghi, P.Moreau, C.Bonniot: "Theoretical Predictions of a High Velocity Premixed Turbulent Flame", presented at Levich Birthday Conference, Oxford, July, 1977
4.71 R.Borghi, P.Moreau: Acta Astron. **4**, 321–341 (1977)
4.72 F.C.Lockwood: Combust. Flame **29**, 111–122 (1977)
4.73 P.A.Libby, K.N.C.Bray: AIAA J. **15**, 1186–1193 (1977)
4.74 P.A.Libby, K.N.C.Bray: Combust. Flame, to appear (1980)
4.75 D.B.Spalding: Chem. Eng. Sci. **26**, 95–107 (1971)
4.76 D.B.Spalding: In Ref. 4.14, pp. 833–843
4.77 K.N.C.Bray: "Equations of Turbulent Combustion I. Fundamental Equations of Reacting Turbulent Flow", University of Southampton Report AASU No. 330 (1973)
4.78 K.N.C.Bray: "Equations of Turbulent Combustion II. Boundary Layer Approximation", University of Southampton Report AASU No. 331 (1973)

4.79 B.E.Launder, A.Morse, W.Rodi, D.B.Spalding: "The Prediction of Free Shear Flows – A Comparison of Six Turbulence Models" NASA Free Shear Flows Conf., Virginia, NASA Rept. No. SP-311 (1972)
4.80 K.N.C.Bray: "Kinetic Energy of Turbulence in Flames", in *Numerical Methods for Investigation of Flow Fields with Chemical Reactions* (AGARD Conf. Proc. 164, NATO, Paris, 1975) pp. II 2-1 to II 2-20
4.81 I.Gokalp: Acta Astron. **6**, 847–860 (1979)
4.82 V.B.Librovich, V.I.Lisitzyn: AIAA J. **15**, 227–233 (1977)
4.83 P.A.Libby, K.N.C.Bray: "Counter-Gradient Diffusion in Premixed Turbulent Flames". American Institute of Aeronautics and Astronautics, 18th Aerospace Sciences Meeting, Pasadena, Calif., Jan. 1980
4.84 G.Tsatsaronis: Combust. Flame **33**, 217–239 (1978)
4.85 K.N.C.Bray, J.B.Moss: Combust. Flame **30**, 125–131 (1977)
4.86 M.Champion, K.N.C.Bray, J.B.Moss: Acta Astron. **5**, 1063–1077 (1978)
4.87 J.B.Moss, K.N.C.Bray: "A Statistical Model of NO Formation in Premixed Turbulent Flames", in Ref. 4.20, pp. 315–320
4.88 J.B.Heywood, J.A.Fay, L.H.Linden: AIAA J. **9**, 841–850 (1971)
4.89 G.A.Lavoie, J.B.Heywood, J.C.Keck: Combust. Sci. Tech. **1**, 313–326 (1970)
4.90 C.Bonniot, R.Borghi: Acta Astron. **6**, 309–327 (1979)
4.91 R.Borghi: "Models of Turbulent Combustion for Numerical Predictions", presented at Conference Methodes de Prévision pour les Ecoulements Turbulents, Institut von Karman de Dynamique des Fluides, Rhode Saint Genèse, Belgium, Jan. 1979
4.92 M.Champion: "Premixed Turbulent Combustion Controlled by Complex Chemical Kinetics". Submitted to Combustion Science and Technology, 1979
4.93 C.du P.Donaldson: "On the Modeling of the Scalar Correlations Necessary to Construct a Second Order Closure Description of Turbulent Reacting Flows", in *Turbulent Mixing in Nonreactive and Reactive Flows*, ed. by S.N.B.Murthy (Plenum, New York 1975) pp. 131–162
4.94 E.S.Fishburne, A.K.Varma: "Investigations of Chemical Reactions in Turbulent Media", paper presented at Sixth International Colloquium on Gasdynamics of Explosions and Reactive Systems, Stockholm, Sweden, 22–26 August 1977
4.95 D.J.Kewley: "A Model of the Supersonic HF Chemical Laser Including Turbulence Effects on the Chemistry", in *Gasdynamic and Chemical Lasers*, ed. by M.Fiebig, H.Hugel (DFVLR-Press, Köln-Porz 1976) pp. 212–223
4.96 D.T.Pratt: In Ref. 4.31, pp. 1339–1351
4.97 J.J.Wormeck, D.T.Pratt: In Ref. 4.6, pp. 1583–1592
4.98 P.G.Felton, J.Swithenbank, A.Turan: "Progress in Modelling Combustors", University of Sheffield, Dept. of Chemical Engineering and Fuel Technology, Rept. HIC 300 (1978)
4.99 S.H.Starner, R.W.Bilger: Combust. Sci. Technol. **21**, 259–276 (1980)
4.100 A.E.Smart, B.Jones, N.T.Jewell: "Measurements of Unsteady Parameters in a Rig Designed to Study Reheat Combustion Instabilities" AIAA Paper No. 76-141, AIAA 14th Aerospace Sciences Mtng. Washington D.C., 1976
4.101 F.E.Marble, S.M.Candel: In Ref. 4.10, pp. 761–769
4.102 R.Borghi, D.Dutoya: In Ref. 4.10, pp. 235–244
4.103 J.B.Moss: "Simultaneous Measurements of Concentration and Velocity in an Open Premixed Turbulent Flame", Combust. Sci. Technol. **22**, 115–129 (1979)

5. The Probability Density Function (pdf) Approach to Reacting Turbulent Flows

E. E. O'Brien

This chapter deals only with the probability density function (pdf) method and within that constraint is limited entirely to the equation of evolution of the pdf as obtained from the conservation laws of a reactive flow system. The subject is new and currently being developed, hindered mainly by a lack of solutions to the turbulent transport problem, and it is impossible at this time to bring it into sharp focus in every detail. In Sect. 5.1 one can find a short historical view of its application to reactive flows, followed in Sect. 5.2 by a derivation of the pdf equation for a typical system. Some closures of the equation have been proposed (Sect. 5.3) but few have been experimentally tested. Nevertheless applications of the method have been attempted and are described in Sect. 5.4. In Sect. 5.5 an effort is made to summarize the present state of research in this field in the hope that it might serve as a spur towards clarification of the many hazy contours in this outline.

5.1 Strategy and Early Developments

Chemically reactive turbulent flows are difficult to predict for the simple reason that neither turbulent transport nor chemical kinetics, the two important phenomena which form their base, are themselves adequately understood. In this chapter we discuss a method of description of such flows which attempts to connect, as efficiently as possible, combined chemical kinetics and turbulent transport to the existing state of knowledge of kinetics on the one hand and turbulent transport on the other. We shall not concern ourselves with important questions of the adequacy of either turbulence theory and approximation or the reliability of chemical kinetic information. We shall focus instead on a recently developed method well suited to combine these two fields and we shall attempt to discern its potential in turbulent reacting flow research, by emphasizing those terms for which approximations are needed and by reviewing what has so far been achieved through its use.

We adopt the common and reasonable assumption that the chemical kinetic scheme describing the reaction set under consideration is known and valid at every instant and spatial location. Thus, it is supposed that there is no disruption by the turbulence of the chemical exchange processes at the molecular level. Nevertheless, even for statistically stationary reactive flows, turbulence may alter the average concentrations of species, the average enthalpy, and the average pressure which are to be found at a given location as compared to what they would been had the flow been laminar, or turbulence of a different character (different intensity, length scales, energy spectrum, etc.). It accomplishes this through the somewhat obscure mechanisms by which a random velocity field transports mass, momentum, and energy in the presence of molecular diffusion effects, and a goal of pdf methods is to assist in exposing the nature of these mechanisms.

It is a more delicate matter to determine what kind of turbulent transport information to adopt as given. Experience has shown that turbulence modeling techniques, which are often adequate to describe Reynolds stresses and other energy-containing turbulence structures, do not fare well when applied to reacting flows [5.1]. One difficulty seems to be in the description of turbulent microscale mixing and, indeed, in a lack of understanding of the details of the mixing mechanisms at this scale [5.2] even when the reaction can be considered to have no dynamical influence on the turbulence (Sect. 1.9).

It might be thought that those closure theories of turbulent transport which express a general relationship between moments will remain useful when reactive fields are studied. For example the quasi-normal hypothesis [5.3], or its more acceptable relative the eddy damped quasi-normal theory [5.4], relates fourth-order moments to second moments as though the variables were normally distributed. Unfortunately almost all of these moment closures have implicitly or explicitly been developed for variables with nearly Gaussian distributions. It is well known that velocity components almost always are nearly Gaussian and, in regions not close to turbulent-nonturbulent interfaces, passive nonreacting scalar fields occasionally are. Reacting species almost never are (Fig. 3.4) unless the reaction is slow compared to a characteristic turbulence eddy turnover time [5.5], which is a time scale of the order of the lifetime of an energy-containing eddy of the turbulence. Thus it seems unlikely that this rich literature on closure approximations will be directly useful.

One form in which turbulence information can be used in reactive flow studies is to require that in the limit of very slow reaction one recovers an existing closure from nonreacting scalar transport theory. In doing this, an important constraint is nonnegativity of the scalar field. In the following pages the reader will note that where approximations or closures are employed for scalar field quantities their form has almost always been taken from those used successfully in the literature on passive scalar transport. By the same token the influence of reaction-generated expansion on the turbulence is often assumed to be adequately handled by using Favre averaging (Sect. 1.9) plus the assumption that Favre-averaged variables can be approximated by closures of the same form as used for turbulence without such effects. As indicated in Chap. 1, one should not understate the dangers in such a strategy; undoubtedly, as measurements in reactive turbulent flows become more sophisticated they will clarify the interaction between turbulence and the reaction processes. In the next section we make specific remarks on Favre averaging.

The pdf method of description, which is the focus of this chapter, was introduced into reactive flow research soon after the Second World War by *Hawthorne* et al. [5.6]. It had been widely used prior to that as an essential building block in the theory of probability by *Doob* [5.7] and a decade and a half earlier, the pdf of a velocity component was one of the earliest turbulence measurements made [5.8]. *Bilger* [5.9] gave a brief but useful description of some of the important properties of scalar pdfs (see also Chap. 3). In the same references, one can find a description of important shear flows in which the converted scalar has a distribution far from Gaussian, even in the absence of reaction. This property, which is especially important near the boundaries of free shear flows, is one of the reasons why the more familiar moment theories of turbulent transport are inadequate to describe reactive flows. A more direct reason can be found if one looks at the chemical production term discussed in detail in Chap. 1. Using the notation developed there, and assuming a global, second-order, forward reaction with an Arrhenius rate expression, the rate of production $\dot{w}_i$ of species i can be written in the form

$$-\dot{w}_i=\varrho^2 k_f Y_i Y_j \exp(-T_a/T), \tag{5.1}$$

where α in the preexponential factor has been taken as zero. *Bilger* has presented the form of $\overline{\dot{w}_i}$ when it is expanded about the mean states using conventional averaging and shows

$$-\overline{\dot{w}_i}=\bar{\varrho}^2 k_f \bar{Y}_i \bar{Y}_j \exp(-T_a/\bar{T})\{1+\overline{\varrho'^2}/\bar{\varrho}^2+\overline{Y_i'Y_j'}/\bar{Y}_i\bar{Y}_j+2\overline{\varrho' Y_i'}/\bar{\varrho}\bar{Y}_i$$
$$+T_a/\bar{T}[\overline{Y_i'T'}/\bar{Y}_i\bar{T}+\overline{Y_j'T'}/\bar{Y}_j\bar{T}+(T_a/2\bar{T}-1)\overline{T'^2}/\bar{T}^2]+\ldots\}, \tag{5.2}$$

where the continuation is for moments of third and higher order and the prime indicates a fluctuation about the mean.

The term $\dot{w}_i$ typically occurs in a conservation equation of the type (1.17) [5.10]

$$\varrho\frac{\partial Y_i}{\partial t}+\varrho\boldsymbol{u}\cdot\nabla Y_i-\nabla\cdot\varrho D_i\nabla Y_i=\dot{w}_i, \tag{5.3}$$

where $\boldsymbol{u}$ is the advecting velocity field and D_i the coefficient of molecular diffusion of species i. In Chap. 1 it is pointed out that because typically $T_a/\bar{T}\gg 1$ the expansion leading to (5.2) is restricted to uninterestingly small temperature fluctuations. However, even in the case of constant density ($\varrho'=0$), isothermal ($T'=0$), and statistically homogeneous flows ($\overline{\boldsymbol{u}\cdot\nabla Y_i}=0$), the direct average of (5.3), which yields linear terms in $\bar{Y}_i$ on the left-hand side, leads, through (5.2), to an infinite series of moments on the right. To make matters worse it has been shown in the case of very rapid reactions in weak turbulence [5.11] that the rms fluctuations in concentration are larger than the mean and any truncation of (5.3) should probably retain terms of all orders in the moment expansion.

This situation is reminiscent of the closure problem in turbulence dynamics where the direct interaction approximation [5.12] and related models have addressed the need to retain moments of all orders in a series expansion of the velocity correlation. There is an extra difficulty for reactive flows, where it is important to retain strict bounds on the concentration such that its mass fraction Y_i satisfies $0 \leqq Y_i \leqq 1$. Presently, there does not exist a version of the direct interaction approximation that retains such bounds although some much simpler low-order moment truncations which do so have been proposed [5.13, 14]. Again, as indicated in Chap. 1, when one turns to the complete problem of variable density, variable temperature shear flows the outlook for even crudely successful modeling by the use of a moment approach to (5.3) seems grim indeed.

By contrast, it was shown many years ago [5.15] that for initially unmixed reactants the closure difficulty associated with $\dot{w}_i$ can be avoided in the limit of fast chemistry when the Lewis number is unity and the reactants dynamically passive. That is, when the coefficients of molecular diffusivity D_i and heat diffusivity are taken as equal for all species in the reaction and $\boldsymbol{u}$ is reaction independent. In this case $\dot{w}_i$ can be eliminated and a simple mixing operator [the left-hand side of (5.3)] acts on sums and differences of weighted species concentrations. The details which are expounded in Chap. 3 can be found in *Bilger*'s [5.9] review, which also extends the analysis with the help of further assumptions. Take, as an example, the case of turbulent convection of a two-species irreversible reaction for which the mass-based, stoichiometric ratio of species 1 and 2 is unity and their molecular weights are equal. One finds

$$(\partial/\partial t + \boldsymbol{u}\cdot\nabla - 1/\varrho\nabla\cdot\varrho D\nabla)\chi = 0\,, \tag{5.4}$$

where

$$\chi = Y_1 - Y_2\,.$$

It is here that one of the virtues of the pdf description first became apparent. To obtain both Y_1 and Y_2 from a solution of (5.4), the flame sheet model assumes

$$Y_1 = 0,\; Y_2 = -\chi \qquad \chi \leqq 0 \tag{5.5a}$$

and

$$Y_1 = \chi,\; Y_2 = 0 \qquad \chi > 0\,. \tag{5.5b}$$

Hence if one can obtain the pdf of χ which satisfies (5.4) and appropriate boundary and initial conditions, one can with the help of (5.5) immediately deduce the pdf of both Y_1 and Y_2 and, consequently, all of their moments. Originally it was simply assumed [5.16] that χ had a Gaussian distribution. In that case Y_1 and Y_2 are both semi-Gaussians, and one can easily express the above results in moment form. Modern statistical mechanics has devised efficient methods to transform sets of differential equations like (5.4) into an evolution equation for the pdf of the dependent variables. In the next section we display the procedure for doing that for a particular reactive system which will display most of the characteristics one finds in these equations.

Some researchers, recognizing that Gaussians are unlikely, especially in free shear flows, but wanting to avoid using the yet incompletely explored evolution equations for the pdf, have combined several kinds of distributions chosen pragmatically to represent a particular reactive flow. For example, if a uniform distribution of reactant B exits from a jet into an ambient atmosphere of A, it is reasonable to assume that at any point the pdf of B will consist of a delta function $c_1\delta(Y_B=0,\boldsymbol{x})$ representing the fluid outside the jet, a continuous distribution $c_2P(Y_B,\boldsymbol{x})$ which might be, for example, a truncated Gaussian, and a delta function $c_3\delta(Y_B=1,\boldsymbol{x})$ to represent contributions from the core fluid which has not yet participated in the reaction. The relative weighting of c_1, c_2, and c_3 at any point is then to be determined and normalization is guaranteed by the constraint $c_1+c_2+c_3=1$. Interesting numerical results have been obtained by such modeling [5.17]. They represent a use of the pdf which has been discussed in Sect. 3.1 and 4.5 and will not be further pursued in this chapter.

5.2 Derivation of Single-Point pdf Equations

There are two, often employed methods for deriving the pdf evolution equation for a given reactive flow problem. One, using the method of functionals, was first introduced into turbulence theory over a decade ago [5.18, 19], and into reactive flow studies some years later [5.20, 21]. From an equation for the probability density functional [5.22], which is both linear and closed, one can deduce an unclosed n-point pdf equation (the equation for the n-point pdf contains the $n+1$ point pdf). The equation for the probability density functional of the turbulent velocity has been studied [5.23], as has that for the concentration of a reacting species in a turbulent flow [5.24, 25] but there seems little prospect that a broad range of applicable results can be obtained at present by such a direct approach.

The n-point pdf equation was also derived by *Lundgren* [5.26] using a more efficient method which produces directly the pdf equation from the partial differential equations which define the conservation laws of the system. This approach has since been used by many authors [5.27–29] and is the one we shall develop here. It should be pointed out that others [5.30, 31] have applied the pdf technique to equations other than the conservation equations for mass, momentum, and energy by making stochastic assumptions at that level. For example *Kuznetsov* and *Frost* assumed that both the turbulence and the scalar field obey Langevin equations and proceed from there.

The technique of derivation outlined below is valid for any set of equations, but for the purpose of demonstration, we shall deal only with its application to a set of conservation laws for a specific chemical flow system.

We suppose that NO, O_3, and NO_2 react according to the chemical kinetic scheme specified below and that no other reactions are important. We further ignore, for this problem, any role played by either density or enthalpy fluctuations. These can easily be included by direct extension of the method and in fact they have been included in formal presentations of the pdf equations and are available in the literature [5.20]. The role of density fluctuations is of special concern and not well understood at present. We address it separately later in the context of even simpler chemical kinetics.

The scheme, which is taken to be

$$NO + O_3 \xrightarrow{k_1} NO_2 + O_2$$

$$NO_2 + O_2 \xrightarrow{k_2} NO + O_3 \,,$$

where k_1 and k_2 are constant reaction rates, is described by the evolution equations

$$\left(\frac{\partial}{\partial t} + \boldsymbol{u} \cdot \nabla - \nabla \cdot D_i \nabla\right) Y_i = \dot{w}_i \,, \qquad i = 1, 2, 3 \,, \tag{5.6}$$

where

$$\dot{w}_1 = \dot{w}_2 = -\dot{w}_3 = -k_1 Y_1 Y_2 + k_2 Y_3 \,. \tag{5.7}$$

Y_i, $i = 1, 2, 3$ are the mass fractions of NO, O_3, and NO_2, respectively, and the concentration of O_2 is taken as fixed. The fact that the molecular weights of the species differ has been neglected for simplicity.

There are six random variables in (5.6), Y_i and the three components of the velocity field $\boldsymbol{u}$ which satisfy the Navier-Stokes equation for incompressible flow. Equations (5.6) and (5.7) with boundary and initial conditions from a complete description of the problem. One may then develop an equation for the pdf of all six variables at n points at simultaneous time, and it is easy to do so. Such a pdf will be a function of the six variables at each of n space locations, and in the absence of spatial symmetries such as homogeneity, or temporal simplifications such as statistical stationarity, it will be a function also of $3n$ space coordinates and time. More generally, if there are N reactants, three velocity components, and no symmetries in space or time, the full n-point pdf will be a function of $n(3+N)+3n+1$ variables.

It used to be said that a function of one variable can be represented by a column of figures, a function of two variables by a page, a function of three variables by a book, and a function of four variables needs a library! The computer has modified this assessment somewhat, but even for $n=1$ the complete pdf of this problem is a function of 10 variables. This in fact is probably the major drawback of the pdf method. One has traded certain improvements in the description of the problem for a rapid increase in the number of independent variables. It seems likely that it is a method which will function best in analyzing idealized and simplified situations of phenomena which cannot be approached convincingly by moment methods or direct numerical simulation. There are ways to reduce the number of variables but one pays a penalty for them also, as we shall now see as we adopt the view that the velocity field is known stochastically through solution of the Navier-Stokes equation. We proceed to obtain an evolution equation for the marginal pdf of the concentration fields.

Assuming existence and uniqueness, one first defines a "fine grained density"

$$\mathscr{P}(\hat{Y}_1, \hat{Y}_2, \hat{Y}_3; \boldsymbol{x}, t) = \delta[Y_1(\boldsymbol{x}, t) - \hat{Y}_1]\,\delta[Y_2(\boldsymbol{x}, t) - \hat{Y}_2]\,\delta[Y_3(\boldsymbol{x}, t) - \hat{Y}_3]$$

such that $\mathscr{P}(\hat{Y}_1, \hat{Y}_2, \hat{Y}_3; \boldsymbol{x}, t)\,d\hat{Y}_1 d\hat{Y}_2 d\hat{Y}_3$ is the probability that at $\boldsymbol{x}$ and t, Y_1 will be in the range $\hat{Y}_1 < Y_1(\boldsymbol{x}, t) < \hat{Y}_1 + d\hat{Y}_1$ while Y_2 is in the range $\hat{Y}_2 < Y_2(\boldsymbol{x}, t) < \hat{Y}_2 + d\hat{Y}_2$ and Y_3 is in the range $\hat{Y}_3 < Y_3(\boldsymbol{x}, t) < \hat{Y}_3 + d\hat{Y}_3$. In the definition of fine grained density the variables with "hats", such as $\hat{Y}_1$, are non-random quantities whereas those without, such as $Y_1(\boldsymbol{x}, t)$, are the random fields whose pdf's are sought. The introduction of the fine grained density can be considered a purely mathematical device but one can also attach some physical meaning to it as follows.

Consider the random scalar field $Y(\boldsymbol{x}, t)$ and denote its fine grained density by

$$\mathscr{P}(\hat{Y}; \boldsymbol{x}, t) = \delta[Y(\boldsymbol{x}, t) - \hat{Y}].$$

$\mathscr{P}(\hat{Y}; \boldsymbol{x}, t)$ is clearly a function of the variable $\hat{Y}$ and a functional of the random field $Y(\boldsymbol{x}, t)$, since it depends on the entire set of values of Y. Consider a fixed value of the variable $\hat{Y}$, say $\hat{Y}_F$, and fix also the space-time point $(\boldsymbol{x}, t)$ of interest. Then $\mathscr{P}(\hat{Y}_F; \boldsymbol{x}, t)$ is zero if $Y(\boldsymbol{x}, t) \neq \hat{Y}_F$ in any one realization of the random field Y and is infinite if $Y(\boldsymbol{x}, t) = \hat{Y}_F$. For a given realization it is also true that $\mathscr{P}$ has all the required properties of a pdf including normalization since $\int \mathscr{P}(\hat{Y}; \boldsymbol{x}, t)\,d\hat{Y} \equiv 1$ by definition of the delta function. Note also that the nth moment of $\hat{Y}$ is by direct calculation $Y^n(\boldsymbol{x}, t)$ for all n.

In other words the fine grained density is a device by which each realization of the random field is written in a pdf manner. As we show later, the connection to the pdf of normal usage comes about on taking the ensemble average of $\mathscr{P}(\hat{Y}; \boldsymbol{x}, t)$ which is, at each space-time point for each ensemble member, a delta function and zeroes. The justification for obtaining mostly smooth, continuous pdfs from such a "spiky" field of realizations was first given by *Stratonovich* [5.32]. More recently *Brissaud* and *Frisch* [5.27] discuss such an entity considered as a measure on Y space. It is always to be interpreted in the sense of a distribution. For example, by the method to be outlined below, we can show that passive scalar transport defined by the equation

$$\frac{\partial \phi}{\partial t} + \boldsymbol{u} \cdot \nabla \phi = D \nabla^2 \phi \tag{5.8}$$

has as the corresponding evolution equation for its fine grained density

$$\frac{\partial \mathscr{P}}{\partial t} + \boldsymbol{u} \cdot \nabla \mathscr{P} + \frac{\partial}{\partial \hat{\phi}} (D \nabla^2 \phi \mathscr{P}) = 0 . \tag{5.9}$$

Equation (5.9) is to be interpreted in the sense that for any infinitely differentiable test function $\psi(\phi)$ (a "good" function of *Lighthill*'s [5.33] terminology)

$$\int d\hat{\phi} \psi(\hat{\phi}) \left[\frac{\partial \mathscr{P}}{\partial t} + \boldsymbol{u} \cdot \nabla \mathscr{P} + \frac{\partial}{\partial \hat{\phi}} (D \nabla^2 \phi \mathscr{P}) \right] = 0 . \tag{5.10}$$

Integration of this last equation by parts using (5.8) shows that it is indeed true for all such test functions.

To return to the reactive flow problem (5.6) we may write

$$\mathscr{P}(\hat{Y}_1, \hat{Y}_2, \hat{Y}_3 ; \boldsymbol{x}, t) = \prod_{\alpha=1}^{3} \delta[\hat{Y}_\alpha - Y_\alpha(\boldsymbol{x}, t)]$$

from which we obtain

$$\frac{\partial \mathscr{P}}{\partial t} = \sum_{\alpha=1}^{3} \frac{\partial}{\partial t} \delta[\hat{Y}_\alpha - Y_\alpha(\boldsymbol{x}, t)] \prod_{\substack{\beta=1 \\ \beta \neq \alpha}}^{3} \delta[\hat{Y}_\beta - Y_\beta(\boldsymbol{x}, t)]$$

or

$$\frac{\partial \mathscr{P}}{\partial t} = \sum_{\alpha=1}^{3} \frac{\partial \delta}{\partial Y_\alpha} \frac{\partial Y_\alpha}{\partial t} \prod_{\substack{\beta=1 \\ \beta \neq \alpha}}^{3} \delta[\hat{Y}_\beta - Y_\beta(\boldsymbol{x}, t)] .$$

Hence

$$\frac{\partial \mathscr{P}}{\partial t} = - \sum_{\alpha=1}^{3} \frac{\partial Y_\alpha}{\partial t} \frac{\partial \mathscr{P}}{\partial \hat{Y}_\alpha} .$$

Inserting (5.6), we find

$$\frac{\partial \mathscr{P}}{\partial t} = \sum_{\alpha=1}^{3} (\boldsymbol{u} \cdot \nabla Y_\alpha - \nabla \cdot D_\alpha \nabla Y_\alpha - \dot{w}_\alpha) \frac{\partial \mathscr{P}}{\partial \hat{Y}_\alpha}$$

or

$$\frac{\partial \mathscr{P}}{\partial t} = - \sum_{\alpha=1}^{3} \boldsymbol{u} \cdot \nabla Y_\alpha \frac{\partial \mathscr{P}}{\partial \hat{Y}_\alpha} - \sum_{\alpha=1}^{3} (\nabla \cdot D_\alpha \nabla Y_\alpha + \dot{w}_\alpha) \frac{\partial \mathscr{P}}{\partial \hat{Y}_\alpha}$$

and finally

$$\frac{\partial \mathscr{P}}{\partial t} + \boldsymbol{u} \cdot \nabla \mathscr{P} + \sum_{\alpha=1}^{3} \frac{\partial}{\partial \hat{Y}_\alpha} (\dot{w}_\alpha \mathscr{P}) = - \sum_{\alpha=1}^{3} \nabla \cdot D_\alpha \nabla Y_\alpha \frac{\partial \mathscr{P}}{\partial \hat{Y}_\alpha} . \tag{5.11}$$

The term on the right-hand side represents the molecular diffusion phenomenon and both looks and is awkward. On the other hand the chemical production term is remarkably simple since $\dot{w}_\alpha$ is a known function of Y_1, Y_2, and Y_3 [see (5.7)]. This can be seen more clearly by writing

$$\dot{w}_\alpha \frac{\partial \mathscr{P}}{\partial \hat{Y}_\alpha} = \frac{\partial}{\partial \hat{Y}_\alpha} \{\dot{w}_\alpha(Y_1, Y_2, Y_3) \delta[\hat{Y}_1 - Y_1(\boldsymbol{x}, t)] \delta[\hat{Y}_2 - Y_2(\boldsymbol{x}, t)] \delta[\hat{Y}_3 - Y_3(\boldsymbol{x}, t)]\}$$

and therefore

$$\dot{w}_\alpha \frac{\partial \mathscr{P}}{\partial \hat{Y}_\alpha} = \frac{\partial}{\partial \hat{Y}_\alpha} [\dot{w}_\alpha(\hat{Y}_1, \hat{Y}_2, \hat{Y}_3) \mathscr{P}(\hat{Y}_1, \hat{Y}_2, \hat{Y}_3; \boldsymbol{x}, t)] .$$

The molecular diffusion term in (5.11) can be written in many alternative ways. To simplify the presentation, take the molecular diffusivities D_α as constants. *Lundgren* [5.26] was the first to deal with a diffusion term when he wrote the one-point pdf equation for the turbulent velocity field. His notation, which has been widely adopted, can be demonstrated in our problem by writing the obvious equality

$$D_\alpha \nabla^2 Y_\alpha \frac{\partial \mathscr{P}}{\partial \hat{Y}_\alpha} = D_\alpha \lim_{\boldsymbol{x}' \to \boldsymbol{x}} \nabla_{\boldsymbol{x}'}^2 Y'_\alpha \frac{\partial \mathscr{P}}{\partial \hat{Y}_\alpha} ,$$

where the prime indicates either a point $\boldsymbol{x}'$ different from $\boldsymbol{x}$ or a function defined at the point $\boldsymbol{x}'$.

On noting that

$$Y'_\alpha = \int \hat{Y}'_\alpha \delta[\hat{Y}'_\alpha - Y_\alpha(\boldsymbol{x}', t)] d\hat{Y}'_\alpha ,$$

Lundgren's form follows

$$D_\alpha \nabla^2 Y_\alpha \frac{\partial \mathscr{P}}{\partial \hat{Y}_\alpha} = D_\alpha \lim_{x' \to x} \frac{\partial}{\partial \hat{Y}_\alpha} \nabla^2_{x'} \int \hat{Y}'_\alpha \mathscr{P}_2(\hat{Y}'_\alpha, \hat{Y}_1, \hat{Y}_2, \hat{Y}_3) d\hat{Y}'_\alpha, \tag{5.12}$$

where $\mathscr{P}_2(\hat{Y}'_\alpha, \hat{Y}_1, \hat{Y}_2, \hat{Y}_3)$ is a two-point pdf defined by

$$\mathscr{P}_2(\hat{Y}'_\alpha, \hat{Y}_1, \hat{Y}_2, \hat{Y}_3; \boldsymbol{x}, \boldsymbol{x}', t) = \delta[\hat{Y}'_\alpha - Y_\alpha(\boldsymbol{x}', t)] \prod_{\beta=1}^{3} \delta[\hat{Y}_\beta - Y_\beta(\boldsymbol{x}, t)].$$

Equation (5.12) reveals one of the grave difficulties with a pdf description of the diffusion process. In trying to write an equation for the pdf at point $\boldsymbol{x}$ we find that the diffusion term has introduced a dependence on the two-point pdf and therefore precludes the possibility than the equation we obtain will be closed. In fact it is necessary to employ the probability density functional if one wishes to escape this difficulty [5.21]. An analogous difficulty arises in a moment description where attempts to write an equation for the intensity, say $\overline{\psi^2}$, of the passive scalar ψ produces a term of the form $D\,\overline{\nabla\psi \cdot \nabla\psi}$, where $\nabla\psi$ depends not on single-point values of ψ but on joint values of ψ at neighboring points. The analogy goes even deeper as we shall see later when, under certain assumptions, the scalar microscale λ_s [5.34], which connects $\overline{\psi^2}$ and $\overline{\nabla\psi \cdot \nabla\psi}$, also appears in the pdf formulation.

Another form of the diffusion term, due to *Fox* [5.35], is perhaps more interesting. He has accomplished a decomposition of it into a part which represents transport in physical space and a part which represents transport in concentration or configuration space. He writes

$$\frac{\partial^2 Y_\alpha}{\partial x_k \partial x_k} \frac{\partial \mathscr{P}}{\partial \hat{Y}_\alpha} = \frac{\partial}{\partial x_k}\left(\frac{\partial Y_\alpha}{\partial x_k} \frac{\partial \mathscr{P}}{\partial \hat{Y}_\alpha}\right) - \frac{\partial Y_\alpha}{\partial x_k} \frac{\partial^2 \mathscr{P}}{\partial x_k \partial \hat{Y}_\alpha}$$

or

$$\nabla^2 Y_\alpha \frac{\partial \mathscr{P}}{\partial \hat{Y}_\alpha} = -\nabla^2 \mathscr{P} - \frac{\partial Y_\alpha}{\partial x_k} \sum_{\beta=1}^{3} \frac{\partial^2 \mathscr{P}}{\partial \hat{Y}_\alpha \partial \hat{Y}_\beta} \frac{\partial Y_\beta}{\partial x_k};$$

therefore

$$D_\alpha \nabla^2 Y_\alpha \frac{\partial \mathscr{P}}{\partial \hat{Y}_\alpha} = -D_\alpha \nabla^2 \mathscr{P} + D_\alpha (\nabla Y_\alpha \cdot \nabla Y_\beta) \frac{\partial^2 \mathscr{P}}{\partial \hat{Y}_\alpha \partial \hat{Y}_\beta}. \tag{5.13}$$

As we mentioned earlier, it is evident that the marginal pdf we seek, $P(\hat{Y}_1, \hat{Y}_2, \hat{Y}_3; \boldsymbol{x}, t)$, is simply the statistical average of the fine grained density $\mathscr{P}(\hat{Y}_1, \hat{Y}_2, \hat{Y}_3; \boldsymbol{x}, t)$ where the average is taken over all realizations of the random functions u, Y_1, Y_2, and Y_3. That is,

$$P(\hat{Y}_1, \hat{Y}_2, \hat{Y}_3; \boldsymbol{x}, t) = \langle \mathscr{P}(\hat{Y}_1, \hat{Y}_2, \hat{Y}_3; \boldsymbol{x}, t) \rangle, \tag{5.14}$$

where the notation $\langle\ \rangle$ represents an ensemble average. We note that the element of probability space Ω, which is usually denoted by ω, has been systematically excluded from this entire development. The interested reader should consult a standard reference on stochastic processes or *Lumley*'s [5.36] monograph on statistical methods used in turbulence.

When (5.14) is applied to (5.11) the form of the resulting equation depends on whether (5.12) or (5.13) is used to represent molecular diffusion. In the former case we find

$$\frac{\partial P}{\partial t}+\langle \boldsymbol{u}\cdot\nabla\mathscr{P}\rangle+\sum_{\alpha=1}^{3}\frac{\partial}{\partial\hat{Y}_\alpha}(\dot{w}_\alpha P)=-\sum_{\alpha=1}^{3}D_\alpha\frac{\partial}{\partial\hat{Y}_\alpha}\left(\lim_{\boldsymbol{x}'\to\boldsymbol{x}}\nabla^2_{\boldsymbol{x}'}\int\hat{Y}'_\alpha P_2 d\hat{Y}'_\alpha\right), \quad (5.15\mathrm{a})$$

where $P_2(\hat{Y}'_\alpha, \hat{Y}_1, \hat{Y}_2, \hat{Y}_3; \boldsymbol{x}', \boldsymbol{x}, t)=\langle\mathscr{P}_2(\hat{Y}'_\alpha, \hat{Y}_1, \hat{Y}_2, \hat{Y}_3; \boldsymbol{x}', \boldsymbol{x}, t)\rangle$. When (5.13) is used the result is

$$\frac{\partial P}{\partial t}+\langle \boldsymbol{u}\cdot\nabla\mathscr{P}\rangle+\sum_{\alpha=1}^{3}\frac{\partial}{\partial\hat{Y}_\alpha}(\dot{w}_\alpha P)=\sum_{\alpha=1}^{3}D_\alpha\nabla^2 P+\sum_{\beta=1}^{3}\sum_{\alpha=1}^{3}D_\alpha\frac{\partial^2}{\partial\hat{Y}_\alpha\partial\hat{Y}_\beta}$$
$$\langle(\nabla Y_\alpha\cdot\nabla Y_\alpha)\mathscr{P}\rangle. \quad (5.15\mathrm{b})$$

In (5.15b) the first term on the right-hand side is spatial diffusion of the pdf while the second represents diffusion in concentration space of a complicated quantity which nevertheless can be given an interpretation in terms of traditional turbulence entities [5.37]. For example, for a single species reaction it can be written

$$D\frac{\partial^2}{\partial\hat{Y}^2}\langle(\nabla Y)^2\mathscr{P}\rangle=D\frac{\partial^2}{\partial\hat{Y}^2}\langle(\nabla Y)^2\mathscr{P}\int d\hat{Z}\delta[\hat{Z}-\nabla Y(\boldsymbol{x},t)]\rangle,$$

since the δ function integral is unity. Therefore, interchanging averaging and integration, we find

$$D\frac{\partial^2}{\partial\hat{Y}^2}\langle(\nabla Y)^2\mathscr{P}\rangle=D\frac{\partial^2}{\partial\hat{Y}^2}\int d\hat{\boldsymbol{Z}}\,\hat{\boldsymbol{Z}}^2 P_{Y\boldsymbol{Z}}(\hat{Y},\hat{\boldsymbol{Z}};\boldsymbol{x},t), \quad (5.16)$$

where $P_{Y\boldsymbol{Z}}(\hat{Y},\hat{\boldsymbol{Z}};\boldsymbol{x},t)$ is the single-point joint pdf of Y and $\boldsymbol{Z}$, the gradient field. Using Bayes theorem in the form

$$P_{Y\boldsymbol{Z}}(\hat{Y},\hat{Z};\boldsymbol{x},t)=P_{\boldsymbol{Z}|Y}(\hat{Z}|\hat{Y};\boldsymbol{x},t)P(\hat{Y};\boldsymbol{x},t)$$

we finally obtain

$$D\frac{\partial^2}{\partial Y^2}\langle\nabla Y)^2\mathscr{P}\rangle=\frac{1}{2}\frac{\partial^2}{\partial\hat{Y}^2}[\langle\varepsilon(\boldsymbol{x},t|\hat{Y})\rangle P(\hat{Y};\boldsymbol{x},t)], \quad (5.17)$$

where $\langle\varepsilon(\boldsymbol{x},t|\hat{Y})\rangle$ is the conditional dissipation of scalar fluctuations at $(\boldsymbol{x},t)$, conditioned on the value of Y at $(\boldsymbol{x},t)$. Analytically it is defined by

$$\langle\varepsilon(\boldsymbol{x},t|\hat{Y})\rangle = 2D \int d\boldsymbol{Z}\, \boldsymbol{Z}^2 P_{\boldsymbol{Z}|Y}(\hat{\boldsymbol{Z}}|\hat{Y};\boldsymbol{x},t),$$

which should be contrasted with the unconditioned statement for the scalar dissipation

$$\langle\varepsilon(\boldsymbol{x},t)\rangle = 2D \int d\boldsymbol{Z}\, \boldsymbol{Z}^2 P_{\boldsymbol{Z}}(\hat{\boldsymbol{Z}};\boldsymbol{x},t).$$

Physically $\langle\varepsilon(\boldsymbol{x},t|\hat{Y})\rangle$ can be obtained by extracting from the ensemble of measured values of the scalar dissipation function ε, at a given space-time point $(\boldsymbol{x},t)$, only those taken when the concentration field at the same point $(\boldsymbol{x},t)$ is at a fixed value, say $\hat{Y}_{\mathrm{F}}$. An ensemble average over this subclass of realizations produces the conditioned expected dissipation function $\langle\varepsilon(\boldsymbol{x},t|\hat{Y}_{\mathrm{F}})\rangle$ which can, of course, also be obtained for other values of $\hat{Y}$ and for other points $(\boldsymbol{x},t)$. It is a quantity of fundamental importance in turbulent reactive flow studies.

Spalding [5.38], *Williams* [5.39], and *Bilger* [5.40] have all proposed formulas for the rate of production of chemical species in a diffusion-limited situation which include the rate of dissipation of the scalar. In particular *Bilger* required the joint pdf of a passive scalar and its gradient. His subsequent assumption of statistical independence in the form

$$P_{Y\boldsymbol{Z}}(\hat{Y},\hat{\boldsymbol{Z}};x,t) = P_Y(\hat{Y};\boldsymbol{x},t)\,P_{\boldsymbol{Z}}(\hat{Z},\boldsymbol{x},t)$$

is equivalent to replacing the conditioned dissipation in (5.17) by the unconditioned, a step for which theoretical support has been found only for high Reynolds number homogeneous turbulence [5.83].

An examination of (5.15a) or (5.15b) also shows that another difficulty has appeared in the marginal pdf equation. The term representing convection by the turbulent velocity field $\langle\boldsymbol{u}\cdot\nabla\mathscr{P}\rangle$ is now nonclosed; it is not expressed in terms of P. This is entirely a consequence of using the marginal pdf $P(\hat{Y}_1, \hat{Y}_2, \hat{Y}_3;\boldsymbol{x},t)$ to describe the system. If one had retained instead the full pdf $P(\boldsymbol{u}, \hat{Y}_1, \hat{Y}_2, \hat{Y}_3;\boldsymbol{x},t)$, which is related to P by

$$P(\hat{Y}_1, \hat{Y}_2, \hat{Y}_3;\boldsymbol{x},t) \equiv \int P(\boldsymbol{u}, \hat{Y}_1, \hat{Y}_2, \hat{Y}_3;\boldsymbol{x},t)\,d\boldsymbol{u},$$

the convective term would be in the closed form $\hat{\boldsymbol{u}}\cdot\nabla P(\hat{\boldsymbol{u}}, \hat{Y}_1, \hat{Y}_2, \hat{Y}_3;\boldsymbol{x},t)$ where $\hat{\boldsymbol{u}}$ is now merely a coefficient (see, for example, [5.20]).

The nonclosure of molecular diffusion and turbulent transport terms is a characteristic of all such pdf equations for reactive flow systems. If one attempts to include the velocity field, as above, for the purpose of closing the transport terms, one finds that the pressure terms of the Navier-Stokes equation are also unclosed in a pdf description. To appreciate this consider the one-point pdf equation for an incompressible turbulent field as obtained by *Lundgren* [5.26]. He found that, if $P_{\boldsymbol{u}}(\hat{\boldsymbol{u}};\boldsymbol{x},t)=\langle\delta[\hat{\boldsymbol{u}}-\hat{\boldsymbol{u}}(\boldsymbol{x},t)]\rangle$, then

$$\frac{\partial P_{\boldsymbol{u}}}{\partial t}+\hat{\boldsymbol{u}}\cdot\nabla P_{\boldsymbol{u}}=-\frac{\partial}{\partial\hat{\boldsymbol{u}}}\lim_{\boldsymbol{x}'\to\boldsymbol{x}}\int\left[-\left(\frac{1}{\varrho}\nabla_{\boldsymbol{x}'}p'\right)+(\nu\nabla_{\boldsymbol{x}'}^{2}\boldsymbol{u}')\right]P_{\boldsymbol{u}'\boldsymbol{u}}(\hat{\boldsymbol{u}}',\hat{\boldsymbol{u}})d\boldsymbol{u}', \tag{5.18}$$

where $P_{\boldsymbol{u}'\boldsymbol{u}}$ is the joint pdf of velocity at two-points. It is understood that the pressure p' in (5.18) can be written in terms of the velocity field through use of the incompressibility condition. That is,

$$-\frac{1}{\varrho}\nabla_{\boldsymbol{x}'}^{2}p'=\frac{\partial^{2}u_i'u_j'}{\partial x_i'\partial x_j'}. \tag{5.19}$$

We can also note that the molecular diffusion term for momentum is entirely analogous to the one we have previously obtained for mass diffusion, (5.15a).

Finally, since the conservation equations all have the form

$$\frac{\partial\chi}{\partial t}=\mathscr{L}_{\boldsymbol{x}}(\chi),$$

where $\mathscr{L}_{\boldsymbol{x}}$ is a space operator pertinent to the particular quantity χ being considered, it is clear from the derivation of the pdf equation, which adds together all terms like $\partial\chi/\partial t$, that the nonclosed expressions for the various scalar and vector quantities of the conservation equations are simply added together linearly. Thus the difficulties we have exposed so far are indeed the only ones which will arise in the equation for the pdf of reacting scalars in incompressible turbulence.

We postpone to Sect. 5.3 descriptions of closure proposals that have been suggested or used for the diffusion, turbulent transport, and pressure terms in a pdf formulation and we now consider the role of variable density which has so far been ignored. Since the density ϱ can generally be represented [5.10] by the equation

$$\frac{\partial\varrho}{\partial t}=-\varrho\nabla\cdot\boldsymbol{u}-\boldsymbol{u}\cdot\nabla\varrho, \tag{5.20}$$

there is no difficulty in formally including such effects in a pdf equation [5.20]. See also the discussion in Sect. 1.9.

For the simplest possible case of the reaction of a single species we shall derive the single-point joint pdf equation for the marginal density defined by $\mathscr{P}(\hat{\varrho},\hat{Y})=\delta[\hat{\varrho}-\varrho(\boldsymbol{x},t)]\,\delta[\hat{Y}-Y(\boldsymbol{x},t)]$, where ϱ and Y satisfy the conservation equations (5.20) and

$$\varrho\frac{\partial Y}{\partial t}+\varrho\boldsymbol{u}\cdot\nabla Y-\nabla\cdot\varrho D\nabla Y=\dot{w}. \tag{5.21}$$

The fine grained density $\mathscr{P}$ for this problem can easily be shown by the method previously described to satisfy [5.37]

$$\frac{\partial \hat{\varrho}\mathscr{P}}{\partial t}+\nabla\cdot \boldsymbol{u}\hat{\varrho}\mathscr{P}=\frac{\partial}{\partial\hat{\varrho}}[\hat{\varrho}^2(\nabla\cdot\boldsymbol{u})\mathscr{P}]-\nabla\cdot\varrho D(\nabla Y)\frac{\partial\mathscr{P}}{\partial\hat{Y}}-\frac{\partial}{\partial\hat{Y}}(\dot{w}\mathscr{P}). \tag{5.22}$$

Some use has been made of Favre averaging of the pdf in combustion flows [5.41, 42] and (5.22) is in a convenient form to investigate this method further.

We use *Bilger*'s [5.41] definition

$$\tilde{P}(\hat{Y};\boldsymbol{x},t)=\frac{1}{\bar{\varrho}(\boldsymbol{x},t)}\int\hat{\varrho}d\hat{\varrho}\langle\mathscr{P}\rangle,$$

average (5.22), and integrate over $\hat{\varrho}$ to find

$$\frac{\partial}{\partial t}\bar{\varrho}\tilde{P}+\nabla\cdot\bar{\varrho}\langle\boldsymbol{u}\mathscr{P}\rangle=-\nabla\cdot\int\left\langle\varrho D\nabla Y\frac{\partial\mathscr{P}}{\partial\hat{Y}}\right\rangle d\hat{\varrho}-\frac{\partial}{\partial\hat{Y}}\int\dot{w}\langle\mathscr{P}\rangle d\hat{\varrho}. \tag{5.23}$$

The terms on the left-hand side of this expression are analogous in form to those obtained assuming constant density, but on the right-hand side it can be seen that Favre averaging does not reduce the diffusion term to a recognizable form, and this remains true even if one computes the *Lundgren* [5.26] or *Fox* [5.35] representations as was done earlier. Since the diffusion terms must be approximated even in the constant density case by making a suitable closure, one might argue that a precise analogy for this term is not a crucial consideration (Sect. 1.8). A potentially more serious difficulty arises with the production rate term, the last of (5.23). Here it is evident that a simple reduction to the constant density form will only occur when the production rate $\dot{w}$ is a linear function of $\hat{\varrho}$. In this case, and only in this case, will the last term be expressible in terms of $\tilde{P}$. It will become

$$-\frac{\partial}{\partial\hat{Y}}[S(\hat{Y})\tilde{P}],$$

where $\dot{w}=\hat{\varrho}S(\hat{Y})$. For complex reactions it is unlikely that $\dot{w}$ will have such a simple dependence on density but unless it does the rationale for using a pdf formulation is lost, since the production rate term will no longer be closed in an equation for $\tilde{P}$.

An adequate description of turbulent reactive flows which includes variable density effects is clearly a long way off for any formulation. Even moment methods have received little attention [5.21, 43], and that mostly formal. Current work with pdfs where variable density is important (e.g., [5.42]) seems to assume a linear dependence of $\dot{w}_i$ on $\hat{\varrho}$ in (5.3), and the same form of closure for terms involving $\tilde{P}$ are used for the pdf P in incompressible flows. *Bilger*

[5.9, 41] also argued for the usefulness of Favre averaging in combustion flows. In the next section we survey the closure approximations that have been proposed for the single-point pdf equations. Then we return to a brief description of multipoint pdf equations and the closure approximations suggested for them.

5.3 Closure Approximations

5.3.1 One-Point pdf Description

It was shown in the previous section that only two kinds of terms are unclosed in a single-point pdf description of turbulent reactive flows using the marginal pdf $P(\hat{Y}_1, \hat{Y}_2, \hat{Y}_3; \boldsymbol{x}, t)$. We shall discuss each in turn. Since the term which includes the production rate $\dot{w}$ is not among those to be closed, it is simpler to begin the discussion by considering a single nonreacting scalar being convected by a turbulent field $\boldsymbol{u}$. *Lundgren*'s form of the pdf equation can then be written

$$\frac{\partial P}{\partial t} + \langle \boldsymbol{u} \cdot \nabla \mathscr{P} \rangle = -D \frac{\partial}{\partial \hat{Y}} \lim_{\boldsymbol{x}' \to \boldsymbol{x}} \nabla_{\boldsymbol{x}'}^2 \int \hat{Y}' P_2(\hat{Y}', \hat{Y}) d\hat{Y}', \tag{5.24}$$

where, by analogy to the preceding section, $\mathscr{P}$ is the fine grained density $\delta[\hat{Y} - Y(\boldsymbol{x}, t)]$ and $P = \langle \mathscr{P} \rangle$, $P_2 = \langle \delta[\hat{Y} - Y(\boldsymbol{x}, t)] \, \delta[\hat{Y}' - Y(\boldsymbol{x}', t)] \rangle$.

The convective term in (5.24) disappears altogether for incompressible, homogeneous fields so let us examine the equation in that situation and consider possible ways to represent the unclosed molecular diffusion term on the right-hand side. We can say at the outset that no altogether satisfactory closure yet exists. Nevertheless it is the term which must be adequately modeled if any process depending heavily on molecular scale mixing is to be represented. Chemical reaction is certainly of this class unless the reactants are premixed before reaction is initiated and subsequent molecular diffusion of thermal energy is not crucial, or unless the reaction is so slow that total mixing can occur before significant reaction takes place. In the latter case one has reaction in a well-stirred homogeneous reactor, the situation often assumed for the measurement of chemical kinetic rates but seldom realized in flows of practical interest.

The first technique of closure, used by *Lundgren* [5.26] for the velocity field and *Dopazo* [5.37] for a scalar field, is to rewrite P_2 in a form where its dependence on P is explicit. Using Bayes' theorem, and assuming homogeneity and incompressibility as mentioned above, we write (5.24) as

$$\frac{\partial P(\hat{Y}, t)}{\partial t} = -D \frac{\partial}{\partial \hat{Y}} \left[\lim_{\boldsymbol{x}' \to \boldsymbol{x}} \nabla_{\boldsymbol{x}'}^2 \int \hat{Y}' P(\hat{Y}' | \hat{Y}; |\boldsymbol{x} - \boldsymbol{x}|, t) d\hat{Y}' P(\hat{Y}, t) \right], \tag{5.25}$$

where $P(\hat{Y}'|\hat{Y})$ is a conditional pdf and the integral in (5.25) is just the conditional expected value of $\hat{Y}'$ given $\hat{Y}$, say $E(\hat{Y}'|\hat{Y})$. Then we may write

$$\frac{\partial P(\hat{Y},t)}{\partial t} = -D\frac{\partial}{\partial \hat{Y}}\left[\lim_{\boldsymbol{x}'\to\boldsymbol{x}} \nabla^2_{\boldsymbol{x}'} E(\hat{Y}'|\hat{Y};|\boldsymbol{x}'-\boldsymbol{x}|,t)d\hat{Y}'P(\hat{Y},t)\right]. \tag{5.26}$$

We shall see later, when discussing multipoint formulations, that a conditional expected value is a form in which all nonclosures in a pdf description can be put. They have an even more general importance in turbulence theory, as shown by *Adrian* [5.44].

One can use well-established mean square estimation theory to approximate $E(\hat{Y}'|\hat{Y})$ since, [5.45], $\int\int[\hat{Y}'-g(\hat{Y})]^2P_2(\hat{Y}',\hat{Y})d\hat{Y}'d\hat{Y}$ is minimized when $g(\hat{Y})=E(\hat{Y}'|\hat{Y})$. The only version of this theory so far used has included the additional constraint that $g(\hat{Y})$ be a linear function of $\hat{Y}$. When that is the case it can be shown [5.46] that

$$E(\hat{Y}'|\hat{Y}) = \bar{Y}(\boldsymbol{x}',t)+\varrho(r,t)\frac{\sigma_{Y'}(\boldsymbol{x}',t)}{\sigma_Y(\boldsymbol{x},t)}[\hat{Y}-\bar{Y}(\boldsymbol{x},t)], \tag{5.27}$$

where $r=|\boldsymbol{x}'-\boldsymbol{x}|$ and ϱ, σ, and $\bar{Y}$ are the usual correlation coefficient, variance, and mean of $Y(\boldsymbol{x},t)$. Using homogeneity and the definition of the scalar microscale as

$$\lambda_s^2 = -2\left(\frac{\partial^2\varrho}{\partial r^2}\right)^{-1}_{r=0},$$

one finds that (5.26) reduces to

$$\frac{\partial P}{\partial t} = \frac{6D}{\lambda_s^2}\frac{\partial}{\partial \hat{Y}}[(\hat{Y}-\bar{Y})P(\hat{Y},t)]. \tag{5.28}$$

Equation (5.28) is quite analogous to *Lundgren*'s [5.26] velocity field pdf equation and it shares the same feature that it is only apparently closed. In fact the microscale λ_s depends on $\varrho(r)$ which is a two-point property and its value cannot be obtained by solving for $P(\hat{Y},t)$ in (5.28). *Bonniot* and *Borghi* [5.42] have discussed some practical assumptions for λ_s. In the form (5.28) the pdf equation is analogous to the intensity equation in the moment form, as we mentioned earlier.

For a more complete discussion of the properties of this closure, which is often misleadingly called the "conditionally Gaussian approximation"; see *Dopazo* and *O'Brien* [5.46]. We label it the linear mean square estimation closure (LSME) and we list some more recent findings on its nature.

Although under this closure an initially Gaussian distribution will remain so, *Frost* [5.47] asserted that other initial distributions do not asymptotically approach the Gaussian as one would expect for simple homogeneous diffusion. This is easy to demonstrate from (5.28) [5.48] if we denote the initial distribution $P(\hat{Y},0)$ by $P_0(\hat{Y})$ and the factor $6D\lambda_s^{-2}$ by β and use the fact that $\bar{Y}$ is time independent. It follows that the solution to (5.28) is of the form

$$P(\hat{Y},t)=\mathrm{e}^{\beta t}P_0[\bar{Y}+(\hat{Y}-\bar{Y})\mathrm{e}^{\beta t}]$$

and hence the initial shape of the profile is never relaxed. In particular if the initial concentration field is binary, that is, if the distribution is simply two delta functions, say at $\hat{Y}=0$ and $\hat{Y}=1$, then the subsequent evolution remains two delta functions approaching symmetrically from two sides the value $\bar{Y}=0.5$ but only reaching it at infinite time. This somewhat pathological case nevertheless reveals an important difficulty with the LSME closure; it fails to model microscale mixing adequately, a potentially serious flaw for its application to reactive flows (Sect. 5.1), and several authors have suggested other more pragmatic approaches which allow two initially binary fields to evolve into continuously distributed fields.

One direct measurement of $E(\hat{Y}'|\hat{Y})$ has been made by Tavoularis in a wind tunnel with a grid generating a homogeneous turbulence and an asymmetric mean temperature. The experiment and the data obtained have been discussed by *Dopazo* [5.37] and, at least in this configuration, give support to the LSME closure. *Bezuglov* [5.49] reported a similar success with measurements taken on the axis of a heated jet.

Frost [5.47] was the first to suggest an alternative approximation which does allow two initial delta functions to relax and produce a continuous distribution of intermediate concentrations. He assumed

$$\frac{\partial P(\hat{Y},t)}{\partial t}=\beta\left\{\int\limits_{\hat{Y}}^{\infty}P(\hat{Y}'')d\hat{Y}''\int\limits_{0}^{\hat{Y}}P(\hat{Y}')d\hat{Y}'-\frac{P(\hat{Y})}{2}\cdot\left[\int\limits_{0}^{\hat{Y}}(\hat{Y}-\hat{Y}')P(\hat{Y}')d\hat{Y}'+\int\limits_{\hat{Y}}^{\infty}(\hat{Y}''-\hat{Y})P(\hat{Y}'')d\hat{Y}''\right]\right\}, \quad (5.29)$$

where β is merely a mixing rate obtained from small-scale arguments of *Kolmogorov* [5.50]; it is equivalent to D/λ_s^2 in the LSME closure. The first term represents the generation of fluid with concentration $\hat{Y}$ by the coalescence of two parcels of fluid with concentrations respectively less than and greater than $\hat{Y}$. The other two terms represent loss of $\hat{Y}$ by the mutual coalescence of a parcal of fluid with concentration $\hat{Y}$ with a parcel whose concentration is either less than or greater than $\hat{Y}$. Other authors have obtained somewhat different forms but with quite similar ad hoc arguments [5.28, 37]. *Frost* [5.47] has used this approximation to compute the structure of a turbulent flame jet using the Burke and Schumann scheme. The flow is far too complicated to serve as a model by which to evaluate his closure. He asserts that with its use all initial distributions tend to the Gaussian. That is certainly not true of *Pope*'s form

[5.28] and neither his nor *Dopazo*'s [5.37] has been adequately investigated. We shall leave it to the reader to obtain further information on these approaches from the cited references.

Dopazo [5.37] has also investigated the possibility of using classical iteration methods to generate a von Neumann series for the pdf equation (5.24) and thereby put some order into the closure approximations. The results are formal and not encouraging since the lowest terms in the approximation introduce unphysical features such as negative diffusivities.

For the usual situation in which there is more than one scalar in a turbulent reacting flow the integral in the diffusion term to be closed is of the form [see (5.15a)]

$$\int \hat{Y}'_\alpha P_2(\hat{Y}'_\alpha, \hat{Y}_1, \hat{Y}_2, \hat{Y}_3; \boldsymbol{x}', \boldsymbol{x}, t) d Y'_\alpha$$

which, using Bayes' theorem, becomes

$$[\int \hat{Y}'_\alpha P_2(\hat{Y}'_\alpha | \hat{Y}_1, \hat{Y}_2, \hat{Y}_3) d\hat{Y}'_\alpha] P(\hat{Y}_1, \hat{Y}_2, \hat{Y}_3) .$$

If one now attempts to approximate the conditional pdf $P_2(\hat{Y}'_\alpha | \hat{Y}_1, \hat{Y}_2, \hat{Y}_3)$ by linear mean square estimation as before, one finds that the resulting pdf equation for P does not yield the correct equations for even the lowest order moments of the $\hat{Y}_1$, $\hat{Y}_2$, and $\hat{Y}_3$ fields [5.37] unless one also assumes that the expected value of $\hat{Y}'_\alpha$ depends only on $\hat{Y}_\alpha$ and on neither of the other two concentrations. This drastic assumption is the only one so far used in discussing multispecies diffusion [5.46] and remains a weak point in the application of pdf methods to reactive flows.

Let us now consider the convective term $\langle \boldsymbol{u} \cdot \nabla \mathscr{P} \rangle$ in (5.15). It is unclosed because it contains the fine grained density $\mathscr{P}$ rather than the marginal pdf P. One is used to approximating the analogous term $\langle \boldsymbol{u} \cdot \nabla \phi \rangle$ in moment formulation of turbulent scalar transport theory; it is the only unclosed term arising when an ensemble average of (5.8) is taken.

We note, first of all, that if $\mathscr{P}(\hat{\phi})$ is the fine grained density for $\phi(\boldsymbol{x}, t)$ then

$$\boldsymbol{u} \cdot \nabla \phi = \boldsymbol{u} \cdot \nabla \int \hat{\phi} \mathscr{P}(\hat{\phi}) d\hat{\phi}$$

and

$$\langle \boldsymbol{u} \cdot \nabla \phi \rangle = \int \langle \boldsymbol{u} \cdot \nabla \mathscr{P} \rangle \hat{\phi} d\hat{\phi} . \tag{5.30}$$

Therefore, it should generally be possible to propose a closure for $\langle \boldsymbol{u} \cdot \nabla \mathscr{P} \rangle$ which reproduces any specified moment closure for $\langle \boldsymbol{u} \cdot \nabla \phi \rangle$. This is a fortunate situation since a great deal of research has gone on in the last decade to find more sophisticated methods of modeling convection in turbulent shear flows [5.51], and it may be possible to incorporate progress in this direction into the pdf formulation. However, as yet only two approximations for this term have been used. By far the most common is the direct use of the eddy diffusivity concept [5.30] to express $\langle \boldsymbol{u} \cdot \nabla \mathscr{P} \rangle$ in terms of P.

For an incompressible $\boldsymbol{u}$ field

$$\boldsymbol{u} \cdot \nabla \mathscr{P} = \nabla \cdot \boldsymbol{u} \mathscr{P}$$

and if

$$\boldsymbol{u} = \bar{\boldsymbol{u}} + \boldsymbol{u}'$$

and

$$\mathscr{P} = \langle \tilde{\mathscr{P}} \rangle + \tilde{\mathscr{P}}' = P + \tilde{\mathscr{P}}'$$

then

$$\langle \boldsymbol{u} \cdot \nabla \mathscr{P} \rangle = \bar{\boldsymbol{u}} \cdot \nabla P + \nabla \cdot \langle \tilde{\boldsymbol{u}}' \tilde{\mathscr{P}}' \rangle$$

or

$$\langle \boldsymbol{u} \cdot \nabla \mathscr{P} \rangle = \bar{\boldsymbol{u}} \cdot \nabla P - \nabla \cdot \boldsymbol{\kappa} \nabla \mathscr{P} . \tag{5.31}$$

In (5.27) $\boldsymbol{\kappa}$ is the eddy diffusivity tensor which could be the same as the classical one defined by

$$\nabla \cdot \langle \boldsymbol{u} \phi' \rangle = -\boldsymbol{\kappa} \nabla \cdot \bar{\phi}$$

since the two definitions are compatible according to (5.30). The representation of turbulent transport of reactants by an eddy diffusivity has been examined from the viewpoint of fundamental stochastic processes in the absence of molecular diffusion [5.52] and is found to be rigorous in some limited circumstances.

The only other closure method applied to the convection term is due to *Dopazo* [5.53] who essentially applied an LSME closure to it. This was possible since for incompressible flow, defining

$$P = \left\langle \prod_{i=1}^{3} \delta[\hat{u}_i - u_i(\boldsymbol{x}, t)] \delta[\hat{\phi} - \phi(\boldsymbol{x}, t)] \right\rangle ,$$

then

$$\langle \boldsymbol{u} \cdot \nabla \mathscr{P} \rangle = \nabla \cdot \boldsymbol{E}(\hat{\boldsymbol{u}} | \hat{\phi}) P(\hat{\phi})$$

and one can apply the approximation to $\boldsymbol{E}(\hat{\boldsymbol{u}} | \hat{\phi})$ in the same manner as was done for $E(\hat{\phi}' | \hat{\phi})$ earlier. It is not easy to see why such expected values should be well represented by such an approximation in the case of mixed velocity and scalar fields, and the status of this closure remains unconfirmed by experiment.

To complete the discussion of single-point closure approximations we mention here the assumption used by *Lundgren* [5.54] to close the pressure term in the Navier-Stokes equation for nonhomogeneous turbulence. As discussed earlier the contribution of pressure fluctuations to the evolution of the velocity single-point pdf is through a two-point pdf, since the pressure at a point depends on the velocities in the neighborhood of the point. After extracting the mean fields the pressure contribution can be written [5.54]

$$\frac{\partial}{\partial \hat{u}}\left\{\frac{1}{4\pi}\int \frac{\partial}{\partial \boldsymbol{x}}\frac{1}{|\boldsymbol{x}-\boldsymbol{x}'|}\left(\hat{\boldsymbol{u}}'\cdot\frac{\partial}{\partial \boldsymbol{x}'}\right)^2[P_2(\hat{\boldsymbol{u}}',\boldsymbol{u})-P(\hat{\boldsymbol{u}}')P(\hat{\boldsymbol{u}})]d\boldsymbol{x}'d\hat{\boldsymbol{u}}'\right\}. \tag{5.32}$$

The proposal to close this term is based on the Boltzmann type of approximation assuming that the main role of the pressure fluctuations is to randomize the velocity distribution. Thus it can be represented by a relaxation model (Krook model in kinetic theory) whereby (5.32) is replaced by the simple expression

$$-\frac{1}{\tau}(P-P_{\mathrm{G}}),$$

where τ is a relaxation time, taken to be

$$\tau \approx 5(\sigma_{\mathrm{u}})^{-2}\left(\varepsilon+\frac{3}{2}\frac{d\sigma_{\mathrm{u}}^2}{dt}\right).$$

The function P_{G} is the Gaussian distribution to which pressure fluctuations are supposed to drive the velocity distribution, so that

$$P_{\mathrm{G}}=(2\pi\sigma_{\mathrm{u}}^2)^{-3/2}\exp[-(\hat{\boldsymbol{u}}-\bar{\boldsymbol{u}})^2/2\sigma_{\mathrm{u}}^2],$$

where $\sigma_{\mathrm{u}}(\boldsymbol{x},t)$ and $\bar{u}(\boldsymbol{x},t)$ are the root mean square velocity fluctuations and average velocity, respectively, and ε is the turbulent dissipation rate.

It is clear that a quantity like ε cannot come from a one-point theory. In a manner analogous to λ_{s}, which arise in the approximation to the diffusion terms, it must be obtained from some other description. The Lundgren closures have been checked by *Srinivasan* et al. [5.55] in a very interesting application to the numerical solution of turbulent Couette flow and they seem to perform well in that case for several assumptions on the nature of ε. We shall return to these calculations later in the discussion of boundary conditions.

There are two other sets of numerical computation of closed single-point pdf equations. One is due to *Dopazo* [5.53] who calculated the evolution of a temperature pdf along the centerline of an axisymmetric heated jet and obtained relatively good agreement with experiments in the self-preserving region by using closures of the type described here. The other is the earlier mentioned study by *Bonniot* and *Borghi* [5.42] who have computed the evolution of the pdf in turbulent combustion in an idealized homogeneous combustor and obtained plausible results for the influence of residence time and degree of mixedness for both one- and two-dimensional distributions. A related numerical work is that of *Frost* [5.47] using a pair of Langevin equations for $\boldsymbol{u}$ and Y as a basis for the single-point pdf of a turbulent diffusion-controlled flame.

The results of these computations seem to confirm that the single-point pdf method can give qualitatively useful results at least in simple enough geometries and with the addition of some information from the moment closure descriptions of turbulence. From a theoretical point of view the state of a single-point description is unsatisfactory because, in fact, it cannot be closed within the pdf formalism. In the next section we describe attempts to close the pdf equations at the n-point level, $n \geqq 2$, where it can be made self-contained.

5.3.2 The Multipoint Descriptions

Using a fine-grained density method similar to that in Sect. 5.2 but, for example, generalizing the delta function to n points in space, one can easily derive the n-point pdf for a system of reactants obeying specified mass conservation equations. As a simple example, single species conservation can be given a two-point description by beginning with the two point fine grained density

$$\mathscr{P}_2(\hat{Y}, \hat{Y}'; \boldsymbol{x}, \boldsymbol{x}', t) = \delta[\hat{Y} - Y(\boldsymbol{x}, t)]\delta[\hat{Y}' - Y(\boldsymbol{x}', t)] .$$

The meaning of $\mathscr{P}_2$ is entirely analogous to the one-point fine grained density ϱ developed in Sect. 5.2 and the two-point pdf is likewise defined analogously as

$$P_2(\hat{Y}, \hat{Y}'; \boldsymbol{x}, \boldsymbol{x}', t) = \langle \mathscr{P}_2(\hat{Y}, \hat{Y}', \boldsymbol{x}, \boldsymbol{x}', t) \rangle .$$

The most general form of the n-point equation can be found in *Ievlev* [5.20] who has managed to incorporate variable density, N reactive species, thermal production, and electromagnetic effects in his description. We shall discuss only the constant density, N reactive species case which is described by the following n-point equation:

$$\begin{aligned} &\frac{\partial P_n}{\partial t} + \sum_{j=1}^{n} \frac{\partial}{\partial x_k^{(j)}}(P_n u_k^{(j)}) + \sum_{m=1}^{N} \frac{\partial}{\partial \psi_{(m)}}(P_n \dot{w}_{(m)}) \\ &= -\sum_{j=1}^{n} \frac{\partial}{\partial u_k^{(j)}}(P_n B_k^{(j)}) - \sum_{m=1}^{N} \frac{\partial}{\partial \psi_{(m)}^{(j)}}(P_n C_{(m)}^{(j)}) . \end{aligned} \tag{5.33}$$

Only the two summations on the right-hand side are unclosed. In (5.33) P_n represents the n-point joint pdf for the velocity and N species at each point; it also depends on $3n$ space coordinates and time. $\psi^{(j)}_{(m)}$ and $\dot{w}^{(j)}_{(m)}$ are, respectively, at the jth point, the concentration and production rate of the mth species and the two unclosed quantities $B^{(j)}_k$ and $C^{(j)}_{(m)}$ are defined as

$$B^{(j)}_k = \lim_{\boldsymbol{x}^{(n+1)} \to \boldsymbol{x}^{(j)}} \left[\nu \frac{\partial^2 E(u^{(n+1)}_k | A_n)}{\partial x^{(n+1)}_q \partial x^{(n+1)}_q} - \frac{1}{\varrho} \frac{\partial}{\partial x^{(n+1)}} E(p^{(n+1)} | A_n) \right]$$
$$C^{(j)}_{(m)} = \lim_{\boldsymbol{x}^{(n+1)} \to \boldsymbol{x}^{(j)}} \left[D_{(m)} \frac{\partial^2 E(\psi^{(n+1)}_{(m)} | A_n)}{\partial x^{(n+1)}_q \partial x^{(n+1)}_q} \right]. \tag{5.34}$$

The superscript $(n+1)$ means the quantity is evaluated at the $n+1$th point (that is why these terms are unclosed in an n-point description). The symbol A_n denotes the set of variables $(x^{(\alpha)}, \psi^{(\alpha)}_{(m)})$ for all n points and $E(A|B)$, as before, is the conditional expected value of variables A given the values of variables B. Through incompressibility the pressure $p^{(n+1)}$ can be written in terms of $\boldsymbol{u}^{(n+1)}$ (5.19) and $\boldsymbol{u}$ satisfies the constraint equations [5.56]

$$\int u^{(j)}_k \frac{\partial p_n}{\partial x^{(j)}_k} d\boldsymbol{u}^{(j)} = 0 \qquad j = 1, \ldots, n. \tag{5.35}$$

A careful reading of (5.33) and (5.34) will indicate how direct an extension they are of the single-point equations and unclosed terms (5.15) and (5.18). Note in particular that the three closures needed are for conditional expected velocity, conditional expected pressure, and conditional expected concentrations as in the single-point case. Again, if one assumed $\boldsymbol{u}$ known, the unclosed $B^{(j)}_k$ term would be replaced by terms of the type $\langle \boldsymbol{u}^{(j)} \cdot \nabla_{\boldsymbol{x}^{(j)}} \psi_{(m)} \rangle$ in an equation for the marginal pdf obtained by integrating p_n over all $\boldsymbol{u}$.

Recalling that the number of independent variables associated with P_n is $n(3+N)+3+1$, we can see that these formal equations are of little practical use except for demonstrating the structure of the pdf formulation and, possibly, for investigating homogeneous fields at the two-point level ($n=2$) for low values of N. There are a few existing studies of the $n=2$ equation. *Lundgren* [5.56] studied the case $N=0$, $n=2$. His closure was obtained by assuming a Gaussian-like relationship between the three-point distribution function and those of lower order. There is a loose resemblance to the "cumulant discard hypothesis" of *Millionshchikov* [5.3] which was subsequently shown to be seriously unphysical [5.57, 58] in that it led to a negative scalar intensity spectrum and a negative energy spectrum function for isotropic turbulence in the energy containing wave numbers, after a time scale of evolution approximately equal to one eddy turnover time. In *Lundgren*'s [5.56] case no lack of realizability has been demonstrated. His closure leads to a value of $C=1.29$ for the universal constant in the Kolmogorov spectrum $E(K) = C\varepsilon^{2/3} k^{-5/3}$ for values of k in the inertial subrange. Such a value is reasonably close to the experimental number $C = 1.44 \pm 0.06$ obtained by *Grant* et al. [5.59] and should be compared to *Kraichnan*'s [5.60] prediction of $C=1.77$ from his Lagrangian history direct interaction theory.

From the point of view of turbulent reacting flow the Lundgren formalism has several specific drawbacks beyond the high dimensionality associated with any pdf method. The major one is that his closure requires the random fields to be close to Gaussian in a sense not yet quantitatively precise. As we mentioned earlier, reactive fields are known to be far from Gaussian when the reaction time scale is of the same order as the mixing time or shorter and in free shear layers. In such cases the Lundgren closure would appear to be unreliable. For the $n=2$, $N=0$ case *Fox* [5.61] has also proposed closures which are similar to those of Lundgren, namely an equation is written which connects the three-point pdf with those of lower order. The details can be found in his papers referenced in the one cited above. There are two major differences between his and the previously described work of Lundgren. One is that the so-called coincidence property is explicitly retained by Fox. This property of the multipoint pdf is the most difficult constraint to satisfy. It requires that the n-point pdf reduce properly to the $n-1$ point pdf when any two of its points of definition are made to coalesce. Lundgren's closure does not satisfy such a condition except for Gaussian fields. *Fox* [5.62] has also simplified the complicated pressure fluctuation terms by using a version of the mean value theorem to approximate their contribution to the pdf. If this approximation turns out to give accurate results, it may be an important step towards making the pdf equation accessible to numerical computation.

Ievlev's n-point closure [5.20, 63] is of quite a different character. It does not invoke any near-Gaussian assumption and may be better suited to describing reactive flows. It does not seem to have received serious attention in this regard, possibly because its format is both incomplete and intricate. *Kuo* [5.64] has investigated the closure at the two-point level and applied it to statistically homogeneous two-species molecular diffusion with and without reaction; the short description which follows is taken from his study.

The formal properties which the Ievlev closure preserves are listed below. They are written for a closure of the P_n equation, $n=2$, but in fact apply at whatever order n the closure is made. If P_2^* is the solution of (5.33) using Ievlev's closure (to be described later), then:

a) P_2^* is normalized for all time if it is normed at $t=0$.

b) P_2^* is real and nonnegative. This property depends on the existence of a unique real solution to (5.33) using Ievlev's closure.

c) The equation for P_1^* obtained by integrating the P_2^* equation is exact.

d) Moment equations can be constructed from the closed equation for P_2^* and are exact up to and including order $n=2$.

e) The coincidence property mentioned above is retained as are the properties of separation and reduction. The former implies the proper behavior of P_n in the limit as one or more points become separated an infinite distance from the others, while the latter specifies the proper relationship between P_{n-1} and P_n [5.56].

We exhibit the nature of the closure in the simplest possible case of molecular self-diffusion of a single species [5.64], which is governed by

$$\frac{\partial \phi}{\partial t} = D\nabla^2 \phi .$$

In the n-point pdf formulation this is not a closed problem in physical space although formal functional solutions have been obtained [5.23, 25] and solutions, asymptotic in time, for the spectrum function of ϕ have been derived assuming certain kinematic features of the large-scale structure of a convecting final-period turbulence [5.65].

For this field the coincidence, separation, and reduction properties mentioned above can be written

Coincidence $\lim_{x' \to x} P_2(\hat{\phi}', \hat{\phi}) = P_1(\hat{\phi}')\delta(\hat{\phi}' - \hat{\phi})$

Separation $\lim_{|x' - x| \to \infty} P_2(\hat{\phi}', \hat{\phi})$
$= P_1(\hat{\phi}')P_1(\hat{\phi})$

Reduction $\int P_2(\hat{\phi}', \hat{\phi}) d\hat{\phi} = P_1(\hat{\phi}')$
$\int P_2(\hat{\phi}', \hat{\phi}) d\hat{\phi}' = P_1(\hat{\phi})$.

The exact pdf equation can be shown, using the method described earlier, to be

$$\frac{\partial P_2}{\partial t} = -\frac{\partial}{\partial \hat{\phi}}[P_2 C(\hat{\phi}, \hat{\phi}')] - \frac{\partial}{\partial \hat{\phi}'}[P_2 C(\hat{\phi}, \hat{\phi}')] , \tag{5.36}$$

where

$$C(\hat{\phi}, \hat{\phi}') = \lim_{x'' \to x}\left[D \frac{\partial^2 E(\hat{\phi}'' | \hat{\phi}, \hat{\phi}')}{\partial x''_g \partial x''_q}\right].$$

Ievlev's closure in this case is to approximate $C(\hat{\phi}, \hat{\phi}')$ by $C^*(\hat{\phi}, \hat{\phi}')$ such that

$$\frac{\partial P_2^*}{\partial t} = -\frac{\partial}{\partial \hat{\phi}}[P_2^* C^*(\hat{\phi}, \hat{\phi}')] - \frac{\partial}{\partial \hat{\phi}'}[P_2^* C^*(\hat{\phi}', \hat{\phi})] \tag{5.37}$$

and the following constraint equations are satisfied:

$$\int C^* P_2^* d\hat{\phi} = \int C P_2^* d\hat{\phi}$$

and

$$\int C^* P_2^* d\hat{\phi}' = \int C P_2^* d\hat{\phi}' .$$

The notation P_2^* is meant to indicate the solution of (5.37) when the approximation C^* is used. He then shows that, with such a C^*, the solution P_2^* will not only have all of the above-mentioned properties but will also minimize the error in estimating the time rate of change of any moment constructed from P_2^*.

The specific proposal made by Ievlev for the form of the approximation C^* is to assume it to be a separable function of $\hat{\phi}$ and $\hat{\phi}'$. The effect of this assumption is not yet clear and the closure must be considered, like the others, to be purely formal. *Kuo* [5.64] has deduced several properties of the closure including its relationship to the previously mentioned ones of *Lundgren* [5.56] and *Fox* [5.61]; he showed that it properly represents molecular mixing. The details can be found in the reference cited.

It should be evident that a great deal more needs to be known about the multipoint description before it can be considered as a workable tool in turbulent reactive flow research. From the theoretical viewpoint it is a far more satisfactory formulation than the single-point description, because it is amenable to more orderly closures (see end of Sect. 5.3.1). It may also be possible to bring powerful spectral methods into reactive flow studies via this approach, although that has not yet been done.

In the next section we return to the one-point formulation to discuss several of the difficult problems, and strategies for handling them, which have arisen in trying to assign boundary conditions to the pdf equation. Attempts to understand this aspect are just beginning in reactive flow studies and it is impossible at present to do more than merely summarize these efforts. Kinetic theory had been confronted with similar problems; it is likely that reactive flow studies using the pdf should look to that theory for some accumulated wisdom.

5.4 Applications of the pdf Method

In this section we review various applications of the single-point pdf method to turbulence and turbulent transport of reactive and nonreactive species. Consistent with the intent of this chapter, attention still is restricted to studies which have dealt with evolution equations for the pdf, bypassing the large literature on pdf measurements of velocity and scalar quantities and on those useful predictive methods which assume the shape of the pdf.

A paper addressing many of the difficulties which arise in nonhomogeneous turbulence is that of *Srinivasan* et al. [5.55], who adopted the *Lundgren* [5.26] closure applied to turbulent Couette flow between parallel plates and obtained numerical solutions using the discrete ordinate method and finite differences. As mentioned earlier, single-point closure requires additional information about the pressure relaxation time τ and the dissipation function ε. In this paper two models of ε are investigated. The first is the traditional assumption of an exact balance between dissipation and production which can be manipulated to express ε as a function of the mean velocity and distance from the wall. The second form for ε is a semi-empirical equation due to *Jones* and *Launder* [5.66], and τ is taken proportional to ε, as proposed by Lundgren.

When either of these forms is appended to the pdf equation a closed system results, which requires only the specification of boundary conditions. For statistically steady Couette flow, advantage can be taken of the symmetries with respect to the stream direction and the coordinate parallel to the boundaries by integrating the pdf over the velocity components in those directions, so as to deal with only a reduced, two-dimensional pdf, one depending on the transverse location and the normal velocity component.

Numerical integration of the equations then becomes relatively straightforward, and we shall not present the details here. However, serious and interesting problems arise with the boundary conditions, problems which will reoccur in all reactive flow studies involving spatial coordinates. The authors have described these difficulties in considerable detail in their paper. Here we summarize briefly the nature of the problem since it appears to be a fundamental one, not yet resolved.

The boundary condition on velocity at a solid surface at rest is usually taken as $\boldsymbol{u}\equiv 0$ for continuum flows. In terms of the pdf of velocity one can interpret this as either

$$\int \hat{\boldsymbol{u}} P_{\boldsymbol{u}}(\hat{\boldsymbol{u}})\, d\hat{\boldsymbol{u}} = 0 \qquad \text{on the surface} \tag{5.38a}$$

or

$$P_{\boldsymbol{u}}(\hat{\boldsymbol{u}}) = \delta(\hat{\boldsymbol{u}}) \qquad \text{on the surface}\,. \tag{5.38b}$$

In the first case the specification of $P_{\boldsymbol{u}}(\hat{\boldsymbol{u}})$ by (5.38a) is not unique and in the second case there is no precise numerical method for specifying a delta function. Near the wall the *Lundgren* [5.26] model assumption of an LMSE closure seems unlikely to represent the physics correctly. That this closure may be inadequate near a bound on the variable being approximated can be seen by examining (5.27) with fixed values of $\boldsymbol{x}'$ and $\boldsymbol{x}$ and stationary statistics, viz.;

$$E\{\hat{Y}'|\hat{Y}\} = A + B[\hat{Y} - \overline{Y(x)}]\,, \tag{5.39}$$

where $A = \overline{Y(\boldsymbol{x}')}$ and $B = \varrho(\boldsymbol{x}', \boldsymbol{x})\,\sigma_y(\boldsymbol{x}')/\sigma_y(\boldsymbol{x})$. Suppose that there is everywhere a bound on the concentration $Y(\boldsymbol{x}, t)$ of the form $Y \geqq 0$. Considered as a function of $\hat{Y}$, $E\{\hat{Y}'|\hat{Y}\}$ is a straight line. There is no apparent constraint that its value at $\hat{Y}=0$, $A - B\overline{Y}(\boldsymbol{x})$, is necessarily nonnegative, and even less that $P(\hat{Y}'|\hat{Y}=0)=0$ for all $\hat{Y}' < 0$. The measurements of *Tavoularis*, mentioned in Sect. 5.3.1 as support for the LMSE approximation, are taken well away from the boundaries in a homogeneous turbulence without any influence from bounds on the scalar field.

After excluding the laminar sublayer from consideration in their numerical scheme, *Srinivasan* et al. [5.55] used two different strategies to overcome the boundary condition problem. They first took as a boundary condition, at some point in the law of the wall region, that the Reynolds stress be constant and that the turbulent energy be conserved. Both of these-features have been observed experimentally. A sufficient condition for them to hold is that the pdf have zero gradient in such a region and it is this condition which they applied together with a specification of the wall shear stress. In this study, for consistency, since dissipation and production are balanced in the law of the wall region, they used that balance to specify ε. Their numerical results compare favorably with experiments.

The arguments in the previous paragraph apply, at best, to flows without mean pressure gradients. In order to obtain a more general method the authors adopted a second strategy of using at the assumed boundary point a Chapman-Enskog form of the pdf for the fluid leaving the wall. The details will not be presented here. The treatment of the pdf is in a mode suggested by previous work on the Boltzmann equation of kinetic theory and is not entirely successful, in the sense that the numerical results are somewhat sensitive to the location in the logarithmic region of the point of application of the boundary conditions. It should be added that in this case the *Jones* and *Launder* [5.66] equation for ε was adopted, and good agreement with data was obtained.

In general the study seems to support the applicability of the pdf method to geometrically simple flows by presenting a successful numerical technique which uses the pdf rather than its low-order moments and by proposing workable ways to treat the boundary conditions. The extent to which the addition of reactants will complicate the boundary conditions is unclear.

Several combustion flows have been studied numerically in systems which, by the assumption of homogeneity, avoid the imposition of spatial boundary conditions. An early study [5.25] which was later improved [5.46] focused on the temperature field of a second-order, one-step irreversible exothermic reaction in which the entire mass production term $\dot{w}$ is approximated as an explicit function of temperature. No particular difficulty arose with the "boundary" conditions on temperature since the description, based on an LMSE closure, is hyperbolic and a straightforward method of characteristics was applicable [5.21]. See also *Dopazo* [5.67].

A more recent numerical study [5.42], which extends some previous work on pdfs applied to turbulent combustion, includes density fluctuations by the *Favre*-averaging technique and explicitly computes concentration pdf behavior. The model is a homogeneous reactor so, again, no spatial boundary conditions need to be invoked. Turbulent mixing is represented by the development of the microscale λ_s which is taken, following *Corrsin* [5.65] to evolve inversely with the time. The initial state of a pair of unmixed reactants which are representable on a two-dimensional pdf plot as four approximate delta functions or, in numerical reality, peaks corresponding to the existence of each species jointly, individually, or not at all. The Arrhenius reaction rate and thermal conditions were chosen to represent poorly mixed pockets of propane at low initial temperatures. There was no sign of the theoretical difficulty with the LMSE closure as regards molecular scale mixing mentioned in Sect. 5.3, perhaps because strict delta functions cannot be incorporated in a numerical scheme, and the results with and without reaction show a plausible evolution of the pdf. This study also introduces directly into the numerical scheme a way to ensure the proper normalization of the pdf, something which has generally proved difficult to achieve. The pdf equation is put in a flux form in concentration space and the space is partitioned into parallelograms separated by surface elements on each of which the flux of pdf is approximated. The space partition evolves with the evolution of the pdf in order to represent peak concentrations with sufficient accuracy.

There also exists application of the pdf equation to scalar transport in shear flows which exhibit some self-similarity. Most of these deal either with transport of a single scalar or with diffusion-limited flame computations employing the *Burke* and *Schumann* [5.15] scheme to reduce the problem to that of a single passive scalar concentration. In such cases an important role must be played by the approximation of the turbulent term, $\langle \boldsymbol{u} \cdot \nabla \mathscr{P} \rangle$ in (5.15). We first discuss some earlier efforts to tackle inhomogeneous problems and then describe a recent approach to the pdf equation which attempts to recognize some of the difficulties with boundary conditions in free turbulent shear flows.

An early paper avoids the free boundaries of an axisymmetric turbulent heated jet by computing just the centerline evolution [5.53] of the axial temperature pdf and by comparing the results with experimental measurements of temperature moments up to the fourth order. The approximation to the molecular diffusion term is an LSME closure with the scalar microscale λ_s related to the velocity microscale λ through *Corrsin*'s [5.68] equation $\lambda_s^2 = 2\lambda^2 N_{Sc}^{-1}$, where N_{Sc} is the Schmidt number and λ is obtained from measurements. When the assumption of similarity is added to the restriction to centerline behavior, the problem concerns an evolution of the pdf in only one dimension (temperature); the subsequent integration can be done analytically and requires only an initial condition, which is taken as a Gaussian. There is good qualitative agreement between the pdf predictions and measurements.

An early and interesting approach to the evolution of a scalar pdf in the full field of a turbulent jet is due to *Kuznetsov* and *Frost* [5.30]. They used a Langevin equation for the scalar field rather than the conservation equation but they include the radial boundary condition on the scalar by recognizing that its pdf at a point consists of a continuous portion when the point is in the jet and a delta function when it is not. The location of the jet boundary is itself a random function of position so that at any fixed point the complete pdf will consist of a mixture of these two shapes, the delta function becoming predominant at large radial distances from the jet axis. Thus they achieved a kind of conditioning of the scalar pdf and are able to obtain coupled equations for the scalar intermittency function γ and the continuous portion of the pdf. With some restrictive assumptions about the nature of this pdf they were also able to obtain analytical solutions for it in two limits, at the centerline and in the limit $\gamma \to 0$, or far from the centerline. One of the serious difficulties in principle with their formulation is that although they conditioned the scalar field, the turbulent velocity field, which is also highly intermittent in a jet, was not so treated. For example they used an eddy diffusivity model (5.31) to write

$$\langle \boldsymbol{u} \cdot \nabla \mathscr{P} \rangle = \bar{\boldsymbol{u}} \cdot \nabla P - \kappa \nabla^2 P,$$

where $\bar{\boldsymbol{u}}$ is the total mean velocity at a point, not the mean velocity of the jet fluid, while κ is the scalar eddy diffusivity and P is the pdf of the scalar carried by the fluid in the jet.

Conditioning of both velocity and scalar fields has been a well-established experimental technique for many years. More recently *Libby* [5.69] introduced analytical techniques for conditioning intermittent turbulent flows and *Dopazo* [5.70] modified and extended the procedure. A similar activity took place in the study of two-phase flows [5.71] and it is now a routine matter to apply analytical conditioning to the conservation equations of an intermittent turbulent shear flow, assuming the boundaries between turbulence and non-turbulence to be sharp. The same is true of the scalar field distribution if the boundary between regions of existence and nonexistence of the scalar field is also sharp, as is thought to be true of the heated jet, for example. *O'Brien* and *Dopazo* [5.72] applied these techniques to condition the pdf equations for both velocity and scalar fields in free turbulent shear flows. *O'Brien* [5.73] has studied a reactive plume using similar techniques.

Conditioning does not alter the need to close the pdf equations for the molecular diffusion and convective terms, but it makes the existing closures more plausible because it gives the possibility of building into them different physical behavior inside and outside of the shear layer. This is clear if one considers, for example, the LMSE closure, noting that it has the form

$$E\{\hat{Y}'|\hat{Y}\} = A + B(\hat{Y} - \bar{Y}). \tag{5.40}$$

For definiteness take $Y \equiv 0$ outside the jet. At a point which is sometimes imbedded in the jet and sometimes not, $\bar{Y}$ will be a positive number. However, for a fluid element outside the jet, there should be no diffusive mixing of the scalar field since it is in a uniform environment of value $Y=0$. The unconditioned form (5.40) asserts, however, that it will tend to the value $\bar{Y}$, and this is clearly false. If (5.40) is used only for the conditioned scalar fields, then the average concentration of the outside fluid is $\bar{Y}=0$, and the proper behavior of no diffusive mixing is obtained.

The technique of conditioning the pdf equations is too recent to permit one to evaluate how useful it might prove to be. Since the fluid is separated into two parts, one in which the scalar field exists and the other where it does not, it is clear that some extra statements are needed to specify the rate at which the latter is entrained into the former. This is a long-standing and difficult problem in turbulence. The *Dopazo* and *O'Brien* [5.74] study made use of experimental conditioned measurements in the plane wake of a heated plate [5.75] and in a heated turbulent jet [5.76] to derive the profiles of entrainment of mass and generalized momentum. These are then compared with models proposed by other investigators; agreement is qualitatively satisfactory but far from sufficient to define the entrainment process adequately.

Another approach to the description of the scalar pdf in a jet is due to *Frost* [5.47] who, in a jet problem, integrated numerically the equation for a pdf which depends on three variables: concentration, radial coordinate, and axial coordinate. He did not treat the pdf radial boundary condition correctly and his study is mostly of interest for the computational method used and his replacement of the LSME closure by a nonlinear convolution-integral representation discussed in Sect. 5.3. Although the representation seems to have no basis in statistical mechanics, it is used to predict a turbulent diffusion-controlled flame jet which has the qualitative properties one expects.

5.5 Summary

In the decade which has elapsed since methods for obtaining the finite-point pdf evolution equations were presented, some progress has been made in applying them to reactive flow situations. At the very least it can be said that the evolution equation itself is readily obtained for any kinetic scheme and flow system and some physical insight into the role of each of the terms in the equation has been achieved. One can begin to feel as comfortable in configuration space as traditional turbulence researchers do in Fourier space, despite the inability in both cases to construct solutions to posed problems.

The pdf method has formed a strong attachment to the methodology of the older discipline of kinetic theory although there are only formal analogies between its equations and those of turbulence and turbulent transport, and it is not yet certain that the union will be fruitful. The predictions of *Lundgren* [5.56] for the small structure of isotropic turbulence and the calculations of his one-point pdf equations for Couette flow [5.55] give some support to kinetic theory methods in idealized continuum situations. Methods that eschew kinetic theory seem to be based on entirely pragmatic grounds which give little hope for either improving one's understanding of the physics of turbulent transport or extending the methods themselves to new situations. The integral closures for the molecular diffusion are the major example of this class. Designed to produce proper mixing at the molecular level, they appear, at the moment, to be too arbitrary to evaluate even in the context of the problems for which they are introduced [5.28, 37, 47].

Turbulent convection occurs in both the moment formulation and the pdf method with roughly the same degree of complexity, although closure approximations in the moment formulation are far more advanced. It is possible that they can be restated in the pdf form with little difficulty [see (5.30)]. In particular, modelers of turbulent shear flows have recently given a great deal of attention to the role of pressure fluctuations, while these have received only the simplest treatment in the pdf method. It would be interesting and useful to know if the latter could benefit from advances due to the former.

The major advantage of the pdf method for reactive flows is its closed-form treatment of the species production rate, which makes it attractive especially for combustion problems. The concomitant disadvantage is the number of independent variables introduced. This has repercussions not only on the complexity of the chemistry that can be handled but also on the sophistication of the numerical methods needed to obtain solutions. When one adds the necessity to clarify spatial boundary conditions, as was discussed in Sect. 5.3, it is clear that research is urgently needed on numerical simulation of simple shear flows with reaction, or at least scalar transport, to better define the limitations and potentials of the pdf description.

Even higher priority should be given to the careful accumulation of reactive flow data in clearly defined, simple turbulent situations that have served so well the evolution of moment theories of turbulence. Wind-tunnel, grid-generated turbulence with dilute, initially unpremixed reactants displaying a simple and well-understood chemical kinetics is yet to be studied. Ideally, enough data should be taken to evaluate all one- and two-point pdfs of the velocity and scalar fields which occur in the system. With the correct use of symmetry properties it may be possible to approach such a detailed description and thereby to evaluate the closure assumptions that have been discussed in this chapter. Since reaction rate depends only on the concentration levels and not on the spatial dimensions of the concentration fields, it is not clear that there is any universal equilibrium range in general reactive flows, nor is there any region of the pdf known to display similar properties in a wide class of reactive flows. Hence data from a quasi-homogeneous region of turbulence behind a grid may not be easy to generalize. Nevertheless the molecular diffusive process which is one of the major difficulties in the pdf description is determined by small-scale structure, and its modeling may be able to be given a universal form for high Reynolds number turbulence.

Equally important, but undoubtedly more difficult, are measurements of the distributions in carefully defined shear layers, free and bounded, for reactants satisfying simple enough kinetic schemes. Instrumentation problems are severe in such flows and recent attempts to study simple reactive shear flows [5.77–79] have fallen short of the ideal described above. They have not accumulated data that can be used as a direct test of any of the closure proposals discussed here.

Finally in trying to evaluate the present state of the pdf method it is necessary to recall all those approaches, neglected in this chapter, which make use of the pdf but not the evolution of the pdf equation obtained directly from the conservation laws. For example the use of a pdf description is crucial in *Bilger*'s [5.40] theory of diffusion flames, *Libby*'s [5.17] modeling of turbulent flows with fast reactions, the premixed flames of *Bray* and *Moss* [5.80], and *Chung*'s [5.31] kinetic theory analogy for reactive flows. It may be that the full pdf method will, for some time, be too cumbersome for the prediction of practical reactive flows, but since the pdf as a descriptive tool is clearly in active use, it is essential to ground it in a proper framework. Proper formulation of the equation, the boundary conditions, and the constraints which the pdf must satisfy will serve that important purpose.

Two recent papers [5.81, 82] which appeared too late for inclusion in this review, describe turbulent flame theories in which the scalar pdf equations obtained from the conservation laws play an important role. Both papers address many of the problems discussed in this chapter.

Acknowledgments. The manuscript for this article was prepared during a stay at the Laboratoire de Mécanique des Fluides of the Ecole Centrale de Lyon. I wish to thank especially Monsieur J. Mathieu and Mademoiselle G. Comte-Bellot for the hospitality extended to me and to note also the useful discussions on the subject topic with many of my colleagues at that institution. The U.S. National Science Foundation has supported much of the author's work in this area over a number of years and has continued its assistance with Grant ENG7710118 which is concerned with the material of Sect. 5.3.2.

References

5.1 D.B.Spalding: Abstract JC3 Bull. Am. Phys. Soc. **24**, 60 (1979)
5.2 A.Roshko: Dryden Lecture, Paper 76–78, AIAA 14th Aerospace Sci. Meeting, Washington, D.C., January, 1976
5.3 M.Millionshtchikov: C. R. Acad. Sci. SSSR **32**, G16 (1941)
5.4 M.Lesieur, P.L.Sulem: "Les Equations Spectrales en Turbulence Homogene et Isotrope Quelques Resultats Theoretiques et Numerques", in *Turbulence and Navier-Stokes Equations*, ed. by R. Temam (Springer, Berlin, Heidelberg, New York 1976) p. 133
5.5 E.E.O'Brien: "Theoretical Aspects of Turbulent Mixing of Reactants", in *Turbulence in Mixing Operations*, ed. by R. S. Brodkey (Academic, New York, San Francisco, London 1975) p. 21
5.6 W.R.Hawthorne, D.S.Weddell, H.C.Hottel: In *Third Symposium on Combustion and Flame and Explosions Phenomena* (Williams and Wilkens Company, Baltimore 1949) p. 266
5.7 J.L.Doob: *Stochastic Processes* (Wiley, New York 1953)
5.8 L.F.G.Simmonds, C.Salter: Proc. R. Soc. London A**45**, 212 (1934)
5.9 R.W.Bilger: Prog. Energy Combust. Sci. **1**, 87 (1976)
5.10 F.A.Williams: *Combustion Theory* (Addison-Wesley, Reading, Mass. 1965)
5.11 E.E.O'Brien: Phys. Fluids **14**, 1326 (1971)
5.12 R.H.Kraichnan: J. Fluid Mech. **5**, 497 (1958)
5.13 C.duP.Donaldson, G.R.Hilst: Environ. Sci. Tech. **6**, 812 (1972)
5.14 C.-H.Lin: Ph. D. Thesis, State University of New York at Stony Brook, New York (1974)
5.15 S.P.Burke, T.E.W.Schumann: Ind. Engr. Chem. **20**, 998 (1928)
5.16 H.L.Toor: AIChEJ. **8**, 72 (1962)
5.17 P.A.Libby: Combust. Sci. Tech. **13**, 79 (1976)
5.18 A.S.Monin: Prikl. Mat. Mech. **31**, 1057 (1967)
5.19 E.A.Novikov: Sov. Phys. Dokl. **12**, 1006 (1968)
5.20 V.M.Ievlev: Dokl. Akad. Nauk. SSSR **208**, 1044 (1973)
5.21 C.Dopazo: Ph. D. Thesis, State University of New York at Stony Brook, New York (1973)
5.22 E.Hopf: J. Ration. Mech. Anal. **1**, 87 (1952)
5.23 R.M.Lewis, R.H.Kraichnan: Commun. Pure Appl. Math. **15**, 297 (1962)
5.24 C.A.Petty, X.B.Reed, Jr.: AIChEJ. **18**, 751 (1972)

5.25 C. Dopazo, E. E. O'Brien: Acta Astron. **1**, 1239 (1974)
5.26 T. S. Lundgren: Phys. Fluids **10**, 969 (1967)
5.27 A. Brissaud, U. Frisch: J. Math. Phys. **15**, 524 (1974)
5.28 S. B. Pope: Combust. Flame **27**, 299 (1976)
5.29 E. E. O'Brien, R. E. Meyers, C. Benkovitz: In *Third Symposium on Atmospheric Turbulence Diffusion and Air Quality* (American Meteorological Society, Boston 1976) p. 160
5.30 V. R. Kuznetsov, V. A. Frost: Izv. Akad. Nauk. SSSR, Mekh. Zhid. i Gaza **2**, 58 (1973)
5.31 P. M. Chung: Phys. Fluids **16**, 1646 (1973)
5.32 R. L. Stratonovich: Doctoral Dissertation, Moscow University (in Russian) (1965)
5.33 M. J. Lighthill, *Fourier Analysis and Generalized Functions* (The University Press, Cambridge, England 1968)
5.34 G. I. Taylor: Proc. R. Soc. London A**151**, 421 (1935)
5.35 R. L. Fox: Phys. Fluids **16**, 957 (1973)
5.36 J. L. Lumley: *Stochastic Tools in Turbulence* (Academic, New York, London 1970)
5.37 C. Dopazo: Phys. Fluids **22**, 20 (1979)
5.38 D. B. Spalding: In *Thirteenth Symposium (International) on Combustion* (The Combustion Institute, Pittsburgh 1971) p. 649
5.39 F. A. Williams: In *AGARD Conference Proceedings No. 164*, ed. by M. Barrère (AGARD, Paris 1975) Chap. III
5.40 R. W. Bilger: Combust. Sci. Tech. **13**, 155 (1976)
5.41 R. W. Bilger: Combust. Sci. Tech. **11**, 215 (1975)
5.42 C. Bonniot, R. Borghi: 6th Coll. Dynamics of Gases in Explosions and Reactive Systems, Stockholm, 1977
5.43 J. E. Moyal: Proc. Comb. Phil. Soc. **48**, 329 (1952)
5.44 R. J. Adrian: Proc. Symp. Turb. in Liquids, Rolla, Missouri, 1975
5.45 A. Papoulis: *Probability, Random Variables, and Stochastic Processes* (McGraw-Hill, New York 1965)
5.46 C. Dopazo, E. E. O'Brien: Combust. Sci. Tech. **13**, 99 (1976)
5.47 V. A. Frost: Fluid Mech.-Sov. Res. **4**, 124 (1975)
5.48 J.-M. Vignon: Docteur-Ingenieur Thesis, L'Université Claude Bernard de Lyon (in French) (1979)
5.49 V. A. Bezuglov: "Change in the Rate of Transfer of Concentration", in *Proc. 20th Sci. Conf. of Moscow Physicotechnical Institute* (in Russian) (Dolgoprudnyi 1975)
5.50 A. N. Kolmogorov: C. R. Acad. Sci. SSSR **30**, 301 (1941)
5.51 J. N. Gence, J. Mathieu: Von Karman Institute Lecture Series, Brussels, May, 1978
5.52 R. E. Meyers, E. E. O'Brien, L. R. Scott: J. Fluid Mech. **85**, 233 (1978)
5.53 C. Dopazo: Phys. Fluids **18**, 397 (1975)
5.54 T. S. Lundgren: Phys. Fluids **12**, 485 (1969)
5.55 R. Srinivasan, D. P. Giddens, L. H. Bangert, J. C. Wu: Phys. Fluids **20**, 557 (1977)
5.56 T. S. Lundgren: *Statistical Models and Turbulence* (Springer, Berlin, Heidelberg, New York 1972) p. 70
5.57 E. E. O'Brien, G. C. Francis: J. Fluid Mech. **13**, 369 (1962)
5.58 Y. Ogura: J. Fluid Mech. **16**, 33 (1963)
5.59 H. L. Grant, R. W. Stewart, A. Moilliet: J. Fluid Mech. **13**, 237 (1962)
5.60 R. H. Kraichnan: Phys. Fluids **8**, 995 (1965); **9**, 1884 (1966)
5.61 R. L. Fox: Phys. Fluids **17**, 846 (1974)
5.62 R. L. Fox: Bull. Am. Phys. Soc. **22**, 1289 (1977)
5.63 V. M. Ievlev: Izv. Akad. Nauk. SSR, Mekh. Zhidk i Gaza **5** (1970)
5.64 Y. Y. Kuo: Ph. D. Thesis, State University of New York at Stony Brook, New York (1977)
5.65 S. Corrsin: J. Aeronaut. Sci. **18**, 417 (1951)
5.66 W. P. Jones, B. E. Launder: Int. J. Heat Mass Transfer **15**, 301 (1972)
5.67 C. Dopazo: Acta Astron. **3**, 853 (1976)
5.68 S. Corrsin: AIChEJ. **3**, 329 (1957)
5.69 P. A. Libby: J. Fluid Mech. **68**, 273 (1975)

5.70 C.Dopazo: J. Fluid Mech. **81**, 433 (1977)
5.71 J.M.Delhaye, J.L.Achard: "On the Use of Averaging Operators in Two-Phase Flow Modeling", Symposium on the Thermal and Hydraulic Aspects of Nuclear Reactor Safety, Vol. 1, *Light Water Reactors*, ed. by O. C. Jones, Jr., S. G. Burkoff (ASME, New York 1977)
5.72 E.E.O'Brien, C.Dopazo: "Behavior of Conditioned Variables in Free Turbulent Shear Flows", in *Structure and Mechanisms of Turbulence*, Vol. 2 (Springer, Berlin, Heidelberg, New York 1978) p. 124
5.73 E.E.O'Brien: J. Fluid Mech. **89**, 209 (1978)
5.74 C.Dopazo, E.E.O'Brien: "Intermittency in Free Turbulent Shear Flows" in *Turbulent Shear Flows*, Vol. 1, ed. by F. Furst, B. E. Launder, F. W. Schmidt, J. H. Whitelaw (Springer, Berlin, Heidelberg, New York 1979) p. 6
5.75 S.F.Ali: Ph. D. Dissertation, The Johns Hopkins University, Baltimore, Maryland (1975)
5.76 N.K.Tutu: Ph. D. Dissertation, State University of New York at Stony Brook, New York (1976)
5.77 R.G.Batt: J. Fluid Mech. **82**, 53 (1977)
5.78 J.R.Shea: J. Fluid Mech. **81**, 317 (1977)
5.79 R.E.Breidenthal: Ph. D. Dissertation, Cal Inst. Tech., Pasadena, California (1978)
5.80 K.N.C.Bray, J.B.Moss: Acta Astron. **4**, 291 (1977)
5.81 S.B.Pope: Phil. Trans. R. Soc. London A**291**, 529 (1979)
5.82 J.Janicka, W.Kolbe, W.Kollmann: "The solution of a pdf-transport equation for turbulent diffusion flames", in *Proceedings of the 1978 Heat Transfer and Fluid Mechanics Institute*, ed. by C. T. Crowe, W. L. Grosshandler (Stanford University Press, Stanford 1978) p. 296
5.83 R.E.Meyers, E.E.O'Brien: "On the Joint pdf of a Scalar and its Gradient" B.N.L. Report August 1980. B.N.L. Upton, New York

6. Perspective and Research Topics

P. A. Libby and F. A. Williams

In the preceding chapters various aspects of the theory of turbulent reacting flows are discussed in detail. The fundamental features of such flows, some of the relevant practical problems connected with turbulent combustion, approaches appropriate for the limiting cases of nonpremixed and premixed reactants, and the direct probability-density approach to the description of such flows are treated in successive chapters. There results a relatively complete view of the present status of the fundamental theory and its application to somewhat idealized flows. In this chapter we attempt to provide an overview and perspective of the field and call attention to research topics of greatest interest from both practical and fundamental points of view.

6.1 Introductory Remarks

In order to obtain tractable limits of coverage in preceding chapters consideration of some important problems in turbulent combustion is excluded. It is appropriate in this chapter to emphasize the importance and difficulties connected with these problems and to provide entries to the relevant literature. Thus in the next few sections we discuss briefly radiative transfer, turbulence in two-phase flows, and the special features of turbulent reacting flows involving high speeds. Since the theory related to these topics is not well developed, we cannot offer a list of seminal references providing points of departure for future work; however, we give an indication of the most appropriate references.

In the absence of turbulence, radiative transfer and the fluid mechanics of multiphase flows constitute specialized fields with their own complexities, achievements, and challenges. The merging of the successful techniques developed in these fields with those appropriate for the study of turbulence has not been carried out to any significant extent. Thus the literature we cite relates primarily to basic theory of radiative transfer and of multiphase fluid mechanics and provides a starting point for the desirable merger.

We have assumed at the outset of our considerations that the Mach numbers associated with our flows are suitably low. In such flows the internal energy of the gas is much greater than the kinetic energy, whence conservation of energy is expressible in terms of the static enthalpy per unit mass or in terms of the temperature. However, there are applications related to advanced propulsive devices for aircraft and missiles calling for turbulent combustion at high speeds. In the description of such flows conservation of energy is more properly handled in terms of stagnation enthalpy, i.e., in terms of the sum of the enthalpy and the kinetic energy per unit mass. Other modifications are required by phenomena accompanying high speeds. Accordingly, an area of research devoted to extensions of the present analyses to high-speed flows is indicated, and references providing entry into the available literature are given.

Another underlying assumption of the preceding presentation is that effects of buoyancy or acceleration are unimportant. These phenomena introduce terms into averaged and fluctuating forms of momentum conservation, thereby necessitating additional considerations under some conditions. These conditions apply, for example, to premixed flame propagation in gravitational fields and to buoyancy-controlled fires. Reference to existing literature on this problem will be given.

Recently novel approaches to the description of turbulent reacting flows have been proposed on the basis of identification of particular types of coherent structures such as vortices or coherently reacting slabs. Although these approaches currently remain in a primitive state of development, their potential for future development is sufficiently great that reference to some of the relevant literature is appropriate. Therefore, a brief, evaluative discussion of approaches in this category is included.

Perturbation approaches to the description of turbulent reacting flows are not addressed specifically in the preceding chapters. Since there are limiting situations in which perturbation methods are useful and others for which such methods show promise, a brief discussion of the subject is given here. References cited concern most specifically the applicable techniques since some results obtained thereby are referenced in other chapters, mainly in Chap. 4.

In all of the preceding chapters methods of closing the describing equations for predicting characteristics of the flow are presented. Although the difficulties, limitations, and resulting questions concerning the modeling are examined, it is taken for granted that the resulting equations can be solved, at least numerically. However, the numerical analysis of systems of partial differential equations constitutes a rapidly developing field of specialization with considerable difficulties. Accordingly, we consider it appropriate to call attention to this field and to provide some entries to its extensive literature.

We next categorize outstanding problems of the restricted field of turbulent reacting flows dealt with in the bulk of this volume into those pertaining to the chemical behavior and those pertaining to fluid-mechanical behavior. It is emphasized that the crucial unsolved problems that relate to chemical behavior are concerned with the effects of finite-rate chemical kinetics; as a consequence the problems of flame holding, flame-out, ignition and extinction, problems of significant practical importance, are indicated as requiring additional research. The fluid-mechanical problems cannot be identified as clearly. Certainly needed is clarification of effects arising from density inhomogeneities associated with heat release. Specifically, we discuss the need to ascertain the extent to which the models used with reasonable success for the description of constant density turbulence can be applied with little modification. The quest for such clarification is found to suggest a variety of research topics.

6.2 Radiative Transfer in Turbulent Flows

Radiative transfer in turbulent reacting flows may be associated either with radiative interactions involving species in a single phase, typically carbon dioxide and water in high-temperature gases, or radiative interactions involving two-phase flow, typically carbon particles in hot gases. While the former may include visible radiation of chemiluminescent origin, often the principal contribution to its energy content lies in the infrared portion of the spectrum and therefore this gaseous radiation sometimes is termed nonluminous radiation. The latter often possesses a broader spectrum (not of a band type) which extends appreciably into the visible region at high temperatures (>1000 K) and therefore has been termed luminous radiation; a gray-body approximation is better for this luminous radiation than for radiation from gases.

Vincenti and *Traugott* [6.1] provide an excellent introduction to radiative transfer in fluid flows including an extensive bibliography. Important applications of such transfer to turbulent flows have been made in geophysical problems, e.g., those concerned with air-sea interaction and with stratification in the upper atmosphere. *Coantic* [6.2] and *Townsend* [6.3] should be consulted for these applications. Methods for calculating radiative transfer in systems of engineering importance were given by *Hottel* and *Sarofim* [6.4]. A recent review [6.5] update these techniques and exhibit the currently tractable levels of approximation for turbulent reacting flows. Illustrative of the current status of the incorporation of radiative processes in predictive methods for such flows is the recent paper of *Tamanini* [6.6].

To include energy gained and lost by the gas through radiative transfer the fundamental equation of energy conservation for multicomponent mixtures, e.g., (1.10) or (1.16), must be modified by addition of a term $\nabla \cdot \boldsymbol{q}_r$ where $\boldsymbol{q}_r$ is the radiation flux vector describing the radiative addition to conductive and diffusive energy fluxes. The most difficult problem associated with the introduction of radiation is that of expressing $\boldsymbol{q}_r$ in terms of other dependent variables describing the flow.

When applied to a flowing medium, the most general formulation of radiative transfer leads to an expression for $\boldsymbol{q}_r$ involving a multiple integral over the spectrum of the radiation and over all space involving significant emission, absorption, or scattering. The integrand depends in a complex fashion on the local and instantaneous temperature and composition. Thus the time-dependent equation of energy conservation for a fluid leads in this case to a partial differential-integral equation. If this equation is to provide the basis for the calculation of time-averaged quantities within a turbulent medium, we face the need to average in some fashion the multiple integral. Significant simplifications are required if useful results are to be realized; two limiting cases suggest directions such simplifications can take.

In the case of an optically thin medium, the case which appears to be of greatest applicability to turbulent combustion free of solid particles, the probability of an emitted photon being absorbed or scattered by another molecule in the gas is small. As a consequence, $\nabla \cdot \boldsymbol{q}_r$ is a positive quantity dependent on the local, instantaneous density, temperature, and composition. While the dependence on composition is complex [6.5], the proportionality $\nabla \cdot \boldsymbol{q}_r \propto \varrho T^4$ may be justifiable, although variation of emissivity with temperature causes the exponent of T to be questionable. To facilitate analysis further approximations of the type $\nabla \cdot \boldsymbol{q}_r \propto \bar{T}^4$ are prevalent [6.6] where $\bar{T}$ is an average temperature for emission of radiation. The value of $\bar{T}$ is not calculated but is estimated from experiment to be somewhat below typical adiabatic flame temperatures [6.5]. Influences of turbulence on this type of approximation deserve further study [6.5]. For optically thin media calculation of radiation absorbed and emitted by surfaces bounding the flow is relatively straightforward in principle but geometrically complex [6.4–6].

In the opposite case an emitted photon is absorbed in a distance not exceeding the Kolmogorov length and the medium is said to be optically thick. Radiative transfer then becomes essentially diffusive and $\boldsymbol{q}_r \propto k_r \nabla T$, where $k_r \propto T^3$, again with a proportionality factor dependent on composition. This limit appears to be encountered seldom for turbulent reacting flows of interest.

In both of these limiting cases the relative simplicity of the radiative flux vector permits the usual averaging process to be carried out. Correlations between the temperature and mass fractions arise and should be treated, for example along the lines discussed in Chaps. 1, 3, and 4; at present, such correlations usually are neglected. However, even with the simplifications attendant with these limiting cases we must keep in mind that significant radiative transfer requires careful assessment of some of the assumptions found to be effective when radiation is negligible. For example, the use of a conserved scalar in nonpremixed reacting flows as discussed in Chap. 3 cannot apply to the enthalpy when radiation is significant. In the optically thin limit the radiative transfer term introduces an algebraic source term which destroys the analogy among the enthalpy and elements. In the optically thick limit the effective thermal conductivity associated with k_r is likely to cause the Lewis number to differ from unity, again excluding enthalpy from the analogy.

Two-phase flows frequently are involved in practical combustion systems for which radiative transfer is significant since solid or liquid particles in the gas often provide the main emitters and absorbers. Such particles may, for example, be soot which forms in fuel-rich portions of the flow or unburned fuel particles injected into the combustion chamber. They may absorb radiant energy and transfer it by conduction to the surrounding gas or be heated by the gas and radiate energy away. If the number of particles is sufficiently high, particle-particle radiative transfer may be significant while the gases are transparent to radiation.

If the particles are sufficiently small, they may be treated as gaseous species with low molecular diffusivities, and their radiative interactions then appear through $\nabla \cdot \boldsymbol{q}_r$. This is the manner in which soot radiation is analyzed; see, for example, [6.5, 6]. The presence of the particles then may simplify the description of the radiation by making a gray-body approximation more acceptable.

In an alternative description the conservation equations are written for the gaseous phase and the influence of the particles appears fundamentally through complicated boundary conditions. With this description, which is needed if particles are sufficiently large, $\nabla \cdot \boldsymbol{q}_r$ applies only to the gas and the problem of accounting for radiant transfer in the presence of turbulent motion of particles involves additional challenges. Little research has been performed on problems of this type.

6.3 Two-Phase Turbulent Flows

Many combustion flows of practical interest involve multiple phases. For example, solid particles occur as pulverized coal or as soot from carbon formation and droplets occur as injected liquid fuel or form by condensation. Under conditions of turbulent flow such combustion is difficult to analyze from a fundamental point of view and has not been extensively studied. The conservation equations for two-phase flows can be developed from several points of view depending on the size and number density of particles or droplets involved. *Soo* [6.7] provide an extensive treatment of the fluid mechanics of two-phase flow but with little material concerning turbulent phenomena in such flows. *Williams* [6.8, 9] present equations describing spray combustion but without reference to turbulence. *Marble* [6.10] review the theory of dusty gases, two-phase flows with particles having negligible volume and small number densities so that particle-particle interactions can be neglected.

To illustrate the nature of the phenomena which are to be considered in multiphase flows consider a flow involving solid particles of a single composition and size carried in a gas. Assume that the medium may be treated as a dusty gas. In this case the gas and particles are considered as coexisting continua and conservation equations are written for the gas and for the particles. In the equations for the particles a continuous density $\varrho_p(\boldsymbol{x}, t) = n_p(\boldsymbol{x}, t)m$, where m is the mass of a single particle and n_p the number density of particles, is introduced. In addition the continuum velocity of the particles $\boldsymbol{v}(\boldsymbol{x}, t)$ and their temperature $T_p(\boldsymbol{x}, t)$ are defined and complete specification of the solid phase in the mixture. The interaction between the two phases is contained in corresponding terms in each pair of conservation equations; for example, in the equations of momentum conservation the difference between the gas velocity $\boldsymbol{u}(\boldsymbol{x},t)$ and the particle velocity $\boldsymbol{v}(\boldsymbol{x},t)$ leads to a contribution to the momentum of the gas and an opposite contribution to the momentum of the particles. Similarly heat lost by the particles increases the energy of the gas while heat gained by the particles cools the gas. The conservation equations developed along these lines can be averaged in order to derive equations for various statistical quantities, for example, for the mean particle velocities $\bar{\boldsymbol{v}}(\boldsymbol{x})$ and for the mean particle density $\bar{\varrho}_{\mathrm{p}}(\boldsymbol{x})$. Modeling to achieve closure is required.

In most turbulent reacting flows of practical interest the particles or droplets exhibit a range of sizes; particles or droplets due to condensation have various sizes depending on the time and nature of the ambient surrounding them since birth. Particles or droplets being consumed by chemical reactions likewise have various sizes depending on their history. Under these circumstances additional source terms describing interaction between the two phases must be considered [6.8]. More significant is the need to treat particle size as a variable [6.8]; it is possible to approximate a continuous distribution of particle sizes by a small number of families of particles each with a fixed size and with exchange from one family to the neighboring one introduced to account for consumption or particle growth. Each family may be considered to have its separate density, temperature, and velocity. Alternatively, a continuous distribution may be retained [6.8].

It is obvious that the treatment of multiphase turbulent reacting flows from a fundamental point of view involves significant difficulties. Thus it is perhaps not surprising that significant advances have not been made in this field. *Mellor* [6.11] provide a recent review of the state of knowledge for practical systems involving spray combustion and concludes that only qualitative trends can be predicted for emissions by existing techniques. If the presence of condensed phases is significant, then fundamental objections may be raised against existing techniques and further research is warranted.

6.4 Effects of High Mach Number

Some propulsion devices for hypersonic aircraft and missiles involve turbulent combustion with high gas speeds. The most extreme case and one suitable for our discussion arises in ramjets involving combustion at supersonic speeds, so-called *Scramjets* (cf. [6.12]). Such ramjets are of interest for the propulsion of hypersonic aircraft with either commercial, military or space applications and of hypersonic cruise missiles and are therefore the subject of current research and advanced development. In these engines air is slowed and thus compressed by a carefully designed inlet from hypersonic flight speeds to Mach numbers of approximately 2–4. The air then enters a combustion chamber and is mixed with fuel, often selected to be gaseous hydrogen. The products of combustion are expanded in an exhaust nozzle to achieve thrust.

The present state of knowledge of supersonic combustion is given in review articles by *Ferri* [6.12] and *Jones* and *Huber* [6.13]. We call attention to two aspects of such combustion of interest in this perspective. The stagnation enthalpy $h_s = h + 1/2\, V^2$, where h is the static enthalpy per unit mass and V is the instantaneous resultant velocity is the obvious variable to consider relative to energy conservation. With a single diffusion coefficient and Prandtl and Schmidt numbers equal to unity the conservation equation for h_s is (1.10) with h replaced by h_s. Thus much of the methodology for low-speed flows is readily extended to high-speed cases. However, the assumption that the pressure is thermochemically constant must be carefully reviewed since small changes in velocity lead to large changes in static pressure in high-speed flows. In addition we expect the effects of fluctuations in pressure to be considerably greater and therefore possibly more influential.

High speeds make the approaches described in Chaps. 3 and 4, based as they are on notions of near or complete equilibrium and on fast chemistry, less likely to prevail. The flow times are so short that considerable attention must be devoted to ignition and flame holding (cf. [6.14]), phenomena which are difficult to treat from fundamental considerations even for low-speed flows.

These brief observations suggest that a wide range of problems connected with turbulent combustion in high-speed flows are of practical interest and call for attention by both theoretical and experimental means.

6.5 Effects of Buoyancy

Although in our general discussion of the describing equations in Chap. 1, we include body forces in the momentum balance via the terms g_i [cf. (1.19, 43, 50)], developments in the chapters which follow largely ignore the effects of such forces. In many flows the velocities are large enough and the length scales small enough so that body forces may be neglected. However, there are flows of practical interest in which this neglect is inappropriate. Body forces arise in several different ways; the most obvious is associated with buoyancy, i.e., with the effect of gravity on fluid elements of different densities. In flows with curved mean streamlines and density inhomogeneities, e.g., in swirling flows, body forces arise from curvature (cf. [6.15]). Finally, premixed turbulent flames are sometimes accelerated by unsteadiness due to the hydrodynamics of the flow field in which they are embedded; in a frame of reference fixed on the flame such acceleration leads to a body force due to acceleration (cf. [6.16]). To simplify our discussion we shall confine our attention to buoyancy effects although the same general considerations apply to the other sources of body forces.

The most important case in which buoyancy plays a significant role in nonreacting turbulent flows is the atmospheric boundary layer (cf. [6.2, 17]). In such a layer turbulent exchange of temperature and species is either decreased or increased depending on whether the mean temperature distribution corresponds to a stable or unstable atmosphere. Similar phenomena prevail in the ocean where variations in both salinity and temperature can lead to buoyancy effects (cf. [6.18]). In large furnaces and in fires the flow rates and length scales are such that buoyancy can alter the flow and thereby its effective chemical behavior. *Tamanini* [6.6, 19] studied the turbulent, buoyancy-controlled round jet and turbulent wall fires.

Since the influence of buoyancy in turbulent reacting flows appears as an interaction between density inhomogeneities due to heat release and gravity, it is sufficient for our purposes to focus on the effect of gravity in a variable density flow without chemical reaction, e.g., on a vertical two-dimensional jet of a light foreign gas discharging into a quiescent ambient consisting of air. If we orient the x axis in the vertical direction, the body force appears only in the x-wise momentum equation as $-\varrho g$ where g is now the gravitational constant. When this equation is averaged, this term becomes $-\bar{\varrho}g$. If we proceed along the lines discussed in Chap. 1 using Favre averaging to develop a system of equations for the description of the mean values and some of the second-order correlations, e.g., the intensities of the velocity and concentration fluctuations, we find that the $-\bar{\varrho}g$ term in the mean x-wise momentum equation is the only explicit manifestation of buoyancy (cf. [6.20]). However, buoyancy influences significantly the intensities and other second-order correlations indirectly via its effect on the various production terms in the conservation equations and on the mean pressure gradients. In fact one important feature of turbulent flows with significant buoyancy effects is the anisotropy of the velocity fluctuations due to the preferential influence of gravity on the fluctuations in one direction. This anisotropy raises questions concerning the modeling of various terms, e.g., of the dissipation of velocity and concentrations fluctuations (cf. [6.19]).

Thus there appear to be interesting and important modifications of the analyses described in Chaps. 3 and 4 required when body forces arise in turbulent reacting flows. In particular, modifications of the modeling used to achieve closure may be called for in order to account for the preferential direction of the operative body force.

6.6 Approaches Identifying Coherent Structures

A newly emerging class of approaches to the prediction of turbulent reacting flows may be viewed as resting in some manner on the identification of coherent structures within the flows. Background for the introduction of such approaches has been given in Sect. 1.15. Although these methods are in early stages of development and their utility cannot be assessed properly at present, it is of interest here to provide entries into their literature.

Spalding and co-workers [6.21, 22] are studying a method that he terms ESCIMO, standing for engulfment, stretching, coherence, interdiffusion, and moving observer. He identifies parcels of fluid which he calls "folds" and attributes to them both "demographic" and "biographic" aspects.

The biographic aspects are the easiest to understand. Each fold is followed by a moving observer, remains coherent with deterministic physics occurring within it, experiences stretching, chemical reaction, and molecular diffusion. Within each fold time is termed "age" and the biographic aspect involves the calculation of the history of the fold as a function of its age subject to prescribed initial conditions and stretchings which might depend on results of the demographic analysis. *Spalding* emphasizes repeatedly that it is possible to calculate complex chemical kinetics within a fold by exercising appropriate computer routines [6.21].

The demographic aspects, being probabilistic, are more difficult to understand. Conservation equations are written for the fraction of fluid in each age range, accounting for turbulent engulfment of folds and for flow of material in and out of the spatial region investigated. The demographic analysis is performed independently of the biographic analysis and requires its own initial and/or boundary conditions which, however, may be taken to depend on results of the biographic analysis, thereby introducing a coupling of the two aspects.

It is difficult to compare the ESCIMO approach with others because of its heuristic foundation. The equations employed, particularly for the demographic aspects, have not been derived formally from the underlying conservation equations given in Chap. 1. Therefore, it is difficult to ascertain what physical effects have been excluded and to assess the fidelity to be expected from predictions. Further investigation of the basis of the approach seems warranted.

Applications of ESCIMO to simplified flows such as the well-stirred reactor [6.21] are beginning to appear. At the present stage of its development, a variety of approximations are evident. These include performance of the biographic analysis only for one representative fold, transport of the fold at the average local velocity, and assignment of an average strain rate based on average velocity gradients. Nevertheless, certain qualitatively realistic results have been obtained, such as extinction of the stirred reactor if the chemistry is too slow. Thus there is some promise in the approach and further development of it is both desirable and intellectually challenging.

Another approach which may be considered to involve coherent structures is the random-vortex method for numerical modeling of turbulent reacting flows [6.23, 24]. This approach is an outgrowth of random-choice methods that have achieved success in certain gas dynamic problems but its physical basis differs radically from that of the gas dynamic analyses. The general philosophy is to incorporate analytical solutions of simplified subproblems into numerical routines for generating solutions to more difficult problems. Unlike ESCIMO the procedures employed are derived with stated assumptions from the underlying conservation equations and therefore the fidelity of the predictions is assessable more readily.

In the random-vortex method the coherent structures are vortices. They are tracked in a more refined manner than the moving-observer sense of ESCIMO; Lagrangian computations are performed, with each vortex being moved under the influence of every other vortex according to its Biot-Savart interaction. Randomness appears for example in the spatial distribution of vortices. Operator-splitting techniques play an essential role in the analysis. There are difficulties associated with the presence of solid boundaries requiring special procedures for the prescription of vortex generation. Also, for computational manageability, prescriptions are needed for eliminating vortices having relatively small influences so that the number of vortices that must be tracked does not grow beyond computer capabilities.

At present a fundamental deficiency of the random-vortex method is that the flows predicted are strictly two dimensional, even on a time-resolved basis. In other words all vortex lines are taken to be straight and parallel. It is well known that in turbulent flows three-dimensional vortices are significant. This is true at small scales even for those special flows in which the large vortices are predominantly two dimensional. Extension of the random-vortex method to account for three-dimensional vortices is difficult in that new physical phenomena arise, such as the interaction of a vortex line with itself. Because of such notable difficulties the ultimate success of the random-vortex method cannot be predicted, although at present its potential range of applicability appears to be quite limited [6.25]. Nevertheless, promising results have been obtained that qualitatively match well with experimentally observed reacting flows in a two-dimensional, turbulent combustor [6.23]. Progress toward three dimensionality has been achieved, and further development of the method will be quite worthwhile.

6.7 Perturbation Methods Appropriate to Turbulent Reacting Flows

Perturbation methods are playing an increasingly active role in general for studies of reacting flows, as is exemplified by the appearance of a monograph [6.26] devoted to the subject. These methods take advantage of a parameter being small or large to simplify the conservation equations. For reacting flows there are numerous such parameters [6.27] engendering analyses of many limiting cases. Less work based on perturbation methods [6.27] has been devoted to turbulent reacting flows but recently there has been an increase in such applications. A few general comments on such methods have been given in Sects. 1.14 and 1.16 and further amplification has been given for nonpremixed systems in Sect. 3.2.3 and for premixed systems in Sect. 4.4.3.

The expansion parameters that thus far have been found to be most useful differ for nonpremixed and premixed systems. In nonpremixed systems a parameter measuring the departure from equilibrium has been employed [6.28, 29]. The introduction of an expansion in this parameter provides a formal assessment of the accuracy of the equilibrium or fast-chemistry approximations in turbulent diffusion flames. It also provides a means for developing corrections to the fast-chemistry approximation in situations where finite rates are significant.

In premixed systems the parameter that has received the most attention in perturbation approaches is the nondimensional activation energy, termed β at the end of Sect. 1.16. Large values of β identify conditions under which the reaction rates increase rapidly with increasing extent of reaction, a behavior representative of combustion. A large value of β is associated with thin zones of reaction and with the occurrence of well-defined phenomena of ignition and extinction. Asymptotic expansions for large β tend to be more complicated to analyze because they generally produce distinct regions within the flow, some having negligible rates of reaction and others yielding fast chemistry or near-equilibrium conditions. Methods of matched asymptotic expansions then are needed in carrying out the perturbation analysis.

For premixed combustion large β analyses have been performed only within the context of a further expansion, namely small gradients of turbulent fluctuations [6.30, 31]. Specifically, the ratio of the thickness of a premixed laminar flame to the Kolmogorov scale of the turbulence is treated as a small parameter. This leads to a formal specification of conditions appropriate to a wrinkled laminar-flame structure of the turbulent combustion. Turbulent flame-speed formulas for wrinkled laminar flames have been derived [6.30, 31] and influences of flame stretch and the curvature of laminar flamelets on turbulent flame propagation have been calculated by allowing molecular transport coefficients for heat and reactants to differ [6.30]. These analyses which should be viewed as the first in a series of studies illustrate the techniques appropriate for perturbation methods applied to turbulent reacting flows.

It is beyond the scope of the present volume to expound perturbation techniques. Readers must consult the cited references. For future work the large parameter β provides an attractive parameter for expansions in both premixed and nonpremixed situations; with regard to the latter progress has been made in calculation of production rates of pollutants such as NO in turbulent diffusion flames [6.28]. Other parameters among those worthy of further investigation are the ratio of the heat release to the initial thermal enthalpy (often a large quantity), Damköhler numbers measuring departures from equilibrium, and nondimensional quantities involving various scales associated with the turbulence. Perturbation methods cannot be expected to describe successfully the entire range of conditions of interest in turbulent reacting flows but should be considered as means for obtaining well-founded results in suitable limiting cases, often without the necessity of introducing questionable closure approximations. Thus perturbation techniques occupy a unique position and merit further exploitation.

6.8 Numerical Methods Appropriate to Turbulent Reacting Flows

The various methods for the description of turbulent combustion introduced in Chaps. 3–5 lead in general to systems of partial differential equations for the calculation of the key statistical quantities characterizing the velocity field and the state variables. When the approximations attendant with boundary layers are applicable so that the pressure is given, these equations are generally parabolic; in these cases initial data at a particular streamwise station and boundary data are required to complete the formulation. In this regard it should be noted that some models adopted to achieve closure modify the nature of the boundary layer approximation so that the resultant equations are hyperbolic [6.32]. The initial data take on special significance in this case and the boundary data which can legitimately be imposed must be carefully considered. The equations in the most general case of turbulent combustion are elliptic so that boundary conditions must be imposed on a closed contour.

Despite the enormous increases in the speed and capacity of high-speed computers achieved over the past decade there remain relative to flow field calculations definite limitations on the geometric and thermochemical complexity and on the spatial resolution which can be currently handled. As a consequence of such increases and such limitations an extensive literature related to computational fluid mechanics has evolved and continues to grow. *Roache* [6.33] and *Holt* [6.34] provide current books on the field; a journal, the International Journal of Computers and Fluids, is devoted exclusively to computational fluid mechanics while other journals, for example, the Journal of Computational Physics, Computational Methods in Applied Mechanics and Engineering, and the SIAM Journal of Numerical Analysis, regularly contain contributions to the subject. Finally, we note that every two years a meeting identified as the International Conference on Numerical Methods in Fluid Dynamics results in proceedings reporting the most recent developments.

While only a portion of the activity indicated by this discussion deals specifically with turbulent flows, many of the methods developed in computational fluid mechanics apply to the equations for the statistical quantities characterizing turbulent combustion. We cite the extensive literature associated with the numerical solution of such equations to impress the unwary with the difficulties that can be encountered. The replacement of the various partial derivatives in a set of equations by difference representations and the formulation of the resultant equations as a set of algebraic equations are deceptively straightforward; however, problems of convergence, error growth, numerical stability, numerical viscosity, etc., problems carefully considered in the cited literature, can readily lead to essential difficulties, meaningless results, and costly frustration. Our message is that care is indicated.

It is beyond the scope of our discussion to consider the details of the numerical methods appropriate to the numerical analysis of turbulent combustion. However, we do wish to note that one feature of turbulent reacting flows introducing special difficulty relates to the existence in some circumstances of relatively concentrated reaction zones with the consequence that adequate spatial resolution of the flow properties within such zones requires a fine computational mesh and a self-adjusting strategy to locate that mesh. If a large number of dependent variables is computed at each mesh point, the requirement of adequate spatial resolution may severely tax the available computational power.

6.9 Outstanding Problems Related to Chemistry

Problems related to chemistry are defined as those associated with the chemical source term $\dot{w}_i$ or with averages involving this term directly. Extensive discussion of the particular average $\dot{w}_i$ has been given in Chaps. 3 and 4; fundamental difficulties in evaluating this quantity are mentioned first in Chap. 1 [cf. (1.56)]. The presentations in Chaps. 3 and 4 show that under conditions categorized loosely as those of fast chemistry, the chemistry problem can be handled adequately in idealized flows that nevertheless possess engineering interest. For partially premixed situations intermediate between those treated in Chaps. 3 and 4 comparably successful methodologies are not available. Progress may be made for such situations by augmenting the description of the reaction zones to allow for a narrow sheet of premixed combustion on one side of the diffusion-controlled reaction surface. The problem becomes more difficult if some of the important reactions are too slow for such an approximation to be applicable; the opposite limiting case of slow reactions is amenable to analysis only if turbulent fluctuations are small enough to affect $\dot{w}_i$ to a negligible extent. Additional research on partially premixed reactive flows is warranted.

The character of the chemistry problem varies appreciably with the predictive approach adopted. For methods based on prediction of *pdf* evolution as presented in Chap. 5 the chemistry problem that appears directly is relatively benign since concentrations and temperature which occur in the complicated functional dependence are independent variables and two-point quantities are not involved in $\dot{w}_i$. However, as indicated in Chap. 5 these methods can be relatively difficult to pursue in flows of direct practical interest. With asymptotic methods for strongly temperature-dependent chemistry the $\dot{w}_i$ term dictates adoption of a randomly moving frame of reference [6.30, 31]. Methods based on ideas of coherent structures alleviate the chemistry problem again by using moving frames of reference. Irrespective of the approach there are complications associated with the chemistry which call for further study.

One class of problems for which approaches to finite rate chemistry are suggested involves major species which experience fast chemistry but trace species which react at slower finite rates. *Bilger* [6.28], and *Williams* and *Libby* [6.29] analyze this situation for nonpremixed reactants and indicate that significant simplifications ensue. Although there appear to be chemical systems and flow conditions of practical interest that are amenable to such an approach, insufficient exploitation of these ideas has appeared.

An alternative approach to finite rate effects involves the extension of the conserved scalar approach described in Chap. 3 along the lines suggested by *Janicka* and *Kollman* [6.35]. The essential point is that the thermochemical state is assumed to be described by a pair of variables, the conserved scalar and either a second mass fraction or a convenient combination of mass fractions. In this situation the probability density approach involves approximating a two-variable function of the form $P(\xi, Y_N; \mathbf{x})$ where Y_N denotes the second variable. Progress toward this objective might be made by considering stretched laminar diffusion flames within the context of a Damköhler number expansion. Further studies of the most appropriate description of such a pdf are required to complement those reasonably developed for single-variable functions.

Our discussion of these approaches with their obvious limitations suggests the need for additional research. In view of the limitation of present treatments of chemical effects a variety of problems of practical importance cannot be dealt with from a fundamental point of view. Included among such problems are flame holding, blowout, thermal ignition, and total extinction, all problems in which chemical kinetic effects play a central role. As a consequence accounting for the influence of the properties of the turbulence on these phenomena remains largely empirical. We thus see on the one hand the limitations of current predictive methods and on the other the richness of the chemistry problem in turbulent combustion of both fundamental and practical interest. This richness calls for a variety of approaches, from careful treatments of highly idealized situations, to direct probability density approaches such as described in Chap. 5, to modeling efforts implementing the moment methods. In all cases it is important to establish close coordination between theoretical and related experimental efforts.

6.10 Outstanding Problems Related to Fluid Mechanics

It is not possible to be very specific regarding fluid-mechanical problems because of current uncertainties. In the moment method even when pdf shapes are approximated such problems relate to the models used to close the describing equations, i.e., to represent the various terms not immediately related to the primary dependent variables. We refer to mean stresses and fluxes, to pressure-velocity and pressure-scalar correlations and to dissipation terms accounting for the influence of molecularity. Models of these various terms are generally adaptations with little alteration of those forms which are more or less successful in the prediction of turbulent flows with constant density. It is thus important to consider whether chemical reaction and variations in density can influence such models.

Whether closure is effected at the level of the conservation equations for mean quantities or at the level of intensities, fluxes, and stresses, gradient transport assumptions in one form or another are almost universally used in the moment method. Gradient transport is remarkably effective in many turbulent shear flows with constant density despite the absence of a fundamental basis. In turbulent reacting flows there are reasons for additional concern for the validity of gradient transport; in some flows with significant heat release the zones of intense chemical activity are sometimes relatively thin with the consequence that large rates of mean strain arise. Such strains suggest stresses and fluxes not accurately represented by gradient transport. A further concern relative to such transport pertains to the possible influence of mean pressure gradients on transport in turbulent flows with inhomogeneities in density such as arise with heat release. Even in the absence of heat-release or pressure-gradient effects, the chemical source term may significantly influence the description of turbulent fluxes of reacting species through fine-scale interactions.

As indicated in Sect. 4.6 *Libby* and *Bray* [6.36] recently provide a theoretical analysis of premixed turbulent flames orthogonal to the oncoming reactants based on an extension of the Bray-Moss model of premixed combustion. The analysis carefully avoids the assumption of gradient transport in order to provide an assessment of the validity of that assumption in premixed flames. Their results show that for modest amounts of heat release the new theory agrees quite satisfactorily with a conventional theory based on gradient transport. However, as the degree of heat release increases the mean flux of product is directed downstream, i.e., contrary to gradient transport, over greater and greater portions of the reaction zone. The physical explanation is readily found to reside in an interaction between the weak mean pressure drop across the flame and the inhomogeneities in density associated with cold reactants and hot products. Experimental results by *Moss* [6.37] in a premixed turbulent flame provide support for these theoretical predictions.

These results engender skepticism regarding the applicability of gradient transport models in variable density turbulence in general and turbulent reacting flow in particular. Additional research may show that for oblique turbulent flames with premixed reactants and for nonpremixed flames the effect of mean pressure gradients may not be as significant as found for normally oriented premixed flames.

It is worth noting that countergradient fluxes are also predicted in constant density flames by the analysis of *Clavin* and *Williams* [6.31]. Their analysis concerns a laminar flame which is wrinkled by weak turbulence. Perturbation methods are used so that no modeling assumptions are required. It is found that the mean turbulent flux of reactant changes sign in the interior of the turbulent reaction zone. Thus again in an entirely different flow analyzed by an entirely different method countergradient diffusion is revealed.

We also call attention to the modeling of another set of terms, those related to dissipation. See, for example, (1.54). In Chap. 3 and 4 the various dissipation terms which arise in the conservation equations are usually modelled in the same fashion as in constant density turbulence. Thus the dissipation of fluctuations of the conserved scalar ξ is usually taken to be proportional to $\bar{\varrho}\tilde{q}^{1/2}\widetilde{\xi''^2}/l$ where we use Favre averages. However, alternative models for the dissipation terms can be developed for reacting flows; *Libby* and *Bray* [6.38] show by means of a straightforward calculation based on the laminar flamelet picture of reacting surfaces within the reaction zone of a premixed flame that $\tilde{q}^{1/2}$ in the usual model should be replaced by the laminar flame speed, a quantity dependent on the chemical system and the flow conditions. Thus the superficial resemblance of the two models may mask significant implications.

In concluding this section we remark on the modeling of the various terms involving the pressure fluctuations. These terms are generally considered to describe important processes in constant density turbulent flows and considerable effort has been devoted to the development of suitable models for such flows [6.39]. Again, if pressure fluctuations are considered at all in the analysis of turbulent reacting flows, these models are adopted with little modification. The increase in noise associated with combustion engenders skepticism regarding this practice even though estimates indicate that the fraction of the turbulent kinetic energy radiated as sound may be small. There are extreme difficulties in measuring fluctuating pressures even in constant density turbulence so that new techniques are certainly required for investigations in turbulent combustion.

6.11 Concluding Remarks

We have closed this volume with a brief review of several aspects of the subject of turbulent reacting flows, aspects whose inclusion within the previous chapters would unduly complicate the exposition; would violate the spirit of a tutorial discussion since they are at the cutting edge of the subject; and finally while relevant to the subject are beyond the scope of the volume. Our message is that most of these aspects involve questions, difficulties, and current limitations whose resolution is of both fundamental and practical interest and is the source of stimulating research topics. We hope that the cited references provide entries to the relevant literature appropriate to the exploitation of these topics and that the current volume may provide a helpful starting point for future investigations.

References

6.1 W.C.Vincenti, S.C.Traugott: "The Coupling of Radiation Transfer and Gas Motion", *Annual Reviews of Fluid Mechanics*, Vol. 3 (Annual Reviews, Palo Alto 1971)

6.2 M.F.Coantic: "An Introduction to Turbulence in Geophysics, and Air-Sea Interactions", AGARDograph No. 232 (1978)
6.3 A.A.Townsend: J. Fluid Mech. **4**, 361–375 (1958)
6.4 H.C.Hottel, A.F.Sarofim: *Radiative Heat Transfer* (McGraw-Hill, New York 1967)
6.5 J. de Ris: "Fire Radiation – A Review", *Seventeenth Symposium (International) on Combustion* (The Combustion Institute, Pittsburgh 1979) pp. 1003–1016
6.6 F.Tamanini: "A Numerical Model for the Prediction of Radiation-Controlled Turbulent Wall Fires", Ref. 6.5, pp. 1075–1085
6.7 S.L.Soo: *Fluid Dynamics of Multiphase Systems* (Blaisdell, Waltham, MA 1967)
6.8 F.A.Williams: *Combustion Theory* (Addison-Wesley, Reading, Mass 1965) Chap. 11
6.9 F.A.Williams: "General Description of Combustion and Flow Processes", *Liquid Propellant Rocket Combustion Instability*, ed. by D.T.Harrje, NASA SP-194 (U.S. Government Printing Office, Washington, D.C. 1972) pp. 37–45
6.10 F.E.Marble: "Dynamics of Dusty Gases", *Annual Reviews of Fluid Mechanics*, Vol. 2 (Annual Reviews, Palo Alto 1970)
6.11 A.M.Mellor: "Turbulent-Combustion Interaction Models for Practical High Intensity Combustors", Ref. 6.5, pp. 377–387
6.12 A.Ferri: J. Aircr. **5**, 3–10 (1968)
6.13 R.A.Jones, P.W.Huber: Astronaut Aeronaut. **16**, 38–48 (1978)
6.14 J.S.Evans, C.J.Schexnayder, Jr.: "Critical Influence of Finite Rate Chemistry and Unmixedness on Ignition and Combustion of Supersonic H_2-Air Streams", AIAA Paper No. 79-0355 (1979)
6.15 D.G.Lilley: AIAA J. **12**, 219–223 (1974)
6.16 G.D.Lewis: "Centrifugal-Force Effects on Combustion", *Fourteenth Symposium (International) on Combustion* (The Combustion Institute, Pittsburgh 1973) pp. 413–419
6.17 P.Bradshaw, J.D.Woods: "Geophysical Turbulence and Buoyant Flows", in *Turbulence*, 2nd. ed. (Springer, Berlin, Heidelberg, New York 1978) pp. 171–192
6.18 T.R.Osborn, C.S.Cox: Geophys. Fluid Dyn. **3**, 321–345 (1972)
6.19 F.Tamanini: "Algebraic Stress Modeling in a Buoyancy Controlled Turbulent Shear Flow", *Symposium on Turbulent Shear Flows* (University Park, Penna. 1977) pp. 6.29–6.28
6.20 R.W.Bilger: Prog. Energy Combust. Sci. **1**, 87–109 (1976)
6.21 D.B.Spalding: "The Influence of Laminar Transport and Chemical Kinetics on the Time-Mean Reaction Rate in a Turbulent Flame", Ref. 6.5, pp. 431–440
6.22 A.S.C.Ma, M.A.Nosier, D.B.Spalding: "Application of the "ESCIMO" Theory of Turbulent Combustion", AIAA Paper No. 80-0014 (January 1980)
6.23 A.F.Ghoniem, A.J.Chorin, A.K.Oppenheim: "Numerical Modeling of Turbulent Combustion in Premixed Gases", *Eighteenth Symposium (International) on Combustion* (The Combustion Institute, Pittsburgh (to appear 1981).
6.24 A.J.Chorin: J. Fluid Mech. **57**, 785–796 (1973)
6.25 W.T.Ashurst: "Numerical Simulation of Turbulent Mixing Layers via Vortex Dynamics", *Turbulent Shear Flows I* (Springer, Berlin, Heidelberg, New York 1979) pp. 402–413
6.26 J.Buckmaster, G.S.S.Ludford: *Theory of Laminar Flames* (Cambridge University Press, Cambridge) (to appear 1980)
6.27 F.A.Williams: "Current Problems in Combustion Research", in *Dynamics and Modeling of Reactine Systems* (Academic Press, New York, to appear 1980)
6.28 R.W.Bilger: Combust. Sci. Tech. **22**, 251–261 (1980)
6.29 F.A.Williams, P.A.Libby: "Some Implications of Recent Theoretical Studies in Turbulent Combustion", AIAA Paper 80-0012 (1980)
6.30 P.Clavin, F.A.Williams: J. Fluid Mech. **90**, 589–604 (1979)
6.31 P.Clavin, F.A.Williams: "Effects of Lewis Number on Propagation of Wrinkled Flames in Turbulent Flows", *Proceedings of the Seventeenth International Colloquium on Gasdynamics of Explosions and Reactive Systems* (to appear 1980)
6.32 P.Bradshaw, D.H.Furness: J. Fluid Mech. **46**, 83 (1971)
6.33 P.J.Roache: *Computational Fluid Dynamics* (Hermosa Publishers, Albuquerque 1972)

6.34 M. Holt: *Numerical Methods in Fluid Dynamics* (Springer, Berlin, Heidelberg, New York 1977)
6.35 J. Janicka, W. Kollman: "A Two-Variable Formalism for the Treatment of Chemical Reactions in Turbulent H_2-Air Diffusion Flame", Ref. 6.5, pp. 421–430
6.36 P. A. Libby, K. N. C. Bray: AIAA J. (to appear 1980)
6.37 J. B. Moss: Combust Sci. Tech. (to appear 1980)
6.38 P. A. Libby, K. N. C. Bray: Combust. Flame (to appear 1980)
6.39 W. C. Reynolds, T. Cebeci: "Calculation of Turbulent Flows". in *Turbulence*, 2nd. ed. (Springer, Berlin, Heidelberg, New York 1978) pp. 193–229

Subject Index

The page number given in *italics* refer to the basic explanation of the entry. Only introductory citations are given for oft-mentioned subjects.

Cavitation and Inhomogeneities in Underwater Acoustics

Proceedings of the First International Conference Göttingen, Fed. Rep. of Germany, July 9–11, 1979
Editor: W. Lauterborn

1980. 192 figures, 6 tables. XI, 319 pages
(Springer Series in Electrophysics, Volume 4)
ISBN 3-540-09939-5

Contents:
Cavitation. – Sound Waves and Bubbles. – Bubble Spectrometry. – Particle Detection. – Inhomogeneities in Ocean Acoustics. – Index of Contributors.

Hydrodynamic Instabilities and the Transition to Turbulence

Editors: H. L. Swinney, J. P. Gollub

1980. (Topics in Applied Physics, Volume 45)
In preparation

Y. I. Ostrovsky, M. M. Butusov,
G. V. Ostrovskaya

Interferometry by Holography

1980. 184 figures, 4 tables. X, 330 pages
(Springer Series in Optical Sciences, Volume 20)
ISBN 3-540-09886-0

Contents:
General Principles: Interference of Light. Optical Interferometry. Holography. Holographic Interferometry. – Experimental Techniques: Light Sources. Hologram Recording Materials. Setups. Experimental Aspects. – Investigation of Transparent Phase Inhomogeneities: Features of Holographic Interferometry of Transparent Objects. Sensitivity of Holographic Interferometry and Methods of Changing It. Holographic Diagnostics of Plasma. Use of Holographic Interferometry in Gas-Dynamic Investigations. – Investigation of Displacements and Relief: The Process of Interference-Pattern Formation in Holography. Methods of Interpreting Holographic Interferograms when Displacements are Studied. Investigation of Surface Relief. Flaw Detection by Holographic Interferometry. – Holographic Studies of Vibrations: Influence of Object Displacement on the Brightness of the Reconstructed Image – The Powell-Stetson Method. The Stroboholographic Method. Phase Modulation of the Reference Beam. Determining the Phases of Vibrations of an Object.

Turbulence

Editor: P. Bradshaw

2nd corrected and updated edition. 1978.
47 figures, 4 tables. XI, 339 pages
(Topics in Applied Physics, Volume 12)
ISBN 3-540-08864-4

Contents:
P. Bradshaw: Introduction. – *H.-H. Fernholz:* External Flows. – *J. P. Johnston:* Internal Flows. – *P. Bradshaw, J. D. Woods:* Geophysical Turbulence and Buoyant Flows. – *W. C. Reynolds, T. Cebecci:* Calculation of Turbulent Flows. – *B. E. Launder:* Heat and Mass Transport. – *J. L. Lumley:* Two-Phase and Non-Newtonian Flows.

H. Haken

Synergetics

An Introduction

Nonequilibrium Phase Transitions and Self-Organization in Physics, Chemistry and Biology.

'Springer Series in Synergetics'

2nd enlarged edition. 1978. 152 figures, 4 tables. XII, 355 pages

ISBN 3-540-08866-0

Contents:

Goal. – Probability. – Information. – Chance. – Necessity. – Chance and Necessity. – Self-Organization. – Physical Systems. – Chemical and Biochemical Systems. – Applications to Biology. – Sociology: A Stochastic Model for the Formation of Public Opinion. – Chaos. – Some Historical Remarks and Outlook.

Springer-Verlag
Berlin
Heidelberg
New York

Ocean Acoustics

Editor: J. A. DeSanto

1979. 109 figures, 5 tables. XI, 285 pages (Topics in Current Physics, Volume 8)

ISBN 3-540-09148-3

Contents:

J. A. DeSanto: Introduction. – *J. A. DeSanto:* Theoretical Methods in Ocean Acoustics. – *F. R. DiNapoli, R. L. Deavenport:* Numerical Models of Underwater Acoustic Propagation. – *J. G. Zornig:* Physical Modeling of Underwater Acoustics. – *J. P. Dugan:* Oceanography in Underwater Acoustics. – *N. Bleistein, J. K. Cohen:* Inverse Methods for Reflector Mapping and Sound Speed Profiling. – *R. P. Porter:* Acoustic Probing of Space-Time Scales in the Ocean. – Subject Index.

Turbulent Shear Flows I

Selected Papers from the First International Symposium on Turbulent Shear Flows, The Pennsylvania State University, University Park, Pennsylvania, USA, April 18–20, 1977
Editors: F. Durst, B. E. Launder, F. W. Schmidt, J. H. Whitelaw

1979. 256 figures, 4 tables. VI, 415 pages

ISBN 3-540-09041-X

Contents:

Free Flows. – Wall Flows. – Recirculating Flows. – Developments in Reynolds Stress Closures. – New Directions in Modeling.

Turbulent Shear Flows II

Selected Papers from the Second International Symposium on Turbulent Shear Flows, Imperial College London, July 2–4, 1979
Editors: J. S. Bradbury, F. Durst, B. E. Launder, F. W. Schmidt, J. H. Whitelaw

1980. 310 figures. Approx. 410 pages

ISBN 3-540-10067-9

Contents:

Turbulence Models. – Wall Flows. – Complex Flows. – Coherent Structures. – Environmental Flows. – Index of Contributors.